Springer Series in

SOLID-STATE SCIENCES 154

Springer Series in
SOLID-STATE SCIENCES

The Springer Series in Solid-State Sciences consists of fundamental scientific books prepared by leading researchers in the field. They strive to communicate, in a systematic and comprehensive way, the basic principles as well as new developments in theoretical and experimental solid-state physics.

138 **Phase Separation in Soft Matter Physics**
Micellar Solutions, Microemulsions, Critical Phenomena
By P.K. Khabibullaev and A.A. Saidov

139 **Optical Response of Nanostructures**
Microscopic Nonlocal Theory
By K. Cho

140 **Fractal Concepts in Condensed Matter Physics**
By T. Nakayama and K. Yakubo

141 **Excitons in Low-Dimensional Semiconductors**
Theory, Numerical Methods, Applications By S. Glutsch

142 **Two-Dimensional Coulomb Liquids and Solids**
By Y. Monarkha and K. Kono

143 **X-Ray Multiple-Wave Diffraction**
Theory and Application
By S.-L. Chang

144 **Physics of Transition Metal Oxides**
By S. Maekawa, T. Tohyama, S.E. Barnes, S. Ishihara, W. Koshibae, and G. Khaliullin

145 **Point-Contact Spectroscopy**
By Y.G. Naidyuk and I.K. Yanson

146 **Optics of Semiconductors and Their Nanostructures**
Editors: H. Kalt and M. Hetterich

147 **Electron Scattering in Solid Matter**
A Theoretical and Computational Treatise
By J. Zabloudil, R. Hammerling, L. Szunyogh, and P. Weinberger

148 **Physical Acoustics in the Solid State**
By B. Lüthi

149 **Solitary Waves in Complex Dispersive Media**
Theory · Simulation · Applications
By V.Yu. Belashov and S.V. Vladimirov

150 **Topology in Condensed Matter**
Editor: M.I. Monastyrsky

151 **Particle Penetration and Radiation Effects**
By P. Sigmund

152 **Magnetism**
From Fundamentals to Nanoscale Dynamics
By H.C. Siegmann and J. Stöhr

153 **Quantum Chemistry of Solids**
The LCAO First Principles Treatment of Crystals
By R.A. Evarestov

154 **Low-Dimensional Molecular Metals**
By N. Toyota, M. Lang and J. Müller

Volumes 91–137 are listed at the end of the book.

Naoki Toyota
Michael Lang
Jens Müller

Low-Dimensional Molecular Metals

With 135 Figures and 9 Tables

Springer

Professor Dr. Naoki Toyota
Tohoku University, Graduate School of Science
Department of Physics
Aramaki
980-8578 Sendai, Japan
E-mail: toyota-n@ldp.phys.tohoku.ac.jp

Professor Dr. Michael Lang
University of Frankfurt, FB 13 Physics
Physikalisches Institut
Max-von-Laue-Str. 1
60438 Frankfurt, Germany
E-mail: michael.lang@physik.uni-frankfurt.de

Dr. Jens Müller
Max Planck Institute
for Chemical Physics of Solids
Nöthnitzer Str. 40
01187 Dresden, Germany
E-mail: mueller@cpfs.mpg.de

Library of Congress Control Number: 2006936714

ISBN 978-3-540-49574-1

Springer is a part of Springer Science+Business Media
springer.com

Printed in Germany

Typesetting: Digital data supplied by authors
Production: LE-TeX Jelonek, Schmidt & Vöckler GbR, Leipzig
Cover production: Manfred Bender, WMX Design GmbH, Heidelberg
SPIN 10980651 Printed on acid-free paper 57/3100/YL - 5 4 3 2 1 0

To Kimiye, Mihoko, Naochika, Daiju
Claire, Moritz, Nicolas
Tina, Luke

Preface

Almost half a century has passed since the first molecular semiconductor with a high electrical conductivity was found in 1954 in halogen-doped organic complexes. This finding, coinciding with the structural identification of DNA, occurred seven years after the principle of a point-contact field-effect transistor had been invented. Needless to say, these two outstanding innovations of the 20th century have had great impact on the relevant fields of science and technology. Similarly, though on a much smaller scale, the discovery of organic semiconductors has set off in motion a large number of elaborate works by chemists and physicists. In consequence, the research areas of today have developed in depth and width, covering fields ranging from conducting polymers, liquid crystals, synthetic metals, molecular magnets, molecular field-effect transistors to single-molecule devices and more.

This particular monograph is concerned with just one aspect of this vast field, the physics of low-dimensional molecular metals (LDMM). This research field has brought together groups from various disciplines such as organic chemistry, physical chemistry, experimental physics, as well as theoretical chemistry and physics. The progress, which has been achieved in the development of new materials and the understanding of their exciting properties, has been the result of intensive interactions among the members of this interdisciplinary club. On behalf of the physics members, the authors, all having their roots in experimental physics, would like to appreciate the efforts made by the chemical members, particularly, for exploring novel molecules and crystals, otherwise such a club could not have existed.

The age of LDMM physics dawned around the early seventies when the first molecular metal TTF-TCNQ, consisting of radical cation and anion groups, was discovered. Their derivatives have been found to form many different charge-transfer salts with various kinds of radical molecules or metal ions. Since then, a number of interesting ground states and phase transitions has been observed: superconductivity, metal-insulator transitions associated with charge- or spin-density-wave condensates, and other cooperative states leading to charge-, antiferromagnetic- and dielectric order. As well as in ordinary metals, all these phenomena are directly related both to the electronic states near the Fermi level and to relevant interactions. An inherent advantage, particularly in LDMM, is due to the fact that the behavior of the conduction electrons, usually one or two carriers in large unit cells, can be expressed surprisingly well by a simple tight-binding picture leading to simple Fermi surfaces (FS) which are being used in introductory textbooks for low-dimensional electronic systems. This characteristic feature is unique for LDMM and seen neither in ordinary bulk metals nor in compounds containing many complicated electron bands. During the last two decades, the FS and its related phenomena have been extensively studied with the experimental tools of low temperatures, and by applying magnetic fields and pressure. Thus the LDMM physics may occupy a unique position in solid state physics in such a way that physical phenomena can be more directly linked to the experimentally clarified details of the electronic states.

With regard to contemporary solid state physics, the research era for LDMM during the last three decades has overlapped with the enormous progress made in other fields of conducting materials and the related physics: for example, superconductivity and magnetism in heavy-fermion rare-earth-based compounds, quantum Hall effect in artificial semiconductor heterostructures, high-T_c superconductivity in cuprates, supramolecular fullerene C_{60}, and superconductivity in compounds with alkaline metals, metallic carbon nanotubes, to name only a few. On deliberating what these discoveries have in common, we may find that the emergence of these new materials has altered and sometimes overturns our common sense, familiar concepts, or theories. This experience may be condensed into words uttered by H. Kammerlingh Onnes: "Door meten tot weten" (Through measurement to knowledge).[1]

While keeping these notions in mind, this monograph intends to provide an overview of LDMM physics including issues of controversy. The authors refrained from including details of chemistry and theories, as these aspects are outside the scope of their working fields. The intention of this book is to describe the physics as plainly and systematically as possible for students at graduate levels in solid state physics and chemistry courses with the background knowledge of introductory solid state physics/chemistry. The wide spectrum of subjects addressed, together with the comprehensive list of references, may be helpful even to experts in this field and also to researchers working at the frontier of the other contemporary research areas mentioned above.

This monograph consists of seven chapters. Chapter 1 will give an overview of the historical survey for molecular magnets and conductors. After a brief introduction of the basic concepts for charge transfer and molecular orbitals in Chapt. 2, the systematics of structural aspects characteristic to LDMM will be given in Chapt. 3. Chapter 4 will be concerned with the normal-metallic states with characteristic transport, optical, thermal, magnetic and dielectric properties, and follow on to instabilities involving spin- and charge-degrees

[1] K. Mendelssohn, *The Quest for Absolute Zero*, 2nd edn., Taylor & Francis, London, 1977.

of freedom leading to spin- or charge-density-wave condensates, spin-Peierls instability states, and charge ordering. In the remaining sections, many-body effects and various phase diagrams will be described. Chapter 5 will focus on the effects of magnetic fields on conduction electrons, the so-called fermiology, and field-induced electronic phase transitions and/or collective states. The vast range of subjects concerning superconductivity will be described in Chapt. 6 covering the characteristic parameters, isotope labeling, the problem of disorder, superconducting vortex states, and pairing symmetry and interaction. In Chapt. 7, characteristic phenomena, caused by interactions of conduction electrons with localized *d*-spins exemplified in DCNQI and BETS salts, will be introduced. A few words will be added in the Epilogue touching on important problems remaining to be solved and on the direction along which the LDMM physics may trace its developments.

The authors would like to express their sincere gratitude to Y. Muto and T. Fukase, Emeritus Professors of Tohoku University (N.T.), the members of the collaborative research center at Frankfurt University, Forschergruppe 412 (M.L.), and S. von Molnár, Director of MARTECH and Professor of Physics at Florida State University (J.M.).

The authors wish to thank many collaborators and colleagues for enlightening discussions, exchange of ideas, or valuable advice: C. Akinci, H. Aoki, A. Ardavan, S. Blundell, J. Brooks, A. Brühl, E. Coronado, P. Day, M. Dressel, S. Endo, R. Feyerherm, H. Fujiwara, U. Geiser, T. Goto, G. Grüner, S. Hill, Y. Iwasa, D. Jérome, S. Kagoshima, W. Kang, M. Kartsovnik, R. Kato, S. Kawamata, A. Kini, H. Kobayashi, F. Kromer, C. Langhammer, A. Lebed, K. Maki, H. Matsui, R. McKenzie, J. Merino, T. Mori, K. Murata, T. Nakamura, E. Negishi, Y. Nishio, Y. Noda, T. Nojima, T. Osada, V. Pashchenko, B. Powell, K. Removic-Langer, L. Pintschovius, F. Pratt, G. Saito, T. Sasaki, J. Schlueter, D. Schweitzer, J. Singleton, M. de Souza, F. Steglich, Ch. Strack, S. Suellow, T. Sugimoto, U. Sushko, J. Suzumura, T. Suzuki, M. Tanatar, K. Tanigaki, H. Tsuchiya, H. Uozaki, M. Watanabe, Y. Watanabe, B. Wolf, A. Wolter, J. Wosnitza.

The authors would like to thank most warmly Carolyne Agnew for her help in proof reading, Momoko Ishikawa and Atsumasa Hojo for their assistance in revising figures and texts, John Schlueter, Masashi Watanabe and Takehiko Mori for providing figures in Chapts. 1-3, and the publisher for continuous encouragement.

Sendai, Frankfurt, Tallahassee, Dresden
November 2006

Naoki Toyota
Michael Lang
Jens Müller

Contents

List of Symbols and Abbreviations

a, b, c	lattice constants
$\boldsymbol{A}$	vector potential
B	magnitude of magnetic induction $\boldsymbol{B}$
B_e	magnitude of applied magnetic field $\boldsymbol{B}_e = \mu_0 \boldsymbol{H}$
B_{c_1}	superconducting lower critical field
B_{c_2}	superconducting upper critical field
$B_{c_{th}}$	superconducting thermodynamic critical field
B_{irr}	irreversibility field
B_J	exchange field
B_{MB}	magnetic breakdown field
B_{MI}	critical field for metal-insulator transition
B_P	Pauli-paramagnetic limiting field (Clogston-Chandrasekhar limit)
B_{sf}	spin-flop field
$c_{i\sigma}(c_{i\sigma}^{+})$	annihilation (creation) operator for an electron with spin σ at i-th site
$-e$	electron charge ($e > 0$)
E	magnitude of electric field vector $\boldsymbol{E}$
$E(\boldsymbol{k})$	electron energy at $\boldsymbol{k}$
E_T	threshold electric field
E_F	Fermi energy
E_G	energy gap (see also Δ)
E_n	energy of the n-th Landau level
$\boldsymbol{E}_\omega$	electric field vector oscillating at ω
f	frequency
F	dH frequency
$F_{\alpha(\beta)}$	dH frequency of $\alpha(\beta)$ orbit
g	electron-phonon coupling constant (see also λ_{ep}); spin g factor
$\boldsymbol{G}_{pq}$	reciprocal lattice vector
h ($\hbar$)	Planck's constant ($h/2\pi$)
H	magnitude of magnetic field $\boldsymbol{H}$
$\mathcal{H}$	Hamiltonian
J	magnitude of current density $\boldsymbol{J}$; exchange constant between spins
J_T	threshold current density

k	magnitude of wave vector $\boldsymbol{k}$
k_B	Boltzmann's constant
k_F	Fermi wave number
ℓ	electron mean free path
m_0	free electron mass
m^*	effective mass
m_b	band mass
m_c	cyclotron mass
$m_c^{\alpha(\beta)}$	cyclotron mass averaged on $\alpha(\beta)$ orbit
m_{th}^*	thermodynamic effective mass
M	magnitude of magnetization $\boldsymbol{M}$
M_{osc}	oscillatory component of magnetization
n	carrier number density; quantum number
n_i	electron number operator at i-th site
$N(E)$	electronic density-of-states as a function of energy E
$N(E_F)$	electronic density-of-states at E_F
N_A	Avogadro's number
p	pressure; magnitude of electron momentum vector $\boldsymbol{p}$
P	electric polarization
q	magnitude of transferred wave vector $\boldsymbol{q}$
Q	magnitude of specified wave number vector $\boldsymbol{Q}$
R	gas constant
r	harmonic index number
$\boldsymbol{r}$	position vector
r_c	cyclotron radius
R_D	electron scattering-induced reduction factor
R_S	spin splitting-induced reduction factor
R_{SC}	superconductivity-induced reduction factor
R_T	temperature-induced reduction factor
s	inter-layer distance
S	magnitude of spin vector $\boldsymbol{S}$; overlap integral
$S_{\alpha(\beta)}$	S_{extr} for $\alpha(\beta)$ orbit
S_k	$\boldsymbol{k}$-space area
S_r	$\boldsymbol{r}$-space area
S_0	$2\pi/\lambda_B^2$
S_{extr}	extremal area of Fermi surface
S_{ij}	overlap integral between i- and j-th orbitals
t_{ij}	transfer energy (integral) between i- and j-th orbitals
T	temperature
T_ρ	resistance-minimum temperature
T_1	spin-lattice relaxation time
T_{AO}	anion-ordering temperature
T_c	superconducting transition temperature
T_g	glass-transition temperature

T_C	Curie temperature
T_{CO}	charge-ordering temperature
T_D	Dingle temperature
$T_{SDW/CDW}$	spin/charge-density-wave-transition temperature
T_N	Néel temperature
T_{MI}	metal-insulator-transition temperature
T_P	Peierls-transition temperature
U	electron-electron Coulomb interaction
$\boldsymbol{v}$	electron velocity
v_F	Fermi velocity
V	sample volume; electrical voltage
V_{mol}	molar volume
V_{ij}	Coulomb repulsive energy between i- and j-th sites
W	bandwidth
Z	surface impedance
α, β, γ	crystallographic angles
γ	anisotropy parameter; Sommerfeld coefficient of electronic specific heat
δ	charge-transfer ratio
δ_ω	skin depth of electromagnetic wave at ω
Δ_d	degree of dimerization
Δ_{CDW}	charge-density-wave energy gap
Δ_{DW}	density-wave energy gap
Δ_s	spin gap
Δ_{SDW}	spin-density-wave energy gap
Δ	superconducting energy gap
$\tilde{\varepsilon}(\omega)$	frequency-dependent complex dielectric constant
ε_1	real part of $\tilde{\varepsilon}(\omega)$
ε_2	imaginary part of $\tilde{\varepsilon}(\omega)$
Θ	Weiss temperature
Θ_D	Debye temperature
κ	Ginzburg-Landau parameter
λ	superconducting penetration depth; also used: London penetration depth λ_L
λ_B	magnetic length
λ_{DW}	dimensionless coupling constant of density-wave
λ_{ep}	electron-phonon coupling constant
μ_0	permeability of vacuum
μ_b	band mass normalized by m_0
μ_c	cyclotron mass normalized by m_0
μ_B	Bohr magneton
ξ	superconducing coherence length; also used:Ginzburg-Landau coherence length ξ_{GL}

ρ	electrical resistivity
$\rho(\boldsymbol{r})$	electron number-density at $\boldsymbol{r}$
$\rho_\sigma(\boldsymbol{r})$	spin-dependent electron number-density for $\sigma =\uparrow,\downarrow$
$\sigma_{xx(xy,zz)}$	components of resistance tensor
$\tilde{\sigma}(\omega)$	frequency-dependent complex conductivity
σ_1	real part of $\tilde{\sigma}(\omega)$
σ_2	imaginary part of $\tilde{\sigma}(\omega)$
τ	electron relaxation time
ϕ_0	magnetic flux quantum ($= h/2e$)
$\chi(\boldsymbol{q})$	electron response function at $\boldsymbol{q}$
$\tilde{\chi}(\boldsymbol{q})$	renormalized electronic response function at $\boldsymbol{q}$
χ_P	Pauli paramagnetic susceptibility
ω	angular frequency
ω_c	cyclotron frequency
ω_D	Debye frequency

1D	one dimension(al)
2D	two dimension(al)
3D	three dimension(al)
AFM (AF)	antiferromagnet(ic)
AFMI	antiferromagnetic insulator
AMRO	angular-dependent magnetoresistance oscillation
BCS	Bardeen-Cooper-Schrieffer
CD	charge disproportionation
CDW	charge-density-wave
CMD	colossal magnetodielectricity
CO	charge order
COP	charge order-induced polarization
CR	cyclotron resonance
dH	de Haas
dHvA	de Haas-van Alphen
DMFT	dynamical mean-field theory
DW	density wave
EMV	electron molecular vibration
ESR	electron spin resonance
FICDW	field-induced charge-density-wave
FISC	field-induced superconductivity
FISDW	field-induced spin-density-wave
FM	ferromagnet(ic)
FS	Fermi surface
HOMO	highest occupied molecular orbital
JP	Jaccarino-Peter
LDMM	low-dimensional molecular metals
LK	Lifshitz-Kosevich

LUMO	lowest unoccupied molecular orbital
MB	magnetic breakdown
MI	metal-insulator
MO	molecular orbital
NMR	nuclear magnetic resonance
NR	negative resistance
PMM	paramagnetic metal(lic)
QVL	quantum vortex liquid
RKKY	Ruderman-Kittel-Kasuya-Yoshida
SDW	spin-density-wave
SP	spin Peierls
SdH	Shubnikov-de Haas
TVL	thermal vortex liquid
μSR	muon spin rotation

(*chemicals*)

BEDO-TTF	*bis(ethylenedioxy)tetrathiafulvalene*
BEDT-TSF	*bis(ethylenedithio)tetraselenafulvalene*
BEDT-TTF	*bis(ethylenedithio)tetrathiafulvalene*
BETS	BEDT-TSF
DCNQI	*N,N'-dicyanoquinodiimine*
DI-DCNQI	*diiode*-DCNQI
DMe-DCNQI	*dimethyl*-DCNQI
DMET	*(dimethylethylenedithio)diselenadithiafulvalene*
dmit	*4,5-dimercapto-1,3-dithiole-2-thione*
EDO-	*ethylenedioxy-*
ET	BEDT-TTF
MDT-	*methyldithio-*
MEM	*N-methyl,N-ethyl-morpholinium*
TCNQ	*tetracyanoquinodimethane*
TM	general abbreviation for TMTTF and TMTSF
tmdt	*tri(methylene)tetrathiafulvalene-dithiolate*
TMTSF	*(tetramethyl)tetraselenafulvalene*
TMTTF	*(tetramethyl)tetrathiafulvalene*
TTF	*tetrathiafulvalene*

1 Introduction: Historical Remarks

The research on molecular magnetism may be traced back to the 19th century when M. Faraday studied the magnetic susceptibility of a variety of organic molecular substances [1]. Almost all of these substances were found, however, to be non-magnetic except for an effect which was later called molecular diamagnetism. In 1927, shortly after quantum mechanics had been established, W. Heitler and F. London [2] had a monumental success in applying this new theory to the hydrogen molecule, the simplest molecule in nature. Following that model, W. Heisenberg [3] gave a general conception of an exchange interaction.[1] Faraday's results were explained by a quantum mechanical rule based on the Pauli principle stating that any molecular orbital in stable molecules, i.e., non-radicals, is necessarily occupied by spin-up and -down electrons. The main problem in those days was to understand the molecular diamagnetism induced by a shielding current flowing along an aromatic ring in response to an external magnetic field [4].

In contrast to the rapidly growing progress achieved from both sides of experimental and theoretical physics in understanding the transport and magnetic properties of metals, alloys, and inorganic compounds, it took a rather long and sluggish period before the new field of *molecular* conductors and magnets emerged in the 1950 – 1960s.

A brief historical survey of molecular magnets will be given first; for a review article the reader is referred to the detailed works by [5, 6, 7, 8, 9, 10].

In the development of magnetic molecular materials, some theoretical guidelines played an important role both in exploring organic radicals with stable spins using an exchange interaction via chemical bonds and in synthesizing molecule-based ferromagnetic (FM) crystals. In 1967, K. Itoh [11] and E. Wasserman et al. [12] independently verified a FM spin alignment of

[1] In this two-electron system, the bonding (antibonding) orbital is symmetric (asymmetric) with respect to the electron exchange corresponding to a spin singlet $S = 0$ (spin triplet $S = 1$). Therefore, the orbital energy difference $E_0 - E_1 < 0$, where E_0 and E_1 are bonding and antibonding orbital energy, respectively, originates from the difference of the spin states. This dependence of the singlet-triplet energy splitting on the spin state of the electrons can be expressed by the Heisenberg Hamiltonian $\mathcal{H} = -2J\boldsymbol{s}_i \cdot \boldsymbol{s}_j$ with an exchange interaction $J = (E_0 - E_1)/2$. Thus, the origin of J is found both in the Coulomb interaction between electrons and in the Pauli principle.

$S = 2$ in polycarbene, *m-phenylenebis(phenylmethylene)*. The idea was based on Longuet-Higgins's theorem [13] using a topological degeneracy, which is in contrast to the conventional Hund's coupling in partially occupied transition-metal or rare-earth ions. This approach has been used for exploring FM organic polymers. To align spins in a molecular crystal, H. McConnell [14] proposed a theoretical idea that a FM interaction could be induced in-between neighboring radical molecules in which positive and negative spin densities do partially overlap. At last, after a FM interaction had been experimentally confirmed in a calvinoxyl crystal [15], M. Kinoshita et al. succeeded in synthesizing the first organic FM, *p*-NPNN (*p-nitrophenyl nitronyl nitroxide*) with a Curie temperature $T_C = 0.65\,\mathrm{K}$ [16, 17, 18]. Although elaborate works have been devoted so far to synthesizing new molecular magnets, T_C is still low, in the order of several Kelvin at most. The reason is attributed to a weak exchange interaction due to both the spatial distribution of spin densities on the magnetic molecule and the relatively large intermolecular distance.

In parallel to these studies of molecular magnets, highly conducting molecular materials have also been explored. The roots can be traced back to a pioneering work carried out in 1954 by H. Akamatsu, H. Inokuchi and Y. Matsunaga [19] who synthesized cation radical salts by doping bromine into perylene, a condensed polycyclic aromatic hydrocarbon. As shown in Fig. 1.1, it exhibits an extraordinarily low electrical resistivity ρ of the order of $10\,\Omega\,\mathrm{cm}$ at room temperature, which increases with decreasing temperature. The family of perylene-halide complexes provided the first semiconductors among organic molecular materials. Along with these studies, R.S. Mulliken's charge-transfer theory [20] promoted the developments of chemical synthesis towards molecular metals.

In 1960 D. Acker et al. of DuPont [21] synthesized a novel acceptor TCNQ (*tetracyanoquinodimethane*), which consists of highly electron attractive dicyanomethylene groups bonding to *para*-quinon. It is a symmetric, planar molecule and, by accepting an electron, it becomes a stable anion radical of larger aromacity with a 6π electronic state.[2]

A number of synthetic studies were devoted, in the 1960s, to heterocyclic donor molecules in which hetero-atoms like O, S and Se had been used to partially replace the carbon atoms. One of the most outstanding donors, TTF (*tetrathiafulvalene*) was synthesized by the groups of D.L. Coffen [22, 23, 24], F. Wudl [25, 26] and S. Hünig [27], see, e.g. [28]. This molecule, which is symmetric with respect to the central double-bond carbons and also planar

[2] The salts M-TCNQ, with monovalent cation radicals (M) like alkaline elements, form one-dimensional chains weakly coupled to each other. These 1:1 salts are insulators in the strong electron-correlation limit at half filling with one electron at each site. These systems have served as one of the model systems for the one-dimensional Hubbard model.

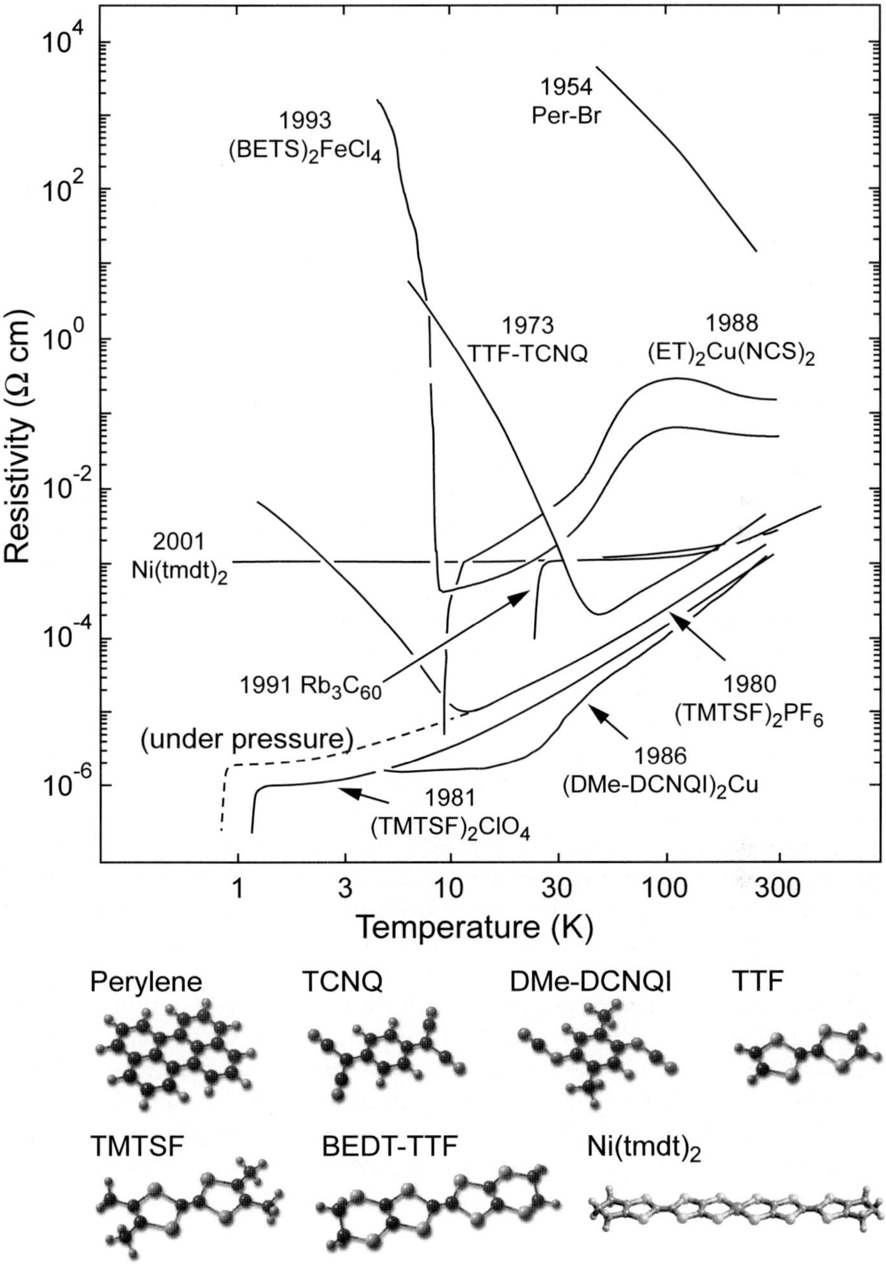

Fig. 1.1. Temperature dependence of the electrical resistivity of epoch-making molecular conductors. The molecular illustrations are by courtesy of M. Watanabe, Tohoku University.

with four sulfur atoms in the five-member rings on both sides, features a high solubility in solvents and a high stability both in its neutral and ionic states.[3]

The first charge-transfer salt, consisting of the above TTF donor and TCNQ acceptor molecules, was the 1:1 compound, TTF-TCNQ. It exhibited low resistivity in the order of $\sim 10^{-3}\,\Omega\,\mathrm{cm}$ at room temperature, which decreases with lowering temperature as shown in Fig. 1.1 [30, 31]. This discovery of an organic molecular metal was a milestone which made a great impact on the chemistry and physics communities, opening the door to the interdisciplinary field of low-dimensional molecular metals (LDMM). From the physics' point of view, the metal-insulator transition around 50 K, as seen in Fig. 1.1, was quite a novel phenomenon, which was soon identified as a Peierls instability [32] in the low-dimensional electron system. Here we note that all the thermodynamic phase transitions occurring in low-dimensional systems are the consequence of three-dimensional couplings. In order to control these couplings or to avoid such low-dimensional instabilities, physicists were motivated to apply pressure, while chemists were encouraged to develop new donors and acceptors potentially with stronger intermolecular couplings.

In 1980, a superconducting transition was observed for the first time in pressurized $(TMTSF)_2PF_6$ (*bis(tetramethyl)-tetraselenafulvalene hexafluoro phosphate*) by D. Jérome and his collaborators [33]. The donor TMTSF, a derivative of TTF, had been synthesized by K. Bechgaard and his chemistry group [34]. As seen in Fig. 1.1, this salt undergoes a metal-insulator (MI) transition around 12 K at ambient pressure due to a low-dimensional instability similar to that in TTF-TCNQ. However, this effect could be suppressed by the application of hydrostatic pressure of about 12 kbar, resulting in a superconducting ground state below $T_c = 0.9\,\mathrm{K}$ [33].[4] Soon after this discovery was made, ambient pressure superconductivity was observed below 1.4 K in $(TMTSF)_2ClO_4$ with the non-centrosymmetric tetrahedral anion [36]. Together with the TMTTF (*bis(tetramethyl)tetrathiafulvalene*) salts with S atoms replacing all the Se atoms in TMTSF, this class of quasi-one-dimensional (quasi-1D) salts have served as model systems for exploring and understanding a variety of physical phenomena at low temperatures and high magnetic fields.

Complementary to these cation radicals, a new acceptor molecule was reported by S. Hünig and his collaborators in 1984 [37]. It was DCNQI (*N,N'-dicyanoquinodiimine*) obtained by substituting the N-CN group for the dicyanomethylene groups in TCNQ, followed by the synthesis of conducting charge-transfer complexes and anion radical salts with monovalent metal cations [38, 39]. In general, this class of salts exhibits a high conductivity similar to that of TTF-TCNQ as seen in Fig. 1.1. In particular, $(DCNQI)_2Cu$ was found to be the first metal with π-d interactions. The interplay of conducting

[3] Chemical details for the derivatives of TTF and their salts are available in [29].

[4] The pressure of ~6 kbar is sufficient to suppress the metal-insulator transition. [35].

π-electrons, with the mixed-valent states of Cu ions providing the localized d-spins, has been extensively studied, see e.g. [40].

The next example for a building block for superconductors was the molecule BEDT-TTF (*bis(ethylenedithio)-tetrathiafulvalene*), abbreviated as ET, that was an extended TTF-derivative with eight S atoms [41]. Both the flatness (except the terminal ethylene groups) and the existence of additional hetero-atoms furnished this donor with a high functionality forming quasi-two-dimensional (quasi-2D) charge-transfer salts with a variety of anions and also with many different stacking patterns and hence various crystal structures. Indeed, charge-transfer salts have been formed with a variety of anions such as linear I_3, tetrahedral ReO_4, polymeric $Cu(NCS)_2$, and thick layered $MHg(SCN)_4$, giving rise to many crystal morphologies denoted as α, β, κ and so on. Among them, the κ-type stacking have provided many superconductors. The first example was κ-$(ET)_2I_3$ with $T_c = 3.6\,K$ found by R. Kato et al. [42, 43], followed by the synthesis of the $Cu(NCS)_2$ salt with higher $T_c = 10.4\,K$ by G. Saito and his collaborators [44, 45]. Soon after that, A. Kini et al. [46] found a higher T_c of 11.6 K in the similar polymeric κ-$(ET)Cu[N(CN)_2]Br$ salt. This material is still the highest-T_c superconductor among LDMM at ambient pressure. Thus the ET family has been a major source for molecular superconductors with an overall number of more than 100. As in the TMTSF and TMTTF salts, the application of pressure has played an important role in some of the ET salts. For example, V.N. Laukhin et al. [47] and K. Murata et al. [48] found a discontinuous upturn of T_c from 1.5 K up to 8 K in β-$(ET)_2I_3$, which was ascribed to the suppression of a lattice modification and an accompanied disordered state. Beside superconductivity, the metallic states in ET salts have been found to be unstable against the formation of magnetic or charge-ordered insulating states due to electron-correlation effects. Likewise, these materials show transitions into anomalous metallic states associated with the low dimensionality as observed in quasi-1D salts.

An asymmetric donor DMET was synthesized by hybridizing ET and TMTSF [49]. Here DMET stands for *(dimethylethylenedithio)diselenadithiafulvalene.* The salts formed with these derivatives such as $(MDT\text{-}TTF)_2AuI_2$ (MDT- being *methylenedithio-*) [50] and $(DMET\text{-}TSF)_2X$ ($X = AuI_2, I_3$) were found to be superconductors with rather low T_c values below 4 K [51]. The next example is $(BEDO\text{-}TTF)_3Cu_2(SCN)_3$ and $(BEDO\text{-}TTF)_2ReO_4{\cdot}H_2O$ with T_c's of about 1 K [52, 53], where BEDO-TTF stands for *bis(ethylenedioxy)tetrathiafulvalene.* Other types of superconductors were found in the salts based on a multi-halogen π-acceptor molecule $M(dmit)_2$ (M = Ni, Pd, Pt) (dmit is *4,5-dimercapto-1,3-dithiole-2-thione*) synthesized by P. Cassoux and his collaborators [54, 55]. In those days, a novel class of superconductors was found in fullerene compounds with alkaline metals yielding high T_c values as shown in the figure for the case of Rb_3C_{60}.[5]

[5] In 1985, H. Kroto et al. discovered the supramolecule C_{60}, called fullerene [56], which had a profound impact on materials chemistry/physics. The many interest-

In further exploring the interplay between magnetism and conductivity as seen in DCNQI, many chemistry groups devoted their efforts to synthesizing magnetic conductors incorporating transition-metal ions or complexes, in which the conducting π-electrons might interact with localized magnetic moments, see e.g. [63]. As a result, a magnetic superconductor $(ET)_4[Fe(C_2O_4)_3]H_3O{\cdot}C_6H_5CN$ synthesized by P. Day, M. Kurmoo and their collaborators [64] was found to exhibit superconductivity at $T_c = 8\,K$ with a Weiss temperature of $-0.2\,K$ due to $S_d = 5/2$ (Fe^{III}). At about the same time, H. Kobayashi et al. synthesized BEDT-TSF (*bis(ethylenedithio)tetraselenaful valene*, abbreviated as BETS) salts with FeX_4 (X = Cl, Br) which have two different crystal morphologies; in the κ-type salt with X = Br, a long-range antiferromagnetic order below $T_N = 2.5\,K$ coexists with superconductivity below 1.1 K [65], while, the λ-type salt undergoes a coupled metal-insulator (MI) and paramagentic-AFM transition as shown in the figure [65, 66, 67]. On the other hand, E. Coronado and his collaborators synthesized $(ET)_3[MnCr(C_2O_4)_3]$, which exhibited the coexistence of a metallic state with FM order below $T_C = 5.5\,K$ [68]. The bimetallic complexes $[Cr^{II}Mn^{III}(C_2O_4)_3]$ form layers of oxalate-bridged hexagonal networks sandwiching the β-type stacking ET layers. Besides ET salts, T. Sugimoto and his coworkers synthesized an asymmetric donor EDT-TTFVO (*ethylenedithiotetrathiafulvalenoquinone-1,3-dithiolemethide*) with a permanent electric dipole, and found the $\pi - d$ coupled metallic state in the $FeBr_4$ salt [69].

The metallic states mentioned so far are the result of intermolecular charge transfer between two different molecular species in a two-component system. Recently A. Kobayashi and her collaborators succeeded in synthesizing a *single-component* system $Ni(tmdt)_2$ which remains metallic down to low temperatures [70, 71, 72] as shown in the figure.

One of the most important targets in the field of molecular conductors has been the increase of the values of the superconducting transition temperature T_c. In this respect, the theoretical article by W.A. Little [73], published in 1964, has played an important role. His idea was that an electron pairing assisted by an excitonic interaction — in contrast to the conventional phonon mechanism — would possibly result in an extraordinarily high T_c. According to Little's model, this requires a one-dimensional conducting polymer attached to which are highly polarizable ligands. Although this theoretical suggestion has been a source of encouragement to synthetic chemists, the idea has remained a dream since in none of the molecular superconductors obtained so far has any signature in favor of this mechanism been found.

---

ing chemical and physical properties have been extensively studied; for example the T_c (19 K in K_3C_{60} [57] and 28 K in Rb_3C_{60} [58, 59]) can be raised up to 33 K in $RbCs_2C_{60}$ [60], see, e.g. [61, 62].

2 Basic Concepts

2.1 Charge Transfer

The prerequisites of forming a conducting material, based on molecular building blocks, are (i) the creation of free charge carriers and (ii) their delocalization throughout the crystal. For the charge-transfer salts described in this monograph, free charge carriers (i) are created by combining a donor molecule D, such as TMTSF, BEDT-TTF (hereafter ET) or their derivatives, with an electron acceptor X. Here D should be an *electron-rich* molecule with a sufficiently low ionization energy I_0 which can be easily oxidized by the acceptor X having the appropriate electron affinity A. Their combination results in a charge-transfer salt $\mathrm{D}_m\mathrm{X}_n$ according to the reaction

$$[\mathrm{D}_m] + [\mathrm{X}_n] \rightarrow [\mathrm{D}_m]^{+\delta} + [\mathrm{X}_n]^{-\delta}, \tag{2.1}$$

with m and n being integers and δ the charge-transfer ratio. This process requires an energy of

$$\Delta E_{CT} = I_0 - A - C < 0, \tag{2.2}$$

where C includes Coulomb-, polarization- and exchange-energy contributions [20]. The number of mobile charge carriers is governed by the charge-transfer process and is independent of temperature. This contrasts with the creation of free charge carriers by doping the system with molecular or atomic *impurities*, whose ionization energies or electron affinities are quite different from those of the host material. As an example, we mention the well-known polyacetylene $(\mathrm{CH})_\mathrm{x}$, a conjugated polymer, which upon doping halogens such as bromine and iodine, becomes conducting [74, 75]. In these materials, the number of carriers can be controlled by the doping level which may vary the conductivity by 10 orders of magnitude.

In forming charge-transfer crystalline solids, the constituent molecules are packed rather densely so that the orbitals may overlap between adjacent molecules. This intermolecular overlap of partially-filled orbitals allows the carriers to delocalize throughout the crystal. These charge-transfer salts have to be distinguished from other classes of molecular conductors such as conjugated polymers or graphite. While in the former, it is essentially an *inter*molecular process leading to the motion of carriers, the conductivity in the latter comes about due to the π-electron system of the extended molecule which provides the path for the carrier motion. (For the *intra*molecular charge transfer in the single-component system $Ni(tmdt)_2$, see Sect. 3.2.3.)

2.2 Molecular Orbitals and Transfer Integrals

As for ordinary inorganic conductors, an understanding of the electronic properties cannot be achieved without a detailed knowledge of their electronic structure. For simple metals such as sodium or potassium, a good starting point for calculating the band structure is to view the conduction electrons as a gas of nearly free fermionic particles whose motion is only weakly perturbed by the periodic potential of the positively charged ions. On the contrary, for some materials such as the transition metals with their partially-filled inner d-shells or the charge-transfer molecular conductors discussed here, the opposite limit is more appropriate. Here the conduction electrons are bound more tightly to the atoms or molecules and the band structure arises from weakly overlapping atomic or molecular wave functions [76, 77]. As a first step in such a tight-binding approach to the band structure of molecular materials, the orbitals of the individual molecules have to be determined. Even for relatively simple molecular systems, the large number of electrons and nuclei per molecule requires stringent simplifications to keep the system tractable.

A most useful approximation is provided by the molecular orbital (MO) method in which the available valence electrons are considered to be delocalized over the whole molecule. The MOs are then obtained by taking a linear combination of atomic s- and p-orbitals of the constituent atoms resulting in σ- and π-orbitals. The number of these molecular orbitals, each of which can accommodate two electrons with antiparallel spins, is equal to the total number of constituent atomic orbitals. While the σ-orbitals are localized along the bonding axis of adjacent atoms, having a nonzero momentum only along this axis (parallel to the molecular plane), the π-orbitals are extended normal to the bonding plane. Due to the lower binding energy of the π-electrons compared to that of the σ-electrons, they tend to delocalize and can be easily excited. As a consequence, the carriers transferred from the donor to the acceptor have a dominantly π-electron character leaving behind a fraction $+\delta$ of π-holes per donor molecular unit $[D_m]$. In most cases, the negatively charged anions $[X_n]^{-\delta}$ adopt a closed-shell configuration and they do not contribute to the electronic properties.[1] For the further treatment it is convenient and, for most cases, sufficient to consider only either the Highest

[1] As an example of a system where the anions do contribute to the electronic properties, we mention the archetype TTF-TCNQ salt [31]. Here the TTF and TCNQ molecules are arranged to form two independent conducting stacks (Sect. 3.2.1). The charge-transfer ratio of $\delta = 0.59$ means that on average each TTF molecule transfers 0.59 electrons to a TCNQ unit (Sect. 4.4.1).

Occupied Molecular Orbital (HOMO) or the Lowest Unoccupied Molecular Orbital (LUMO) of the π-electrons moving in a potential formed by the σ-electrons together with the nuclei and the inner-shell electrons. This forms the basis of the so-called π-electron approximation and can be considered as an extension of the idea of the *frontier orbital theory* for chemical reactions of molecules to form molecular solids, see e.g. [78].

For the typical molecules concerned in this monograph, Fig. 2.1 shows the HOMOs of donors such as TMTSF, BEDT-TTF (ET), and BEDT-TSF (BETS) and the LUMOs for 2-R1,5-R2-DCNQI. The HOMOs on these donors have a strong π character with a much smaller contribution from the $3d_{yz}$-orbitals. Due to B_{1u} symmetry (D_{2h} point group), nodes appear on all the C–S bonds. Since the Slater coefficients of the $3p_z$-orbital wavefunctions in the chalcogen atoms are almost twice as large as that of the C atoms, the electron density is most largely settled on the inner S or Se atoms. On the other hand, the LUMO π^*-orbitals of DCNQI are largest on the N atoms of the cyano group so that DCNQI is expected to be a strong donor ligand.

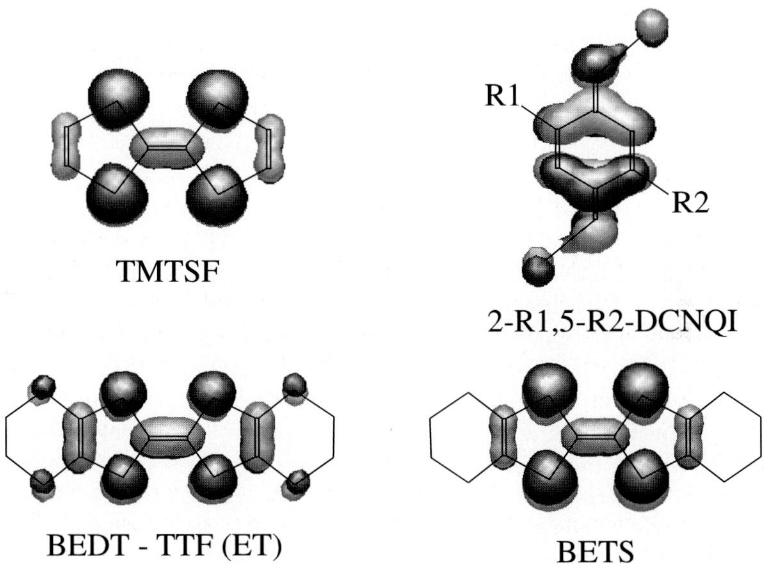

Fig. 2.1. HOMOs of the donor molecules TMTSF, BEDT-TTF (ET), and BEDT-TSF (BETS) and LUMOs of the acceptor 2-R1,5-R2-DCNQI. The heavy and light grays correspond to the in-phase and out-of-phase contributions of the orbital wavefunctions, calculated by MOPAC. By courtesy of T. Mori, Tokyo Institute of Technology.

These molecular orbitals are used for calculations of the intermolecular orbital overlaps [79, 80, 81, 82] (refer to Fig. 3.3 for a schematic visualization in the case of $(TMTSF)_2PF_6$). As shown in Sect. 3.2, all these donor and ac-

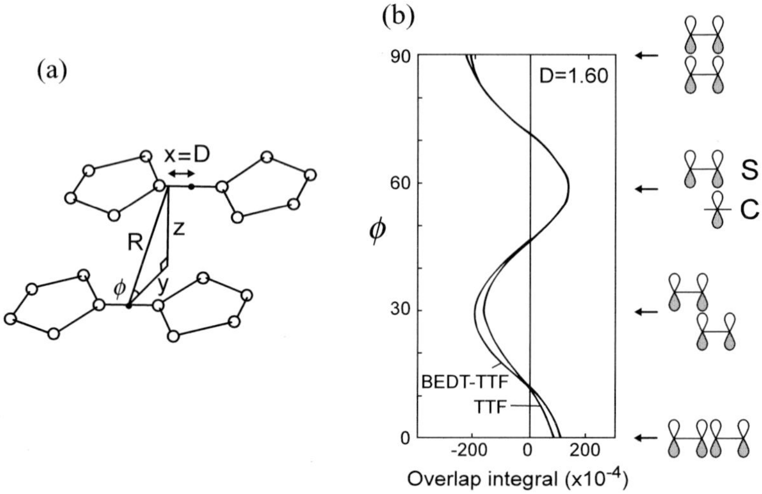

Fig. 2.2. (a) Relative orientation of two TTF donor molecules. D represents the molecular slipping along the long axis, R stands for the distance between the long axes, and ϕ varies from 0°, the side-by-side configuration, to 90°, the face-to-face configuration. (b) ϕ-dependence of the overlap integrals S_{ij} in units of 10^{-4} for the HOMOs between two TTF or ET donors. After [82].

ceptor molecules stack so as to keep both their long axes and molecular planes parallel to each other so that the relative rotation can be safely neglected. The relative configuration of two molecules is represented by a set of cylindrical coordinates (D, R, ϕ) as shown in Fig. 2.2(a). The overlap integrals S change with ϕ; the largest integrals at $\phi = 90^\circ$ correspond to the normal face-to-face stacking, the interaction at $\phi = 0^\circ$ is due to the side-by-side stacking with the π-like coupling of p_z-(and d-) orbitals. These overlaps together with the local maximum at $\phi = 30^\circ$ come from the components of the p_z-orbitals projected parallel and perpendicular to the bonding direction. The peak at $\phi = 60^\circ$ is due to the overlap between wavefunctions of C (central) and S atoms.

With these overlap integrals, the transfer energy t_{ij} can be directly obtained according to the relation of the extended Hückel method,

$$t_{ij} = \frac{1}{2} K (E_i + E_j) S_{ij} , \qquad (2.3)$$

where E_i is the ionization energy at the i-th site and K is a constant usually taken to be 1.75. The conversion factor from S_{ij} to t_{ij} cannot be determined exactly so that it is usually assumed to be about $-10\,\mathrm{eV}$.[2]

Experimental methods have been developed to determine the charge-transfer ratios δ in the crystals. For the TTF-TCNQ crystal, for example,

[2] This ambiguity makes it difficult to discuss band parameters such as band width, Fermi energy, and band mass quantitatively.

X-ray diffuse scattering and diffraction, Raman scattering, and X-ray photoemission spectroscopy played an important role in determining $\delta = 0.59$. In particular, the superlattice formation caused by the quasi-1D Fermi-surface instability provided important insight as will be described in detail in Sect. 4.4.1. In contrast to this non-stoichiometric charge transfer, stoichiometric transfer of $\delta = 1$ occurs in the 2:1 salts D_2X based on TMTTF, TMTSF, ET, BETS, and DCNQI molecules.

As a consequence of the charge transfer, some strain (deformation) is induced in the donor-molecular framework [83]. Figure 2.3 shows four independent bonds in the TTF-skeleton of the ET molecule. The oxidation of ET lengthens the carbon double bonds C=C (a, d) at central and side positions and shortens the S–C bonds (b, c). This effect is used to determine the quantity of charge localized on the donor due to charge-ordering phenomena [84], which will be described in Sect. 4.4.3.

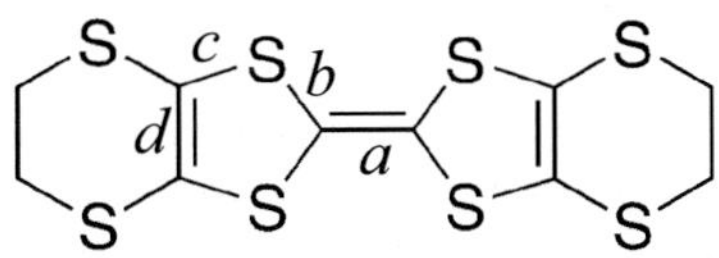

Fig. 2.3. BEDT-TTF (ET) molecule and the chemical bonds; a and d for central and side C=C, b and c for S–C. For $\delta = 0$, $a = 1.343(4)$ Å, $b = 1.756(3)$ Å, $c = 1.760(3)$ Å, $d = 1.333(4)$ Å. For the empirical rule between the charge δ and the strain $(b + c) - (a + d)$, see [83, 84].

3 Structural Aspects

3.1 Molecular Building Blocks

Building blocks in ordinary non-magnetic metals are atoms. Electrons of s or p character contribute to the quasiparticle states at the Fermi surface which support the transport of charges through the crystal. In relatively simple metallic compounds like sodium or potassium, the constituent atoms each provide an outer-shell s-electron, which, in turn, becomes the conduction electron, and can be considered as a gas of nearly free carriers. *Free electrons* are described by the simple Drude model where the dc-conductivity (inverse resistivity) is expressed as

$$\sigma = \rho^{-1} = \frac{ne^2\tau}{m} = ne\mu. \tag{3.1}$$

Here n denotes the free carrier density, e the electronic charge, τ the relaxation time, and m the electron mass. $\mu = e\tau/m$ is the carrier mobility which is a measure of how easily the charge carriers can be accelerated by an applied electrical field. Equation (3.1) is applicable to *conduction electrons* also, when σ denotes the conductivity tensor and n the electron number density depending on the shape of the Fermi surface and direction of acceleration. The electron mass m then has to be replaced by m^*, the effective carrier mass which is inversely proportional to the curvature of the energy band $E(\boldsymbol{k})$, $m^* \propto [\partial^2 E(\boldsymbol{k})/\partial k^2]^{-1}$. In simple metals the Fermi surface — reflecting the anisotropy of the crystal structure — is of spherical shape. This results in a more or less isotropic electrical conductivity.

The organic charge-transfer conductors, on which we focus in this monograph, are chemically more complicated due to the fact that their building blocks are complex molecules rather than atoms. Owing to the large difference between the bonding energies within the molecules (chemical bonds) and the weak intermolecular potentials, the constituent parts of the system may be viewed as semi-rigid bodies rather than points. This is reflected in the corresponding vibrational frequencies which consist of *intra*-molecular motional degrees of freedom of the constituent atoms within the molecules as well as rigid-body motion of the molecular entities themselves, i.e. *inter*-molecular motions or rotations, see Sect. 4.3.1. This is one of the most distinct features intrinsic to crystalline molecular conductors. As will be discussed below, the building-block composition of the present organic materials allows for a vast variety of systems with different physical properties derived from different packing schemes of donor-radical cations [28]. Additionally, the building-block composition results in (i) a distinct anisotropy of the crystal structures which is reflected in highly anisotropic Fermi surfaces, and related physical properties, as e.g. electrical transport. (ii) Relatively soft crystal lattices consisting of weakly-bonded molecular entities which make the physical properties highly susceptible to changes in external parameters such as temperature or pressure.

As described in Chapt. 2, in the present materials, unpaired carriers are created by a partial transfer of charge between the two constituent parts of a charge-transfer complex: an organic electron-donor molecule D is combined with an — inorganic, organometallic, or organic — electron-acceptor complex X, forming a system D_mX_n, where m and n are integers, see (2.1). Organic conductors have been derived from a variety of different organic electron-donor molecules D. Figure 3.1 shows a selection of important organic molecules furnishing charge-transfer complexes, discussed in this monograph. Most of them are derivatives of the archetype TTF molecule. Its combination with the electron acceptor TCNQ in 1973 led to the synthesis of the first quasi-1D organic conductor TTF-TCNQ [30, 31]. This material and its derivatives served for a long time as model systems for exploring the physical properties of quasi-1D conductors.

Organic and organometallic chemistry have provided an enormous number of donor molecules which may serve as building blocks for organic conductors and superconductors. Nearly all the organic superconductors known to date are composed of — both symmetrical and asymmetrical — tetrachalcogenafulvalenes derivatives such as TMTSF, TMTTF, BEDT-TTF, BEDT-TSF, BEDO-TTF, DMET, and MDT-TTF, see Fig. 3.1. In the vast majority of D_mX_n salts, the donor-acceptor molecular ratio $m : n$ is fixed to $2 : 1$, i.e., two donor molecules transfer one electron to the acceptor X. In most cases, the negatively charged anions adopt a closed-shell configuration and thus do not contribute to the electrical conductivity. TMTSF differs from TMTTF in that the four sulfur atoms of the latter have been replaced by selenium. In the same way, BEDT-TSF (or simply BETS) is derived from BEDT-TTF (ET). It is interesting to note that organic superconductors have been synthesized not only by using these symmetric molecules, but also on the basis of asymmetric complexes, as e.g. DMET [49, 50, 85], which is a hybrid of TMTSF and ET, see [86] for an overview. The various synthetic methodologies used in the preparation of tetrachalcogenafulvalenes have been intensively studied, see e.g. [28, 87].

Besides the peculiarities in the vibrational excitations and the pronounced anisotropy stemming from the building-block composition, the carrier density n of organic charge-transfer salts can intuitively be expected to be small

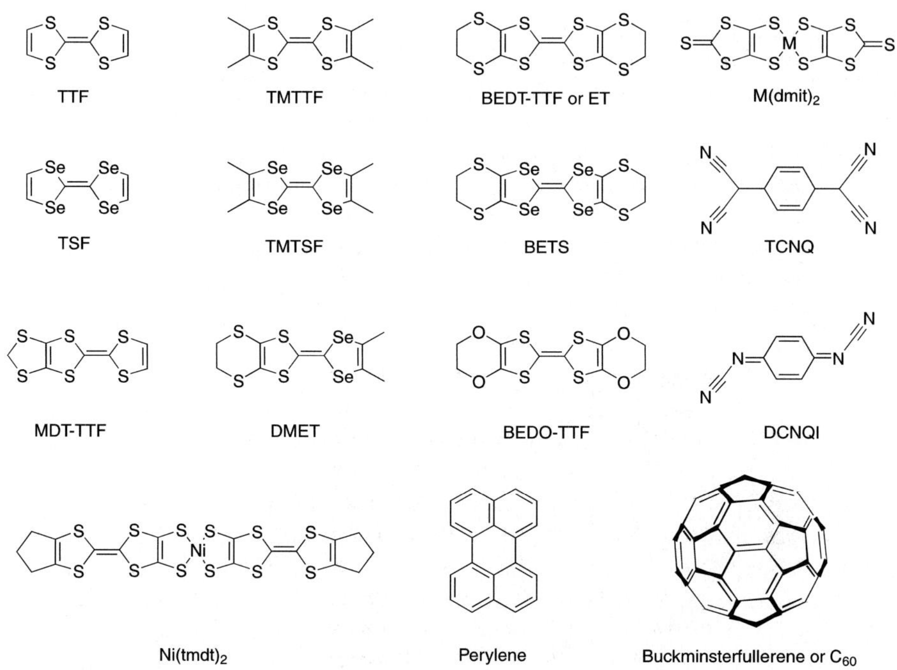

Fig. 3.1. Selected molecules that furnish organic conductors.

since the volume of the organic molecules providing the conduction electrons is much larger than that of an atom in an ordinary metal. The volume of the unit cell divided by the number of formula units contained in the unit cell of some typical $(ET)_2X$ salts is in the order of $V \sim 850 \times 10^{-24}\,\mathrm{cm}^{-3}$, cf. Table 3.1, leading to carrier densities in the order of $n \sim 1.2 \times 10^{21}\,\mathrm{cm}^{-3}$, which is much lower than that of a metal like Cu: $n_{\mathrm{Cu}} \sim 8.5 \times 10^{22}\,\mathrm{cm}^{-3}$. Also the carrier mobility $\mu = e\tau/m^*$ can be expected to be smaller than that of a normal metal because the scattering probability $1/\tau$ is increased due to both the narrow conduction bands, which means a larger density of available momentum states at the Fermi level to scatter into and the large number of vibration modes contributing to the scattering process: the unit cell of the typical salt κ-$(ET)_2Cu[N(CN)_2]Br$ with four formula units ($Z = 4$) contains 236 atoms and the vibrational spectrum consists of 708 phonon branches. These are composed of 576 internal modes of the ET molecules, 132 (low-frequency, translational and librational) intermolecular modes and modes of the polymer-like anion chains including the three acoustic phonon branches [88, 89]. Indeed, the room-temperature intraplane conductivity of κ-$(ET)_2Cu(NCS)_2$, measured e.g. along the crystallographic b-axis, $\sigma_b = 16.7\,(\Omega\mathrm{cm})^{-1}$, is several orders of magnitude smaller than that of Cu, $\sigma_{\mathrm{Cu}} = 6 \times 10^5\,(\Omega\mathrm{cm})^{-1}$.

3.2 Crystal Structures

3.2.1 Quasi-One-Dimensional Compounds

As mentioned above, TTF-TCNQ is the prototype organic charge-transfer salt. It shows a relatively large conductivity down to about 60 K where a sharp transition into an insulating charge-density-wave ground state occurs, see e.g. [90] for a recent review. Figure 3.2 shows the crystal structure of this compound. Donor (TTF) and acceptor (TCNQ) molecules form homologous columnar stacks with interplanar (intrastack) spacings of 3.47 Å and 3.17 Å, respectively [91], so the overlap of the molecular orbitals is largest along the stacking b direction resulting in a one-dimensional electron dispersion. TTF and TCNQ molecular planes are set at an angle of 58.5° to one another. They are oppositely directed by 24.5° and 34.0°, respectively, relative to the axis $c^* = a \times b$, see Fig. 3.2.

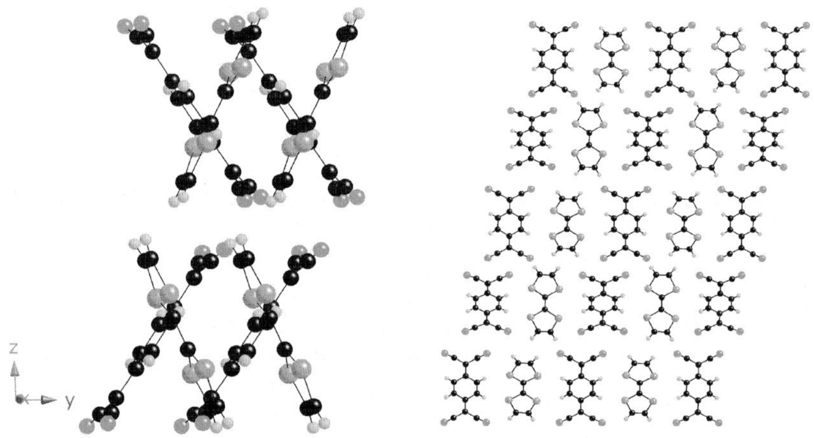

Fig. 3.2. Left: Crystal structure of TTF-TCNQ showing the angular relation between TTF and TCNQ molecules. Right: Top view of the layered structure. For the lattice parameters, see Table 3.1. Courtesy of J.A. Schlueter, Argonne National Laboratory.

In $(TMTSF)_2X$ the charge carriers are generated by electron transfer between two donor molecules TMTSF and an anion X^-, whereas in TTF-TCNQ free carriers are generated in both columns by partial charge transfer between TTF and TCNQ. TMTSF provides the basis for the so-called Bechgaard salts $(TMTSF)_2X$ which form with a variety of inorganic monovalent octahedral or tetrahedral acceptor molecules X. Indeed, it was $(TMTSF)_2PF_6$ where in 1980 superconductivity had been observed for the first time in an organic compound [33]. At ambient-pressure conditions, the material was found to undergo a metal-insulator transition around 12 K which had been identified as a spin-density-wave (SDW) ordering [92]. By the application of hydrostatic pressure of 12 kbar, the SDW instability can be suppressed and superconductivity forms below $T_c = 0.9\,\mathrm{K}$ [34, 33]. A replacement of PF_6 by ClO_4 has resulted in the first and — till now — only member of the quasi-1D salts which becomes superconducting at ambient pressure [36]. Figure 3.3 exemplarily shows the crystal structure of $(TMTSF)_2PF_6$. The lower part of

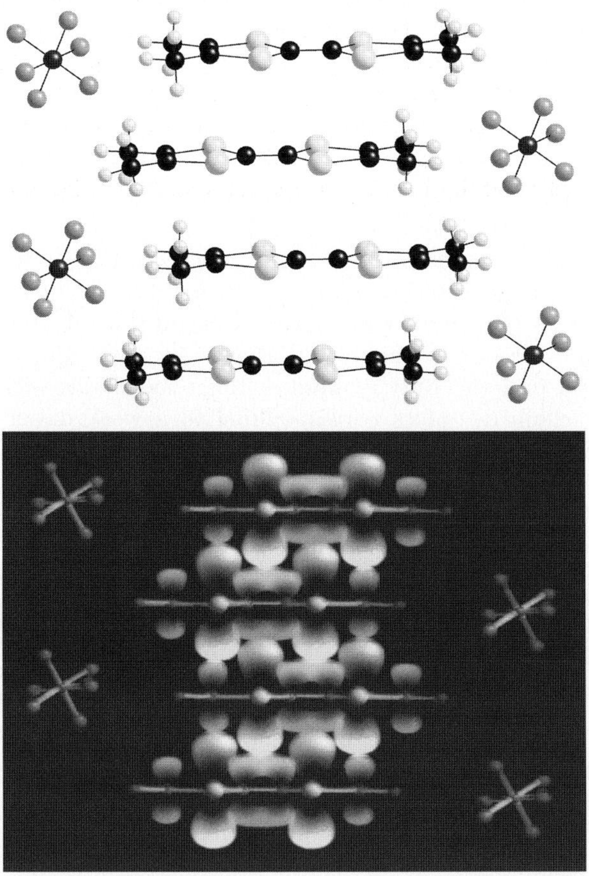

Fig. 3.3. Crystal structure of $(TMTSF)_2PF_6$ viewed from somewhat tilted angle relative to the b direction (top) and schematics showing the π-molecular orbitals of TMTSF (bottom, after [93]). The axis of highest conductivity is the a-direction, where the molecular orbital's overlap is largest, and c, the least-conducting direction, respectively. Courtesy of J.A. Schlueter, Argonne National Laboratory.

Fig. 3.3 schematically shows the molecular orbitals of the TMTSF molecules. As described in Sect. 2.2, π-molecular orbitals, formed from p_z-atomic orbitals extending perpendicularly to the planes of the donor molecules, have low binding energies — much lower than those of the σ-electrons. Hence, they can easily be excited, as can be seen in the charge-transfer process. Accordingly, the charge carriers thus created have a π-electron (hole) character. Prerequisites for high metallic conductivity are both the creation of a sufficiently large number of carriers by charge transfer and their delocalization throughout the crystal, i.e. a sufficiently high mobility. When the crystal is formed, the latter criterion may be fulfilled by a dense packing of the donor molecules in segregated molecular stacks. The most advantageous features of TTF-type donor molecules are that the HOMO has a large amplitude and the same sign on every chalcogen atom, see Fig. 2.1. The spacing between the molecules along the stacking direction is shorter than the sum of the van der Waals radii of the Se atoms having large electron clouds, hence the π-orbitals of the partially-filled outer molecular shells overlap and an electronic band structure is formed. As these bands are formed by the overlap of molecular orbitals of adjacent, more or less flat molecules, it is obvious that their relative angle and overall spatial arrangement give rise to directional dependent electronic properties which can lead to quasi-1D and quasi-2D conductivity behavior.

All members of the $(TM)_2X$ family, where TM stands for TMTSF and TMTTF, are isostructural with triclinic symmetry. From Fig. 3.3, it is clear that the most-conducting direction is along the stacking a-axis and the least-conducting direction the c-axis, where the anions, which are located at inversion centers of the lattice, separate the conducting stacks in the ab-plane. The donor molecules, which are nearly planar and almost perpendicular to the chain axis, are arranged in a zigzag-type manner along the chains with two slightly different intrachain distances corresponding to a weak dimerization. The fairly close Se $\cdots$ Se distance of 3.87 Å along the b-axis, being smaller than the sum of the van der Waals radii of 4 Å, results in a weak inter-chain overlap [92, 90] and thus a weakly 2D electronic character. An important structural aspect of the $(TM)_2X$ compounds is related to the symmetry of the anion X. Where these anions are centrosymmetric, such as the octahedral $X = PF_6$, AsF_6, SbF_6 or TaF_6 complexes, their orientation is fixed in the structure. In contrast, $(TM)_2X$ salts formed with non-centrosymmetric tetrahedral anions such as ClO_4, ReO_4, and BF_4 or triangular NO_3 undergo a structural transition from a disordered high-temperature state to an ordered low-temperature phase [94]. High-pressure experiments on the isostructural family based on the sulfur molecule TMTTF with the same series of anions showed that $(TMTSF)_2X$ and $(TMTTF)_2X$ salts both belong to the same $(TM)_2X$ family, the properties of which can be summarized in a generic phase diagram [90], see Sect. 4.6.

As will be discussed in detail in Sect. 7.1, one way of designing highly conducting *magnetic* salts is to incorporate transition-metal ions or complexes into the crystal structure. This enables the study of the interactions between conducting π-electrons and those localized magnetic moments resulting from partially-filled d-shells of the transition metals. In 1986 S. Hünig and his collaborators synthesized $(DCNQI)_2Cu$, the first quasi-1D metal with π-d interactions (Fig. 7.1), see e.g. [39] for a recent review.

3.2.2 Quasi-Two-Dimensional Compounds

Apparently, the strong tendency of the Bechgaards salts to undergo a metal-insulator transition, inherent to quasi-1D electron systems, counteracts with their ability to become superconducting. In order to achieve ambient-pressure superconductivity in these organic complexes with potentially high transition temperatures, it is necessary to increase their dimensionality. According to this strategy, in 1982 the first metallic compound based on the new electron donor molecule ET [87], cf. Fig. 3.1, was synthesized by G. Saito et al. [95]. The underlying idea was to enhance the overlap between π-orbitals of adjacent molecules by enlarging the π-electron system on each molecule. This has been accomplished by adding rings of carbon and sulfur atoms at the outer ends of the TTF skeleton.

The electronic interactions in the present materials are determined not only by the distance between adjacent molecules but also by their spatial arrangement. In contrast to the Bechgaard salts, where the donor molecules form infinite stacks, steric effects specific to the ET molecules prevent such an infinite face-to-face stacking in the $(ET)_2X$ salts. As a consequence, the side-by-side overlap between π-orbitals of adjacent molecules becomes stronger and, in some cases, comparable to the face-to-face interaction resulting in a quasi-2D electronic structure of the ET-based salts, see Sect. 2.2. The intermolecular electronic interactions are mostly determined by the hetero-atoms' S or Se contacts, both due to their larger size compared to the C atoms and their location at the periphery of the molecule [28]. The combination of ET with the monovalent anion X = $Cu(NCS)_2$ achieved in 1988 led to the discovery of an ambient-pressure superconductor with a transition temperature in the range of 10 K [96].[1]

While all members of the $(TM)_2X$ family share the same crystal structure, the rather loose intra-stack coupling of the $(ET)_2X$ salts gives rise to a variety of polymorphic phases (packing motifs) which are distinguished by Greek characters; the most important amongst them are the α-, β-, κ-, θ-, and λ-phases, cf. Fig. 3.4. For a comprehensive review of the structural properties of these salts, see, e.g. [98, 99, 100]. In some cases, as for instance realized in the compound with the linear anion X = I_3, various structural modifications exist even for the same anion, i.e. α-, β-, κ-, θ-$(ET)_2I_3$, see [101] for a recent discussion of $(ET)_2X$ systems based on these trihalide anions. The α-type structure consists of stacks arranged in a herring-bone pattern. In α-$(ET)_2I_3$, the ET molecules are connected via S $\cdots$ S contacts being shorter than the sum of the van der Waals radii of 3.6 Å. The θ-type structure also consists of linear stacks of ET molecules. The β-type packing is reminiscent of the stacking arrangement found in the Bechgaard salts. However, the smaller inter-stack distances in β-$(ET)_2I_3$, lead to a more two-dimensional electronic

[1] The first ambient-pressure superconductor in this class of materials — the second generation of organic superconductors — was β-$(ET)_2I_3$ [97].

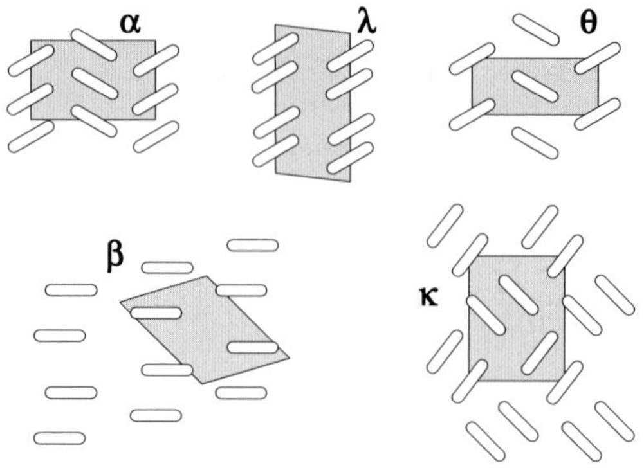

Fig. 3.4. Schematic packing motifs showing the spatial arrangement of the ET molecules in the α-, β-, κ-, θ-, and λ-phases of $(ET)_2X$ viewed along the long axis of the ET molecules (white ellipses). The gray areas represent the unit cells.

structure. The κ-phase is unique in that it does not consist of interacting stacks but rather of interacting dimers formed by two face-to-face aligned ET molecules. Adjacent dimers are arranged almost orthogonal to each other so that the *intra-* and *inter*-dimer interactions are of the same size. This results in a quasi-2D electronic structure with a small in-plane anisotropy. The κ-type compounds with polymere-like anions are of particular interest with respect to their superconducting properties as they exhibit the highest transition temperatures.

The neutral ET molecule $C_6S_8[(CH_2)_2]_2$ [102] is slightly distorted from planar geometry. However, in forming the crystal, apart from the ethylene $[(CH_2)_2]$ groups at the outer ends of the molecules (see Sect. 3.3), the charged ET molecules untwist at their center and become more planar. The planar C_6S_8 skeleton of the ET molecules permits a rather dense packing with a variety of possible packing arrangements. As a result, the interdimer interaction becomes comparable to that within the dimers giving rise to a quasi-2D electronic structure. Besides the intermolecular $S \cdots S$ contacts, i.e. the donor-donor interaction, the donor-acceptor couplings also play an important role for the physical properties of these multilayer systems. The latter interaction is provided by electrostatic forces as a consequence of the charged molecules and the hydrogen bonds joining between the carbon atoms at the donor site and the sulfur, carbon or nitrogen atoms being located at the acceptor site. The relative strength of these different interactions, the conformational degrees of freedom of the ethylene groups (see Fig. 3.12) along with the flexibility of the molecular framework give rise to a variety of different ET complexes [86, 98, 99, 100].

κ-(BEDT-TTF)$_2$X with Polymeric Anions

Despite their complex crystal structure with rather low symmetry, it is convenient to think of the $(ET)_2X$ compounds as layered systems consisting of conducting sheets formed by the ET molecules, which are intersected by more or less thick insulating anion layers, see Fig. 3.5. Prime examples are the κ-phase

$(ET)_2X$ salts with $X{=}Cu(NCS)_2$, $Cu[N(CN)_2]Br$ and $Cu[N(CN)_2]Cl$ which are the most intensively studied and best characterized members of this class of materials. These compounds are of particular interest, not only because of their relatively high superconducting transition temperatures but also owing to certain similarities in their normal- and superconducting-state properties with those of the high-temperature cuprate superconductors [103, 104]. Figures 3.5 and 3.6 display the crystal structures of κ-$(ET)_2Cu(NCS)_2$ and κ-$(ET)_2Cu[N(CN)_2]Z$, respectively. In both cases, the layered structure con-

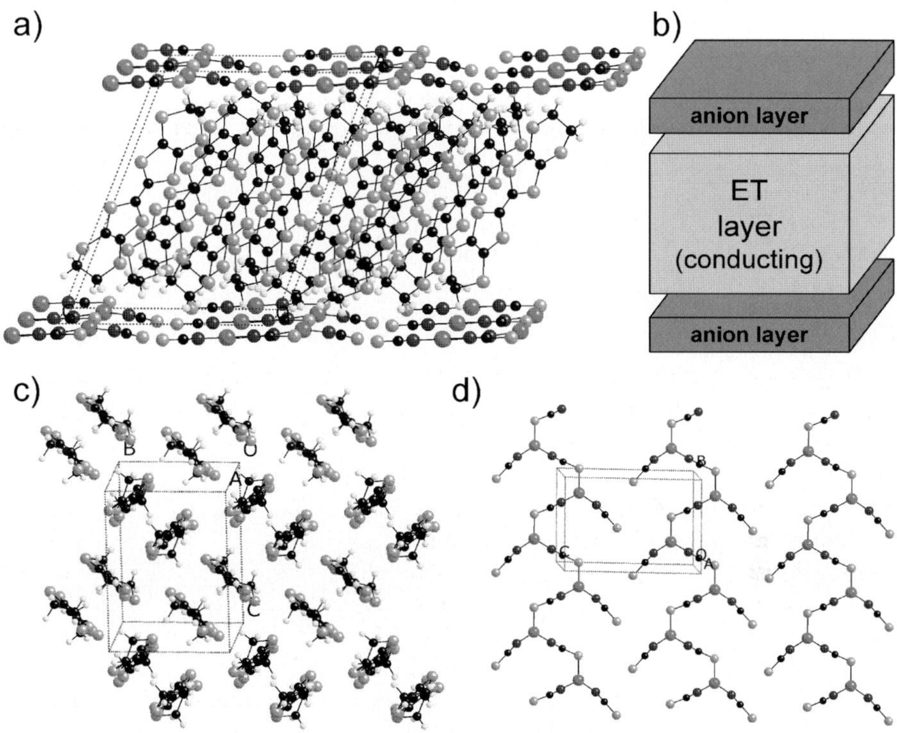

Fig. 3.5. (a): Crystal structure of κ-$(ET)_2Cu(NCS)_2$. The unit cell is indicated by dashed lines. (b): The conducting ET layers are separated by thin insulting anion layers, schematically shown on the right side. (c): Arrangement of the ET molecules, and (d): the anion layer of κ-$(ET)_2Cu(NCS)_2$ viewed along the a^*-direction, i.e. perpendicular to the conducting planes. The a-axis is slightly tilted from the a^*-axis which is normal to the conducting bc-plane. The short distances between the sulfur atoms of different ET molecules result in a good conductivity within the ET layers. Courtesy of J.A. Schlueter, Argonne National Laboratory.

sists of conducting planes with the characteristic κ-type arrangement of the ET molecules (see Fig. 3.5) separated by insulating anion layers. While the crystal structure of κ-$(ET)_2Cu(NCS)_2$ has monoclinic symmetry with two

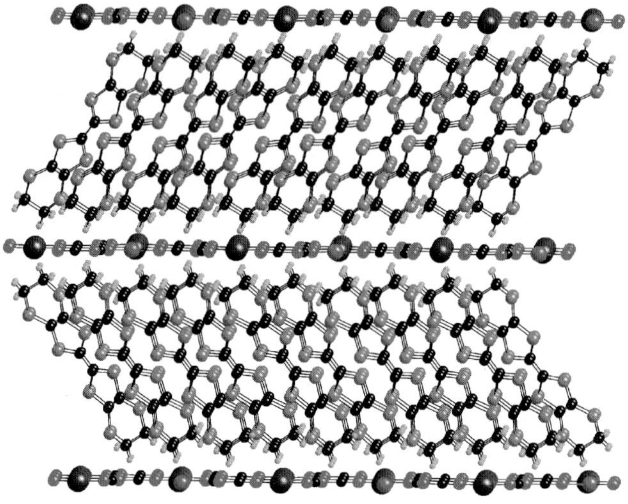

Fig. 3.6. Crystal structure of κ-$(ET)_2Cu[N(CN)_2]Z$ with Z = Br and Cl. The direction perpendicular to the conducting plane is the crystallographic b-axis. The anion layers are parallel to the ac-plane. The polymeric-like anion chains are running along the a-direction. Courtesy of J.A. Schlueter, Argonne National Laboratory.

dimers, i.e. two formula units per unit cell, the κ-$(ET)_2Cu[N(CN)_2]Z$ salts are orthorhombic with a unit cell containing four dimers, see Table 3.1. Due to the particular polymeric arrangement, the anions lack a center of inversion symmetry. Subtle changes in the intermolecular spacing or relative orientation of the ET molecules, as e.g. induced by either external pressure or anion substitution, may significantly alter the π-electron overlap between adjacent molecules. This can have a severe influence on the electronic properties as demonstrated for the κ-$(ET)_2Cu[N(CN)_2]Z$ system for various Z: while the compound with Z = Br is a superconductor with $T_c = 11.6$ K [46], replacement of Br by the slightly *smaller* Cl results in an antiferromagnetic insulating ground state. On the other hand, the application of hydrostatic pressure of only about 300 bar drives the latter system to a superconductor with a T_c of 12.8 K [105, 106, 107], the highest transition temperature found among this class of materials so far.

α-(BEDT-TTF)$_2$MHg(SCN)$_4$

The isostructural systems α-$(ET)_2MHg(SCN)_4$, with M = K [108], Rb [109], Tl [110], and NH_4 [111] are strongly two-dimensional systems due to the thick anion layer which consists of three sub-layers. This is reflected in a large ratio of the electrical conductivity parallel and perpendicular to the layers, $\sigma_{\|}/\sigma_{\perp}$, of about $10^5 - 10^6$ [112]. The thickness of the anion layer of about 7 Å is almost comparable to the length of the donor ET molecules of 11.8 Å measured between the carbon atoms of the ethylene endgroups [86]. The anion layer exhibits a complex structure itself, whereas M^+ and Hg^{2+} ions are embedded between layers of SCN^- units, see Fig. 3.7. This leads to a strong decoupling of the conducting layers.

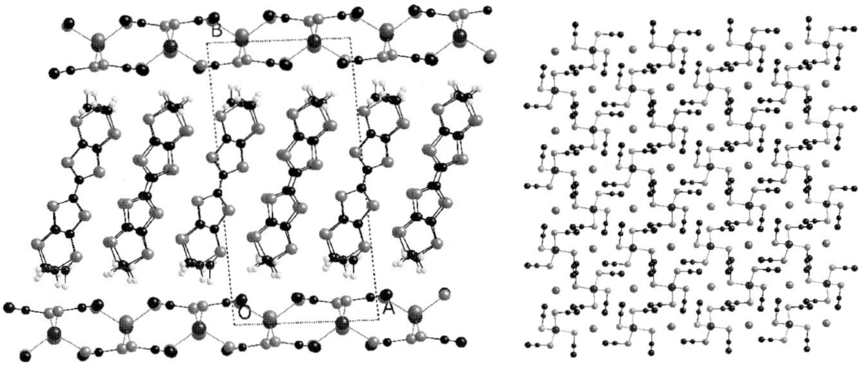

Fig. 3.7. Crystal structure of α-$(ET)_2MHg(SCN)_4$, with M = K, Rb, Tl, and NH_4 (left, note the thick anion layer) and top view on the molecular arrangement of the anion layer (right). Courtesy of J.A. Schlueter, Argonne National Laboratory.

The systems with M = K, Rb, and Tl show a magnetic transition around 8 – 10 K which has been discussed in terms of a spin- or charge-density-wave transition, see Sects. 4.4 and 5.3.2. The system M = NH_4 becomes superconducting below about 1 K [111, 113]. Uniaxial stress perpendicular to highly conducting layers, i.e. along the b^*-axis induces superconductivity in the density-wave compound M = K and increases T_c in the NH_4 system [114, 115]. This behavior is in line with a pressure-induced increase in the interlayer coupling strength, although an exact explanation for the effect might be more complicated.

β''-(BEDT-TTF)$_2$SF$_5$CH$_2$CF$_2$SO$_3$

In the systems discussed so far, the radical anions are generated oxidatively *in situ* during the electrochemical crystallization process which is difficult to control in all details (see below). New possibilities arise from counterions which are introduced already in the form of anions with a well-defined charge state. Such anions can be relatively large and have a discrete rather than a polymeric structure. It is possible to synthesize families of isostructural salts whose structure/physical-properties correlations can be investigated. First promising candidates were systems with organometallic anions $M(CF_3)_4^-$ with M = Cu, Ag, Au [116, 117] which crystallize in the κ-phase and show relatively high T_c values up to 10 K [86]. A problem in studying these systems is that the anion contains solvent molecules whose volatility gives rise to undesirable time dependencies. A further step towards new materials was taken with the discovery of β''-$(ET)_2SF_5CH_2CF_2SO_3$. This system is unique as it is the first superconducting charge-transfer salt ($T_c \sim 5$ K) which is completely organic, i.e. even the anion layer does not contain any metal atoms

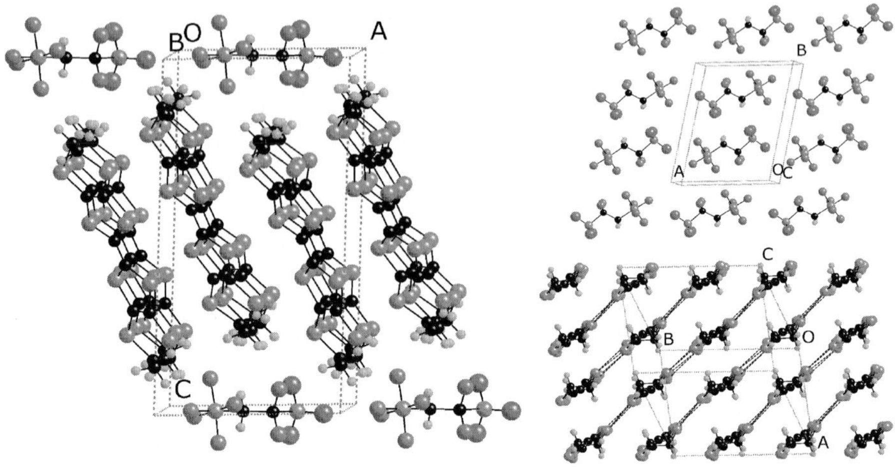

Fig. 3.8. Crystal structure of β''-$(ET)_2SF_5CH_2CF_2SO_3$ (left), structure of the discrete anions (top right), and packing of the electron-donor radical cations within the conducting layer (bottom right, inter-molecular contacts shorter than the sum of the van der Waals radii are indicated by dashed lines). Boxes indicate the unit cell. Courtesy of J.A. Schlueter, Argonne National Laboratory.

[118, 117, 119]. The system is of particular interest as it is a text-book example for a two-dimensional metal, whose physical properties have been studied extensively via transport, magnetic, and thermodynamic measurements [119], see also Sect. 5.3.3. Its layered crystal structure [118] contains tilted stacks of formally $+1/2$-charged cations of the electron donor in a typical β''-type packing[2] except for a doubling of the transverse axis due to the space requirements of the large anion, see Fig. 3.8. The anion layer is relatively thick and consists of discrete complexes $SF_5CH_2CF_2SO_3$.

λ-(BETS)$_2$Fe$_x$Ga$_{1-x}$Cl$_4$

A new class of materials which has recently gained much interest is based on the donor molecule BETS and its combination with the discrete anions MX_4^- (M = Fe, Ga, In, Co, Ni, Mn, ...; X = Cl, Br). The idea behind the examination of BETS conductors was to develop quasi-2D systems where the π-type carriers can interact with $3d$ magnetic moments localized on the anion site as in FeX_4^- and to compare such systems with isostructural materials containing non-magnetic anions such as GaX_4^- [67]. As described in Sect. 7.2.1, two structural modifications have been found: the orthorhombic κ-type structure (Pnma) which results in plate-like crystals and the triclinic ($P\bar{1}$) λ-type

[2] The β''-packing differs from the β-type structure (Fig. 3.4) in a slight tilting of each ET molecule within the stacks.

variant which grows in a needle-like manner [120, 121, 122, 67], see Figs. 7.5 and 7.6. The λ-$(BETS)_2GaCl_4$ salt is a superconductor with $T_c = 6\,K$ [123]. Upon substituting Ga by Fe in λ-$(BETS)_2Fe_xGa_{1-x}Cl_4$, superconductivity becomes continuously suppressed with increasing x [123, 124, 125] and, for $x \geq 0.5$, replaced by an antiferromagnetic insulating ground state. In λ-$(BETS)_2FeCl_4$ and κ-$(BETS)_2FeBr_4$ field-induced superconductivity at high magnetic fields has been observed, see Sect. 7.2.6.

3.2.3 Three-dimensional Compounds

Prototypes of molecular compounds with a 3D electronic structure are the alkali-metal-doped fullerenes A_3C_{60}, where e.g. K_3C_{60} and Rb_3C_{60} are superconductors with T_c values of about 20 K and 30 K, respectively. The discussion of fullerenes is beyond the scope of this monograph. However, similar to the materials discussed so far are single-component molecular metals which also exhibit 3D intermolecular interactions.

Single-Component Molecular Metals

The materials discussed above are based on the intermolecular charge transfer between the donor molecules forming the conduction band and the charge compensating, "electrically inert" anions. In TTF-TCNQ, both the donor and acceptor form conduction bands. One important target in the field of molecular conductors is the realization of a single-component molecular metal,

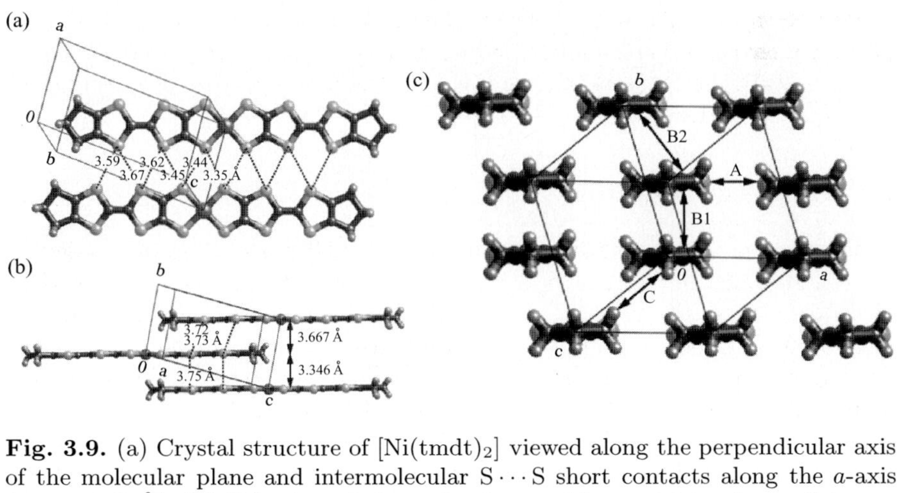

Fig. 3.9. (a) Crystal structure of $[Ni(tmdt)_2]$ viewed along the perpendicular axis of the molecular plane and intermolecular S···S short contacts along the a-axis ($3.35 - 3.67\,\AA$). (b) Side view of the molecular stacking and interplanar distances with S···S short contacts. There are two types of overlap modes, and in both cases almost half of the molecules are overlapped. (c) End-on projection of $[Ni(tmdt)_2]$. After [72].

Table 3.1. Room-temperature crystallographic data of some organic conductors/superconductors including the space group SG, lattice parameters and corresponding angles, unit-cell volume V, number of formular units per unit cell z, as well as the superconducting transition temperature T_c (for superconducting salts or where external pressure induces superconductivity). In the case of quasi-2D salts the lattice parameter perpendicular to the highly-conducting planes is underlined. (S/C)DW denote a (spin/charge)-density-wave type ground state, SC superconductivity, and (AF or CO)I an (antiferromagnetic or charge-ordered) insulating state, respectively, and M a metallic ground state. Where superconductivity can be induced by external pressure, the critical pressure is given in parenthesis.

	SG	a (Å)	b (Å)	c (Å)	α (°)	β (°)	γ (°)	V (Å^3)	z	ground state	T_c (K)	Ref.
TTF-TCNQ	$P2_1/c$	12.298	3.819	18.468		104.46		839.9	2	CDW		[91]
$(TMTSF)_2PF_6$	$P\bar{1}$	7.297	7.711	13.522	83.39	86.27	71.01	713.1	1	SDW	1.1 (6.5 kbar)	[126, 127]
$(TMTSF)_2ClO_4$	$P\bar{1}$	7.266	7.678	13.275	84.58	86.73	70.43	693.6	1	SC	1.2 [§]	[127]
α-$(ET)_2I_3$	$P\bar{1}$	9.211	10.850	17.488	96.95	97.97	90.75	1690.3	2	I		[128]
β-$(ET)_2I_3$	$P\bar{1}$	6.615	9.097	15.291	94.35	95.55	109.75	855.9	1	SC	1.5 (8 at 1 kbar) [*]	[97]
β-$(ET)_2IBr_2$	$P\bar{1}$	6.593	8.975	15.093	93.79	94.97	110.54		1	SC	2.8	[129]
κ-$(ET)_2I_3$	$P2_1/c$	16.387	8.466	12.832		108.56		1687.6	2	SC	3.5	[43]
θ-$(ET)_2I_3$	$Pnma$	10.076	33.85	4.964				1693	2	SC (M) [♯]	3.6	[130]
κ-$(ET)_2Cu(NCS)_2$	$P2_1$	16.248	8.440	13.124		110.3		1688.0	2	SC	10.4	[44, 131]
κ-$(ET)_2Cu[N(CN)_2]Br$	$Pnma$	12.942	30.016	8.539				3317	4	SC	11.2	[46]
κ-$(ET)_2Cu[N(CN)_2]Cl$	$Pnma$	12.977	29.979	8.480				3299	4	AFI	12.8 (300 bar)	[105]
α-$(ET)_2KHg(SCN)_4$	$P\bar{1}$	10.082	20.565	9.933	103.7	90.91	93.06	1997	2	DW [†]		[109]
α-$(ET)_2NH_4Hg(SCN)_4$	$P\bar{1}$	10.091	20.595	9.963	103.65	90.53	93.30	2008.1	2	SC	1.1	[109]
β''-$(ET)_2SF_5CH_2CF_2SO_3$	$P\bar{1}$	9.260	11.635	17.572	94.69	91.70	103.10	1835.5	2	SC	5.3	[118]
κ_L-$(ET)_2Cu(CF_5)_4(TCE)$	$Pnma$	13.169	38.031	8.539				4276.4	4	SC	4 [‡]	[132]
θ-$(ET)_2RbZn(SCN)_4$	$I222$	10.175	43.301	4.65				2047	2	(CO)I		[133]
κ-$(BETS)_2GaBr_4$	$Pnma$	11.773	36.635	8.492				3662.6	4	SC	1	[121]
λ-$(BETS)_2GaCl_4$	$P\bar{1}$	16.172	18.616	6.607	98.38	96.75	112.56	1805.5	2	SC	5.5	[121]
$[Ni(tmdt)_2]$	$P\bar{1}$	6.376	7.359	12.012	90.384	96.688	103.587	534.7	1	M		[72]

[§] Slowly cooled to insure long-range anion ordering.
[*] A high-T_c state can be created by application of pressure of about 1 kbar where T_c increase discontinuously.
[♯] Not all crystals show superconductivity. Some stay metallic down to low temperatures without showing a phase transition.
[†] Superconductivity can be induced by uniaxial pressure applied perpendicular to the planes.
[‡] A minority phase κ_H grows with the same anion exhibiting a T_c value of 9.2 K.

see [72, 134] for a review. Recently, a first example of a crystal containing a neutral transition metal complex and extended-TTF ligands, [Ni(tmdt)$_2$] (tmdt denotes *trimethylenetetrathiafulvalenedithiolate*), exhibiting metallic behavior down to very low temperatures (see Fig. 1.1) has been prepared [70, 71]. Here, the carriers are generated by intramolecular electron transfer between HOMO and LUMO bands of the same molecule. This requires a very small HOMO-LUMO gap ΔE, i.e. small compared with the bandwidth $1/2 \cdot (W_{\mathrm{HOMO}} + W_{\mathrm{LUMO}})$, and sufficiently large intermolecular interactions (especially transverse interactions). These two conditions are satisfied by transition metal dithiolate complexes with extended-TTF ligands. Figure 3.9 shows the crystal structure of [Ni(tmdt)$_2$] which is the first 3D metal composed of single-component planar molecules as revealed by the crystal and band structure analysis, as well as resistivity [70, 71] and dHvA [135] measurements. The resistivity measured along the a-axis revealed metallic behavior down to 0.6 K and a room-temperature conductivity of 400 Scm^{-1} has been found, the latter value being almost a factor of 10 bigger than those for typical ET-based superconductors.

Crystal Growth by Electrochemical Crystallization

The quasi-1D and -2D systems tend to grow in needle- and plate-like shapes, respectively, although the morphology is not an indicator of the dimensionality of the system. For example, crystals of κ-(ET)$_2$Cu[N(CN)$_2$]Br can grow as either blocky crystals or rods, depending on crystallization conditions. The crystallization of organic molecular conductors can be divided into two steps: (1) the organic synthesis of the neutral donor molecules and the preparation of suitable electrolytes containing the desired charge compensating components and (2) the oxidation of the electron donor molecules in the presence of charge compensating anions resulting in the crystallization of cation-radical salts. For step (1) several independent methods have been used for the synthesis of electron-donor molecules, see e.g. [87]. Conversely, to perform step (2), the electrocrystallization technique has proven to be the method of choice for synthesizing high-quality crystals because the crystal growth rate can be controlled. In the following, we will briefly describe a common growth method for the κ-(ET)$_2$Cu(NCS)$_2$ system. For details on the chemical synthesis and crystal growth techniques, the reader is referred to [136, 28, 87, 86] and references cited therein. During the electrocrystallization process, the neutral donor molecules are oxidized to radical cations at a controlled rate. Figure 3.10 shows a schematic sketch and a photograph of a glass cell used to grow the organic superconductor κ-(ET)$_2$Cu(NCS)$_2$. The anode and cathode chambers are separated by an ultra-fine porosity glass frit. Platinum-wire electrodes are inserted into each chamber. An electrolyte solution, containing the desired-charge compensating anion in about a ten-fold excess, is added to each chamber. A neutral donor molecule, such as ET, is added to the anode solution. Typically, the anode solution is saturated with the neutral electron-donor molecule, with some undissolved donor remaining on the anode cell floor. The cell is then sealed under inert-gas atmosphere, such as nitrogen or argon. The platinum electrodes are then connected to a power supply. Generally, the cation-radical salts are grown under constant current conditions, but recent results have indicated that in certain cases, constant voltage can produce superior quality crystals. Crystals are typically grown in a dark, vibration- free environment at constant temperature. Ambient temperature is suitable for the growth of many crystals. However, elevated temperatures are sometimes required to achieve a sufficient concentration of donor molecules in solution. In general, lower crystallization temperatures will grow higher quality crystals, but the temperature must be high enough to achieve a minimum concentration of both electrolyte and neutral donor molecules at the electrode surface. When the radical cations, which are generated at the anode surface, combine with the dissolved counterions, the charge-transfer salts begin to crystallize (because of their lower solubility product) and black crystals begin to grow. Typically, these crystals grow on the anode surface, but it is not uncommon for these crystals to grow on the cell walls. The glass frit serves as a filter preventing the diffusion of the reduction products generated at the cathode to the working electrode. During a constant-current experiment, the voltage between the electrodes remains relatively stable. Crystals are harvested when either the voltage starts to change significantly, the color of the electrolyte changes, or the neutral donor molecules are used up. This process usually takes several days, weeks, or even a few months. Although electrocrystallization is a self-purifying process, clean electrochemical cells

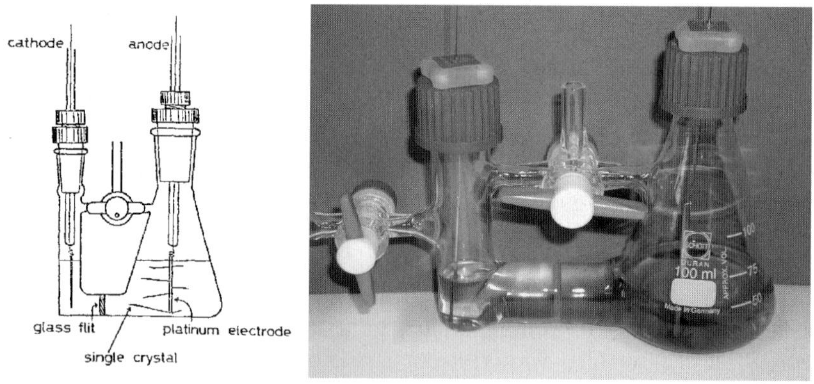

Fig. 3.10. Schematic sketch and photograph of an electrocrystallization cell. As a different design, H-type cells are often used. Single crystals of κ-$(ET)_2Cu(NCS)_2$ are formed at the anode (right electrode). Note the change in color of the solution between the cathode (left) and anode side (right) after growth time of several weeks. The glass frit serves as a filter which prevents the chemical reaction products on the cathode side from diffusing to the anode side.

and electrodes, high-purity inert gases and starting reagents are required to produce the highest-quality crystals. Other factors influencing crystal growth are the identity and concentration of the electrolyte, the surface roughness and type of electrode, as well as the temperature and current density. The optimal conditions can vary from system to system. To grow, for example, high-quality single-crystals of κ-$(ET)_2Cu(NCS)_2$, the following starting material can be used: pre-synthesized ET, KSCN, CuSCN, and 18-crown-6 ether (molar ratio 1 : 4 : 4 : 4 in a 90 % 1,1,2 trichloroethane/10 % ethanol solution). A possible anodal chemical reaction is:

$$ET^0 \longrightarrow ET^+ + e^-$$

$$ET^+ + ET \longrightarrow ET_2^+$$

$$KSCN + CuSCN + 18\text{-crown-6 ether} \longrightarrow (K^+ 18\text{-crown-6 ether})[Cu(NCS)_2]$$

$$ET_2^+ + [Cu(NCS)_2] \longrightarrow \kappa\text{-}(ET)_2Cu(NCS)_2 .$$

The advantage of the electrocrystallization method compared to a direct mixing of the starting material is a "clean" redox reaction where the growth rate of the crystals can be controlled within certain limits by adjusting external parameters such as temperature and current density. However, the reaction of the supporting electrolyte is often difficult to predict.

3.3 Anion Ordering and Glassy Phenomena

In discussing molecular conductors and superconductors, an important issue which should not be overlooked is disorder and its possible implications on the electronic properties, see Sect. 6.3.3 below. In this respect, we have to distinguish between different kinds of imperfections. The *extrinsic* disorder, i.e. impurity concentrations, contaminations or crystal defects can be vastly controlled in the preparation process although some aspects remain puzzling, see e.g. the discussion on the resistivity maximum in Sect. 4.3.1. In a study of the alloy series β-$(ET)_2X$ with $X = (I_3)_{1-x}(IBr_2)_x$ $(0 \leq x \leq 1)$ — where the salts with the two limiting compositions with $x = 0$ and $x = 1$ are superconductors — a clear correlation between the residual-resistivity ratio (RRR) and T_c was found [137]. These experiments show that superconductivity is very sensitive to the induced random potentials which lead to electron localization. The effect of random potentials created by radiation damage effects — resulting in a suppression of superconductivity — has been studied for the Bechgaard salts, β-$(ET)_2I_3$, and very recently, for the κ-$(ET)_2Cu(NCS)_2$ [138]. For more details, the reader is referred to [86] and references therein.

$(TM)_2X$ Salts

However, certain kinds of *intrinsic* disorder are unavoidable and can be of particular importance for experiments attempting to explore superconducting-state properties. The latter type of imperfections concern materials where, by symmetry, certain structural elements can adopt one of two possible orientations which are almost degenerate in energy [139, 140] and correspond to short and long contacts between the Se (S) atoms of the donor TMTSF (TMTTF) molecule and a peripheral electronegative atom of the anion [90]. This can be seen in the $(TM)_2X$ salts with non-centrosymmetric anions such as tetrahedral ClO_4, ReO_4, NO_3, SCN, and FSO_3. As a result, these anions are disordered at room temperature with an equal occupation for both orientations. Upon cooling, entropy is gained by a more or less perfect ordering of the anions, depending on how fast the system is cooled through the ordering temperature T_{AO}. A perfect long-range anion ordering, realized to a good approximation when cooled sufficiently slowly, then introduces a new periodicity of the lattice. Depending on the anion, this can have quite different implications on the electronic properties: for the $(TMTSF)_2ClO_4$ salt, for example, the anion ordering below 24 K is accompanied by a doubling of the periodicity along the *b*-axis, i.e. perpendicular to the stacking axis which leaves the conducting properties almost unaffected. In contrast, the anion ordering for $(TMTSF)_2ReO_4$ opens up a large gap at the Fermi level leading to an insulating ground state. If the compound had been cooled quickly through T_{AO}, a disordered *quenched* state is adopted at low temperatures whose properties are quite different from the *relaxed* state obtained after slow cooling.

The left panel of Fig. 3.11 shows the effect of anion ordering on the salt $(TMTSF)_2ClO_4$. For a slowly-cooled system, the anion-ordering phase transition occurs at $T_{AO} = 24\,K$. This is accompanied by a decrease of the resistivity due to the reduction of scattering by randomly distributed anion potentials. Upon further cooling, superconductivity sets in with $T_c = (1.2 \pm 0.2)\,K$. In contrast, for a crystal cooled rapidly, the high-temperature disordered state becomes quenched and the system undergoes a metal-insulator transition at $T_{MI} = 6.05\,K$. The insulating quenched state has been identified via NMR and ESR measurements [143, 144] as a SDW state with an energy gap, $\Delta_0/k_B T_{SDW} = 3.64$ close to the mean-field value of 3.52 [142]. The phase transition at T_{SDW} has been explored more recently by specific heat measurements [145]. Also from these measurements, a value of $\Delta S = R \ln(4/3)$ has been estimated for the entropy released at the anion-ordering transition, where $R = 8.314\,J/(molK)$ denotes the gas constant, which is below the value of $R \ln 2$ expected for an order-disorder transition with two possible configurations for the ClO_4 tetrahedron in the high-temperature disordered state. This has been interpreted as a generally incomplete ordering at the transition [145]. The right panel of Fig. 3.11 shows that the low-temperature resistivity may change over several orders of magnitude depending on the thermal his-

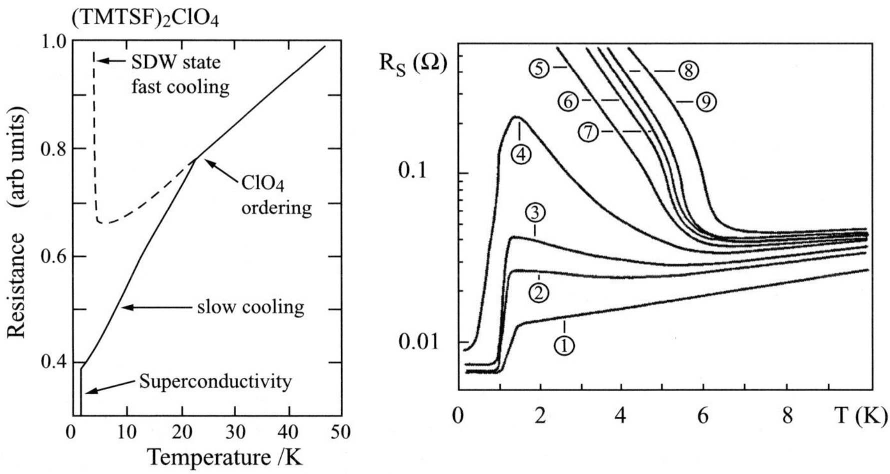

Fig. 3.11. Left panel: Effects of anion ordering in $(TMTSF)_2ClO_4$ at $T_{AO} = 24\,K$, on the resistivity, from [141]. Right panel: Semilogarithmic plot of the resistance vs. temperature in states with various degrees of disorder characterized by T_Q numbered from 1 to 9 (see text), with 1: $T_Q = 0$ (relaxed state), 2: $T_Q = 22\,K$ to 9: $T_Q = 35\,K$; after [142].

tory of the sample. Intermediate states with various degrees of frozen anion disorder have been produced by warming the sample up to 40 K where the high-temperature disorder is fully developed, then slowly cooling to T_Q and later quench cooling from T_Q to liquid-helium temperature. When $T_Q \geq T_{AO}$, the degree of disorder increases with T_Q: an intermediate state with only a partial degree of anion ordering is produced where superconducting and SDW domains coexist in the sample [142]. Specific heat measurements showed that there is no difference between the specific heat anomalies at T_{AO} of slowly-cooled and those of quenched-cooled samples, i.e. the structural transition is independent of the kinetic conditions. From these results, the authors concluded that the reordering transition of the anions is of second order and occurs within the experimental timescales [145]. This is contrary to what would be expected for a glass-like transition. Such a glassy behavior, however, has been observed at the ordering temperatures of the compounds with X = ReO_4 ($T_{AO} = 180\,K$) and FSO_3 ($T_{AO} = 86\,K$). For more details see [146, 86] and references therein.

κ-$(ET)_2X$ Salts

Indications for frozen-in disorder have been also reported for the quasi-2D salts of the κ-$(ET)_2X$ family. In an ac-calorimetry study, a glass-like transition has been found for κ-$(ET)_2Cu[N(CN)_2]Br$ and κ-$(ET)_2Cu[N(CN)_2]Cl$

[147, 148, 149]. The authors observed step-like anomalies in the heat capacity around 100 K, which have been attributed to a freezing out of the intramolecular motions of the ET molecule's ethylene endgroups which represent the most deformable parts of the ET molecules [150]. As shown schematically in Fig. 3.12, the relative orientation of the outer C−C bonds can either be parallel (eclipsed) or canted (staggered). At high temperatures, the ethylene endgroups become disordered due to the strong thermal vibrations. Upon cooling to low temperatures, the endgroups adopt one of the two possible conformations, depending on the anion and the crystal structure, see [86].

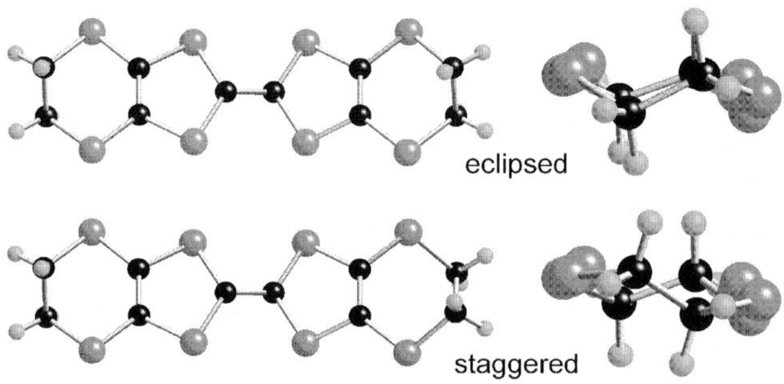

Fig. 3.12. Schematic view of the relative orientations of the ethylene endgroups $[(CH_2)_2]$ of the ET molecule: eclipsed (top) and staggered (bottom) conformation. Right side shows the view along the long axis of the molecule. Courtesy of J.A. Schlueter, Argonne National Laboratory.

In contrast to *thermodynamic phase transitions*, the systems which undergo a *glass transition* are in a metastable state, away from thermodynamic equilibrium. When approaching the glass-transition temperature T_g from above, the dynamics of certain processes (e.g. molecular motions or rotations) slow down so rapidly that thermodynamic equilibrium cannot be reached in these motional or rotational degrees of freedom and a short-range *structural* order, characteristic for T_g, becomes frozen in. "Classic" glass-forming systems are undercooled liquids like e.g. glucose: when cooled down carefully, the temperature of liquids with a melting temperature T_m can be lowered to about 2/3 of T_m before the liquid then suddenly freezes. At the freezing temperature, the specific heat and volume thermal expansion show abrupt jumps, reminiscent of a continuous phase transition. However, the transition in such undercooled liquids is caused by relaxation phenomena and is called a glass transition. The mechanisms of relaxation processes in glasses are rather general in nature so that the basic concepts apply e.g. also for spin-glass systems where below the spin-glass temperature a certain *magnetic* disorder becomes frozen in. Likewise, these basic concepts apply for the crystalline, *glass-like* solids discussed here, where only for certain molecular entities of the compound, the thermodynamic equilibrium state cannot be established below T_g.

The heat capacity, C_p, and volume thermal expansion coefficient, β, of a system are ideal to study glass (or glass-like) transitions, since they allow to distinguish between mere excitations of lattice vibrations and actual changes in the structural order. There are three important features which characterize a glass or glass-like transition: (1) $C_p(T)$ and $\beta(T)$ increase substantially above T_g due to the additional excitation of structural degrees of freedom. (2) T_g depends on the cooling/warming rate $q = dT/dt$, where one possible definition for the glass temperature is

$$-|q| \cdot \left.\frac{d\tau}{dT}\right|_{T_g} \cong 1, \tag{3.2}$$

with τ the relaxation time of the corresponding molecular entities which, in turn, is proportional to the system's viscosity η. [3] It is easy to see that the higher the cooling rate q, the higher the glass-transition temperature T_g. (3) The shape of the glass-transition anomaly in $C_p(T)$ or $\beta(T)$ depends on the thermal history, i.e. a distinct hysteresis is observed: the warming curves show a relatively sharp discontinuity with characteristic under- and overshoot behavior, see Fig. 3.13, whereas the cooling curves are more broadened and lack the under- and overshoot behavior. For details regarding thermodynamic aspects, see e.g. [152, 153] and references therein.

Clear evidence for a glass-like transition in the κ-(ET)$_2$X systems with polymeric anions X = Cu[N(CN)$_2$]Cl, Cu[N(CN)$_2$]Br, and Cu(NCS)$_2$ have been derived from thermal expansion measurements [155, 115, 154]. The outcome of this study confirmed, on the one hand, the above-mentioned specific heat results and clarified on the other hand the nature of the thermal expansion anomalies reported previously [156, 157]. In addition, this study showed that a more complicated glass-like transition also exists for the κ-(ET)$_2$Cu(NCS)$_2$ salt. Figure 3.13 shows the linear thermal expansion coefficients, $\alpha_i(T)$, for the three principal axes of the superconducting system κ-(ET)$_2$Cu[N(CN)$_2$]Br in the vicinity of the glass-like transition (cf. Fig. 4.19 for a wider temperature range). The transition shows all the characteristics expected for a glassy transition discussed above. At $T_g \approx 75$ K step-like anomalies with hysteresis and the characteristic under- and overshoot behavior in the warming curves are observed. The analysis, briefly described below, shows that indeed the conformational degrees of freedom in the ethylene endgroups of the ET molecule cause the observed relaxation phenomena:

[3] We note that a definition of T_g is arbitrary [151] and depends on the measured quantity and measuring technique. The given expression defines T_g as the temperature where the cooling and warming curves, taken at the same $|q|$, in $C_p(T)$ or $\beta(T)$ measurements intersect. It is also important to note that experimentally a transition *region* (i.e. a transformation interval) is observed rather than a well-defined sharp transition.

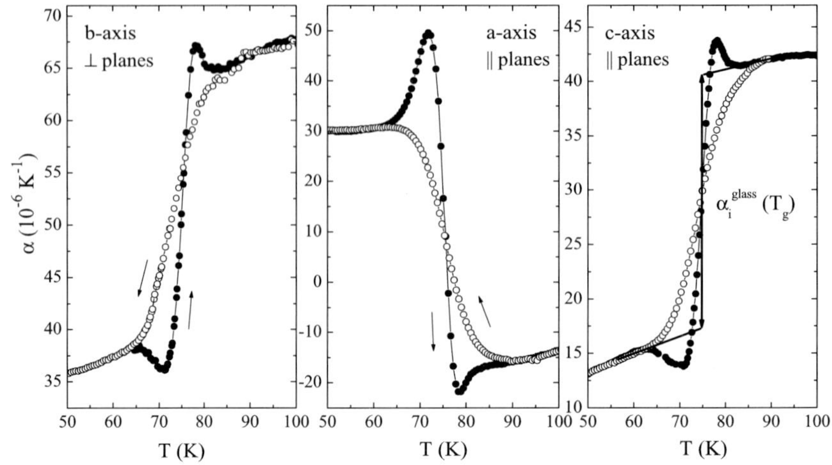

Fig. 3.13. Linear coefficients of thermal expansion, $\alpha_i(T)$, of κ-$(ET)_2Cu[N(CN)_2]Br$ along the three principal crystal axes in the vicinity of T_g. Solid and open symbols represent data taken while warming up and cooling down the sample, respectively. After [154].

for kinetic reasons, the ordering process of the $[(CH_2)_2]$ moieties cannot be completed on lowering the temperature below T_g; instead a characteristic disorder in the orientational degrees of freedom of the ethylene endgroups becomes frozen in. The degree of frozen disorder depends on the transition temperature T_g, which in turn depends on the cooling rate $|q|$.

For κ-$(ET)_2Cu[N(CN)_2]Br$, in early X-ray studies the terminal ethylene moieties have been found to be thermally disordered at room temperature and to prefer the eclipsed conformation at 127 K. At 20 K, the ordering in the eclipsed orientation has been confirmed [158, 159]. In recent high-resolution X-ray experiments, at room temperature, the occupation factor of the eclipsed form refines to $69 \pm 2\,\%$. At 100 K, i.e. slightly above the glass transition, the disorder is already considerably reduced as indicated by an occupation factor of $92 \pm 2\,\%$ of the eclipsed form [160]. That means at T_g the frozen-in disorder amounts to less than about 8 %.

Figure 3.14 shows in more detail the linear coefficient of thermal expansion for κ-$(ET)_2Cu[N(CN)_2]Br$ measured along an in-plane crystallographic direction at temperatures near T_g. The inset on the right side of Fig. 3.14 shows the inverse of the glass-transition temperatures, T_g^{-1}, vs. the cooling rate $|q_c|$ in an Arrhenius plot. The data nicely follow a linear behavior as expected for a thermally activated relaxation time [161, 162]:

$$\tau(T) = \nu_0^{-1} \cdot \exp\left(\frac{E_\mathrm{a}}{k_B T}\right), \tag{3.3}$$

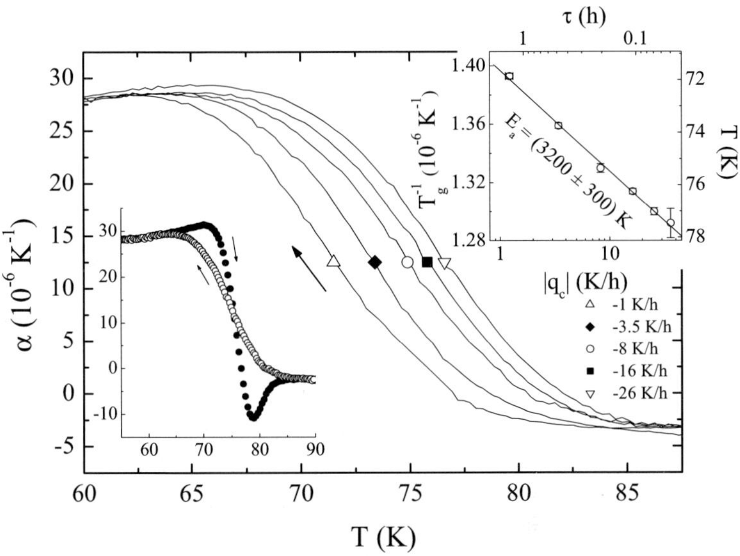

Fig. 3.14. Linear thermal expansion coefficient, α, vs. T measured parallel to the conducting planes of κ-$(ET)_2Cu[N(CN)_2]Br$ in the vicinity of the glass transition defined as the midpoint of the step-like change in α for varying cooling rates q_c. Insets: hysteresis between heating and cooling curves around T_g (left side) and Arrhenius plot of T_{g}^{-1} vs. $|q_c|$ and τ (right side), where $|q_c|$ is the cooling rate and τ the relaxation time. After [155].

where E_{a} denotes the activation energy barrier. The prefactor represents an attempt frequency ν_0. A linear fit to the data of Fig. 3.14 yields $E_{\mathrm{a}} = (3200 \pm 300)\,\mathrm{K}$ [155].

The characteristic activation energy of the $[(CH_2)_2]$ conformational motion (cf. Fig. 3.12) was determined to $E_{\mathrm{a}} = 2650\,\mathrm{K}$ by ^{1}H-NMR measurements [163]. These values have to be compared to 2400 K as estimated from resistivity and ac-calorimetric measurements [164, 148], $(2000 \pm 500)\,\mathrm{K}$ from resistivity measurements of structural relaxation kinetics [165], and 2600 K also from resistivity-relaxation measurements [166]. $E_{\mathrm{a}} = 2660\,\mathrm{K}$ has been found in magnetization measurements for κ-$(D_8$-$ET)_2Cu[N(CN)_2]Br$ [167]. A similar size of the activation energy derived from Fig. 3.14 along with the observation of a mass-isotope shift when replacing the hydrogen atoms in $[(CH_2)_2]$ by deuterium [4] provide clear evidence that the ethylene endgroups are the relevant entities involved in the relaxation process [155, 115]. For κ-$(ET)_2Cu[N(CN)_2]Cl$, a similar analysis of T_g vs. q from thermal expansion data yields $E_{\mathrm{a}} = (2650 \pm 300)\,\mathrm{K}$ for κ-$(ET)_2Cu[N(CN)_2]Cl$ [155] which agrees

[4] T_g shifts to higher temperatures in the deuterated system in good agreement with the larger mass (i.e. larger relaxation time) of the $[(CD_2)_2]$ compared to $[(CH_2)_2]$.

well with (2600 ± 100) K from ^{1}H-NMR [168] and 2700 K from ac-calorimetry [148].

Further information comes from the distinct anisotropy of the glassy features as observed in Fig. 3.13 (note the different scales). The additional contribution to the volume thermal expansion above T_g, $\beta^{\text{glass}}(T) = \sum_i \alpha_i^{\text{glass}}(T)$ where $(i = a, b, c)$, caused by the excitation of the $[(CH_2)_2]$ group's motional/rotational degrees of freedom, is a measure for the pressure dependence of the entropy related to the ethylene endgroup's ordering process, whereas S_{ethy} in turn is a measure of the degree of frozen disorder,

$$\left.\frac{\partial S_{\text{ethy}}}{\partial p_i}\right|_T = -V_{\text{mol}} \cdot \alpha_i^{\text{glass}}(T). \tag{3.4}$$

From Fig. 3.13 one reads for $T = 90$ K (where thermal equilibrium can just be established), $\partial S_{\text{ethy}}/\partial p_i = -1$ J/(molKkbar), $+2.6$ J/(molKkbar), and -1.1 J/(molKkbar) for $i = b$- (perpendicular to the planes), a-, and c-axes, respectively. Consequently, pressure along the b- and c-axes should increase the degree of $[(CH_2)_2]$ order, whereas for uniaxial stress along the a-axis the opposite effect is expected. On combining the uniaxial-pressure coefficients, one finds $\partial S_{\text{ethy}}/\partial p_{\text{hydr}} = +0.5$ J/(molKkbar), i.e. a substantial reduction of the degree of ethylene order upon the application of hydrostatic pressure.[5]

Thus, κ-$(ET)_2Cu[N(CN)_2]Br$ is an ideal system to study the influence of intrinsic disorder on the superconducting properties as the degree of disorder can be varied on the same sample by applying different cooling rates or pressure conditions. The influence of frozen disorder on the electronic properties will be discussed in more detail in Sect. 6.3.3.

[5] In the two-state model, the difference in entropy between the totally ordered and disordered state amounts to $S_{\text{ethy}}^{\text{max}} = Nk_B \ln 2 = 2R \ln 2 = 11.53$ J/(molK) with $N = 2N_A$ as the number of relevant ethylene units per mole. N_A is Avogadro's number and R the gas constant.

4 Normal-State Properties

4.1 Tight-Binding Model

As discussed in Chapt. 2, weakly overlapping π-molecular orbitals provide a useful starting point for estimating the electronic states for molecular materials. As an input for such tight-binding band-structure calculations, the molecular orbitals are usually calculated within the extended-Hückel approximation which uses some empirical molecular parameters [169, 170, 82, 79, 81]. As has been discussed in more detail in [86], the extended-Hückel approximation enables the calculation of the overlap integrals S and the corresponding transfer energies t between the π-orbitals of adjacent molecules. The latter quantity is a measure of the probability for an electron to hop from one molecule to the next. The so-derived extended-Hückel tight-binding bands have the usual cosine energy-dispersion relation

$$E(\boldsymbol{k}) = 2t_a \cos(k_a a) + 2t_b \cos(k_b b) + 2t_c \cos(k_c c), \qquad (4.1)$$

where a, b, and c denote the intermolecular distances along the a-, b- and c- axes, $\boldsymbol{k} = (k_a, k_b, k_c)$ the wave vector of the electron and t_a, t_b, and t_c the directional-dependent transfer energies. For simplicity, a lattice with orthorhombic symmetry has been assumed. This semi-empirical method has been very successfully applied in the case of the present molecular metals, see, e.g. [80, 82, 171, 96, 109, 172, 173, 174].

As the transfer energies are a direct measure of the overlap of the molecular orbitals between neighboring molecules, they strongly depend on the arrangement of the molecules in the crystal structure, i.e. their packing motif. In the case of the $(TM)_2X$ salts, where the flat TM (= TMTSF or TMTTF) molecules form segregated infinite stacks, the overlap of π-orbitals along the stacking axis (a-axis) predominates which results in a quasi-1D conductor. Applying an extended-Hückel tight-binding calculation to the $(TMTSF)_2X$ salts, P.M. Grant estimated the transfer energies $t_a = 0.25\,\mathrm{eV}$, $t_b = 0.025\,\mathrm{eV}$ and $t_c = 0.0015\,\mathrm{eV}$ [175], corresponding to a ratio $t_a : t_b : t_c$ of $1 : 0.1 : 0.006$. Table 4.1 compiles the transfer energies obtained from quantum chemical calculations for various $(TM)_2X$ salts. As a consequence of a structural dimerization, the intra-stack transfer energy alternates between adjacent molecules by an amount $\Delta t_S / t_S$.

Table 4.1. Intra- (t_S) and inter-stack (t_I) transfer energies of selected $(TM)_2X$ salts taken from [146]. The ratio $\Delta t_S/t_S$ indicates the structural dimerization.

	t_S (eV)	t_I (eV)	$\Delta t_S/t_S$
$(TMTSF)_2PF_6$	0.365	0.0262	0.16
$(TMTSF)_2ClO_4$	0.366	0.0216	0.14
$(TMTTF)_2Br$	0.240	0.01	0.13
$(TMTTF)_2PF_6$	0.115	0.0123	0.38

In the vast majority of D_mX_n salts, the donor-acceptor molecular ratio $m : n$ is fixed to $2 : 1$, i.e., two donor molecules transfer one electron ($\delta = 1$) in (2.1) to the acceptor X.[1] In terms of a simple single-electron picture which neglects electron-electron interactions, a partially-filled conduction band and thus metallic behavior is expected. The degree of band filling is determined by the stoichiometry. The conduction band, formed by the donor's HOMO which can accommodate two electrons, is three-quarter-filled by electrons (or quarter-filled by holes) in the above $2 : 1$ compounds with the donor carrying half a hole on average. Due to a structural dimerization in the TMTTF compounds (where it is more pronounced compared to the TMTSF salts) and the κ- and λ-phase $(ET)_2X$ and $(BETS)_2X$ compounds, cf. Fig. 4.2, a dimerization gap may open thereby splitting the conduction band into two parts. The larger the dimerization, the more the two bands split, which is accompanied by a gradual crossover from a three-quarter-filled to an effectively half-filled situation. The effect of strong dimerization on a one-dimensional band structure is visualized in Fig. 4.1. Consequently, according to the above single-particle band picture, metallic behavior is expected for the whole $(TM)_2X$ series. This is, however, in contrast to the observations, yielding a variety of different ground states, dependent on the donor and acceptor molecule.

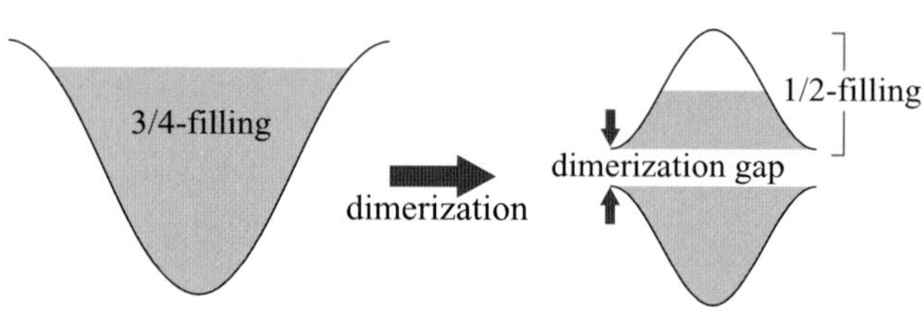

Fig. 4.1. Schematic representation of a one-dimensional band structure without (a) and with (b) dimerization, after [176].

[1] As exceptions to the usual $2 : 1$ stoichiometry we mention the quasi-1D $1 : 1$ compound TTF-TCNQ and $(ET)_4[Ni(dto)_2]$, a 2D metal with a $4 : 1$ composition [177]. For a review on organic conductors with unusual band fillings we refer to the recent publication by T. Mori [178].

To understand the diversity of behaviors in this class of materials, the joint effects of strong electron-electron repulsion and low dimensionality have to be taken into account.

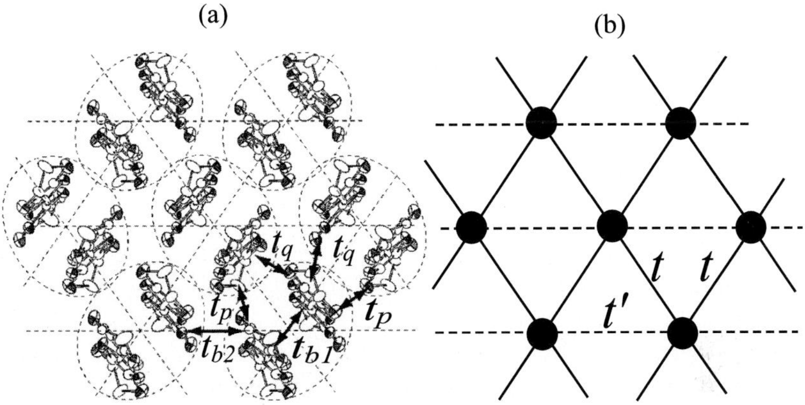

Fig. 4.2. (a) Crystal structure of κ-(ET)$_2$X viewed along the long axes of the ET molecules with the definition of the transfer integrals. (b) For a predominant intradimer hopping t_{b1}, pairs of ET molecules represent dimer units (black dots) which form an anisotropic triangular lattice with $t' \sim t_{b2}/2$ and $t \sim (|t_p| + |t_q|)/2$, after [179].

Figure 4.2 shows the spatial κ-type arrangement of the ET molecules, including the definition of the nearest-neighbor interactions (a) and the resulting effective triangular lattice structure for the dimer model as the effective model (b) [180]. According to various quantum chemical calculations for different anions [96, 181, 182, 183, 150], see Table 4.2, the intradimer-transfer integrals t_{b1} are more than twice as large as the other interdimer integrals t_{b2}, $|t_p|$ and $|t_q|$. As a result, the band structures can be thought of as being composed of bonding and antibonding orbitals of the dimers [180]. Since these bonding and antibonding orbitals are separated by an energy approximately twice this intradimer transfer energy t_{b1}, a mixing between these orbitals can be neglected and the interdimer interaction can be treated as a perturbation [180]. In the further treatment, only the antibonding orbitals on the dimers are considered. Each dimer has two nearest neighbors with interaction energy $t' \sim t_{b2}/2$, and four next-nearest neighbors along the diagonal directions in the unit cell with $t \sim (|t_p| + |t_q|)/2$. Because of the much larger distance between dimers in the vertical direction in Fig. 4.2, hopping along this path can be neglected. The so-derived effective-dimer model has the same co-ordination

Table 4.2. Values of the transfer (hopping) integrals between the ET molecules in κ-(ET)$_2$X for different X as defined in Fig. 4.2. Each dimer has two nearest neighbors with interaction energies $t' \sim t_{b2}/2$ and four next-nearest neighbors with $t \sim (|t_p| + |t_q|)/2$, cf. Fig. 4.2(b). Values are based on quantum chemical calculations using the extended-Hückel approximation and the ambient-pressure crystal structure, unless denoted otherwise.

	t_{b1} (eV)	t_{b2} (eV)	$\|t_p\|$ (eV)	$\|t_q\|$ (eV)	t'/t	Ref.
I_3	0.247	0.088	0.119	0.033	0.58	[181]
$Cu(NCS)_2$	0.23	0.113	0.099	0.033	0.86	[181]
$Cu(NCS)_2$	0.136	0.05	0.041	0.027	0.74	[150]
$Cu(NCS)_2$	0.244	0.026	0.022	0.02	0.62	[182]
$Cu(NCS)_2$ (7.5 kbar)	0.162	0.054	0.058	0.03	0.61	[150]
$Cu(NCS)_2$ (20 kbar)	0.324	0.045	0.027	0.031	0.78	[182]
$Cu[N(CN)_2]Br$	0.224	0.071	0.094	0.04	0.53	[183]
$Cu[N(CN)_2]Br$	0.244	0.092	0.101	0.034	0.68	[181]
$Cu[N(CN)_2]Cl$	0.273	0.104	0.105	0.039	0.722	[99, 100]
$Cu_2(CN)_3$	0.224	0.115	0.08	0.029	1.06	[181]

as the triangular lattice which implies significant magnetic frustration due to the competition between the two hopping terms t' and t.[2]

For some compounds such as β-(ET)$_2$I$_3$ [186, 187], κ-(ET)$_2$Cu(NCS)$_2$ [188], and κ-(ET)$_2$Cu[N(CN)$_2$]Br [183], the band structure has been determined also by first-principles calculations based on the local-density approximation (LDA). Apart from the first attempts [186], which suffered from some crude approximations, these calculations were found to be in accordance with the results of the above semi-empirical approach as regards the overall shape of the bands and the topology of the Fermi surface, see below. However, as has been pointed out in [188], significant differences to the extended-Hückel calculations have been recognized in some parameters such as the bare band masses being larger by a factor of about 2 in the LDA calculations, see also Sect. 4.5.

To understand the diversity in the electronic properties encountered in molecular solids, the Coulomb repulsion between the electrons can play an important role and has to be taken into account, see, e.g. [90, 176]. The basic Hamiltonian to examine such a problem is a tight-binding model based on either the HOMO or LUMO of the relevant molecules together with the

[2] Among the κ-(ET)$_2$X family, the X = $Cu_2(CN)_3$ system is unique in showing a ratio of transfer integrals t'/t close to unity, cf. Table 4.2, favorable for a spin-liquid ground state, cf. [179, 184, 185].

Coulomb interaction. The latter affects not only electrons staying at the same site, which requires an on-site energy U. Coulomb interactions are also of relevance for electrons residing at neighboring sites i and j. This effect is accounted for by the inter-site interaction V_{ij}. The resulting Hamiltonian is the extended-Hubbard model

$$\mathcal{H} = \sum_{<ij>} \sum_{\sigma} (t_{ij} c_{i\sigma}^{\dagger} c_{j\sigma} + h.c.) + \sum_{i} U n_{i\uparrow} n_{i\downarrow} + \sum_{<ij>} V_{ij} n_i n_j, \qquad (4.2)$$

where $< ij >$ denotes the pair of lattice sites i and j and σ the spin index, which can be either $\uparrow$ or $\downarrow$. The simple Hubbard model, i.e. $V_{ij} = 0$, was first used by H. Kino and H. Fukuyama [189] in an attempt to describe systematically the electronic properties of $(ET)_2X$ salts by taking the anisotropic hopping terms t_{ij} into full account. They treated the on-site Coulomb repulsion U within the standard mean-field approximation and adopted the t_{ij} terms from the above-mentioned extended-Hückel model. It has been argued that (4.2) can be applied to a broad range of compounds, with the parameter U varying to some extent depending on the donor and acceptor molecules involved. For a given molecule, however, it is usually assumed to adopt the same value which is typically of the order of the width of the conduction band, i.e., of the order of ~ 0.5 eV. On the other hand, t_{ij} is more sensitive to the molecular arrangement. As for the inter-site Coulomb term V_{ij} a point-charge relation $V_{ij} \propto 1/r_{ij}$ approximately holds, where r_{ij} is the distance between the centers of the molecules at sites i and j. For some compounds it has been found that V_{ij} between neighboring molecules can be as large as $\sim 50\%$ of U [190, 191]. This long-range part of the Coulomb interaction has been shown to be of crucial importance for stabilizing a charge-disproportionated state [192], cf. Sect. 4.4.3. C. Hotta and H. Fukuyama [193] have discussed another parameter — besides the ratio of the effective on-site Coulomb repulsion and bandwidth U/W — to characterize the κ-$(ET)_2X$ salts with quarter-filled hole bands at strong dimerization (the system is simplified to a triangular lattice at half filling), namely the degree of splitting of the antibonding HOMO bands which determines the shape of the Fermi surface. In their model, the degree of band splitting turns out to be the key parameter determining the ground states. The possible importance of nesting is pointed out which leads to the stability of the different (Mott-insulating or spin-density-wave) ground states [193]. In an attempt to provide a unified theoretical picture for the quasi-2D organic conductors with different packing motifs (see Fig. 3.4), a new minimal model based on an anisotropic lattice has been proposed in [194].

At this point, it is instructive to discuss briefly two limiting cases implied in (4.2) for the $D_2^{+1}A^{-1}$ compounds, i.e. an average valence of the donor molecule of $+1/2$ resulting in a quarter-filled band, cf. Fig. 4.3. Given that the electron-electron interaction is larger than the kinetic energy, i.e. for a strongly correlated system, the electrons will favor a situation where they stay apart from each other to reduce their potential energy. At a quarter filling,

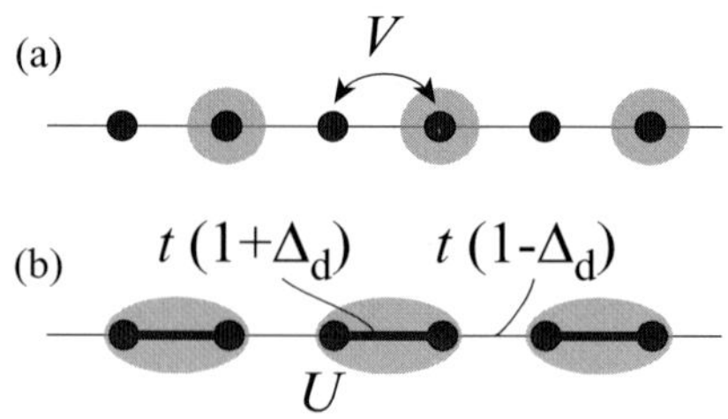

Fig. 4.3. (a) Schematic representation of two limiting cases in a one-dimensional quarter-filled system with strong electron-electron interactions: (a) a charge-ordered (CO) state in the absence of dimerization and (b) a dimer-Mott state for strong dimerization Δ_d. Black dots and grey ellipse represent the lattice sites and localized carriers, respectively. After [176].

the electrons will occupy every other site, as shown in Fig. 4.3(a) for a one-dimensional system, which corresponds to a long-range charged-ordered (CO) state. For such a CO state to be realized, the inter-site Coulomb interaction V_{ij} is of crucial importance [192]. On the other hand, if a lattice dimerization is present, a single electron will occupy the bonding molecular orbital of each dimer, cf. Fig 4.3(b), giving rise to an effectively half-filled band. As a consequence of the strong on-site Coulomb repulsion U, the electrons will reside at the dimer sites, resulting in a dimer-Mott-insulating state. In contrast to the CO state, the dimer-Mott state shows a uniform charge distribution. In the actual charge-transfer salts discussed in this monograph, the situations are less clear. Reasons lie in the moderate strength of the Coulomb energy, being comparable to the kinetic energy, and the actual band structure which is determined by the anisotropy of the hopping terms t_{ij}.

4.2 Fermi Surfaces and Low Dimensionality

Based on the above-mentioned extended-Hückel tight-binding calculations, P.M. Grant has calculated a model band structure for the $(TMTSF)_2X$ compound, see Fig. 4.4. While for an ideal one-dimensional metal the Fermi surface (FS) consists of a pair of parallel sheets at the wave vectors $\pm\, k_F$, where k_F is the Fermi wave vector, the FS sheets of $(TMTSF)_2X$ are slightly corrugated. The finite dispersion along k_b is due to the non-vanishing interstack electron-transfer energy t_b. Since t_b amounts to only about 1/10 of t_a, the FS remains open in the Brillouin zone. Of particular importance for the ground-state properties of low-dimensional metals in general are the topological aspects of the Fermi surface, i.e., their nesting properties. The latter refer to the case that a particular translation vector in the k-space — the so-called nesting vector — can cause large areas of the Fermi surface to map onto each other. In this case, the system can lower its electronic energy by

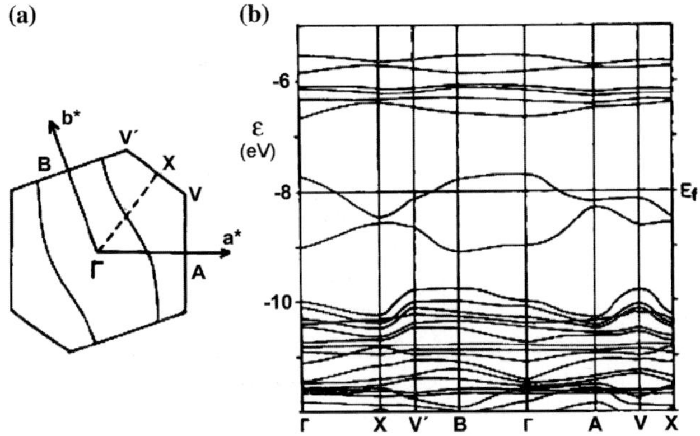

Fig. 4.4. (a) Fermi surface and (b) tight binding band structure calculated by the extended-Hückel scheme for $(TMTSF)_2X$, after [195, 175].

introducing a $Q = 2k_F$ distortion in k-space. In real space, the $2k_F$ modulation corresponds to an additional periodicity. In analogy to the gaps that open at the Brillouin-zone boundary at $\pm\pi/a = \pm q_a/2$ as a consequence of the lattice periodicity a with $q_a = 2\pi/a$, the $Q = 2k_F$ modulation creates band gaps over parts of the FS at $\pm k_F$. Since the energy gain scales with the area of the FS involved, the nesting has to be sufficiently good for this mechanism to work. Depending on the degree of nesting and the strength of electron-electron interactions, the charge- or the spin-density can become modulated and the corresponding states are known as charge- (CDW) or spin-density-wave (SDW) states. Apparently, an ideal one-dimensional metal is inherently instable against such a $Q = 2k_F$ modulation and the ground state of such a system is an insulator.

Figure 4.5 shows the results of semi-empirical band-structure calculations and FS topologies projected on the basal plane of the first Brillouin zone for typical α-, β-, β'', and κ-type $(ET)_2X$ salts. Since, in these calculations, any contribution from an interlayer electron transfer has been neglected, the resulting Fermi surfaces are strictly two-dimensional. For the β-phase salts, such as β-$(ET)_2I_3$, there are two ET molecules per unit cell transferring one electron to the I_3^- anion. As Fig. 4.5 indicates, the FS of this compound is very simple. It is closed within the first Brillouin zone with a shape of a slightly deformed cylinder, reflecting the almost isotropic in-plane electron transfer integrals. On the contrary, the FS of the α-$(ET)_2KHg(SCN)_4$ salt combines 2D closed parts with open sheets of quasi-1D character. The good nesting properties of the latter are believed to be responsible for the low-temperature instability found in the isostructural series α-$(ET)_2MHg(SCN)_4$ with M = K, Rb and Tl. While previous susceptibiliy- [200] and muon-spin-rotation experiments [201] seemed to indicate a SDW transition, more recent high-

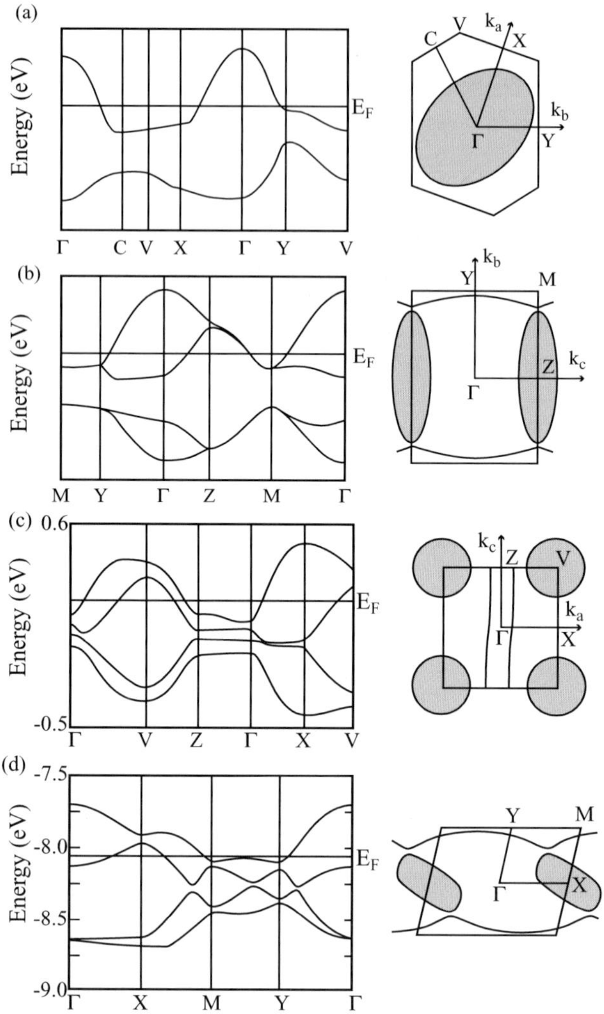

Fig. 4.5. Calculated energy dispersions and Fermi surfaces for (a) β-phase (after [196]), (b) κ-phase [197, 198, 96], (c) α-phase [109], and (d) β''-phase ET salts [199].

field results of different groups appear to be inconsistent with the expected B-T phase diagram for such a SDW state [202, 203]. Their experimental data, however, were found to be quite consistent with a CDW scenario [204, 205], see also Sect. 5.3.2. For the κ-salts, the arrangement of the dimers causes a doubling of the unit cell resulting in two transferred electrons in two HOMOs. Again the resulting FS is very simple resembling a free-electron-like scenario where intersections of the FS with the zone boundary give rise to small energy gaps due to Bragg reflections. The band structures and FS topologies are similar among the various κ-(ET)$_2$X salts [46, 206, 158, 207], except for the degeneracy of the two upper bands along the Z-M zone boundary for the X = I_3 salt. Due to the lack of a center-of-inversion symmetry in the X =

$Cu(NCS)_2$ and $Cu[N(CN)_2]Br$ salts, this degeneracy is lifted and a gap opens at the zone boundary. The resulting FS for the κ-$(ET)_2X$ salts consists of two parts, a closed hole-like orbit (α-pocket FS, see Fig. 5.6(b)) centered at the Z-point and a pair of corrugated sheets.

The FS topology of the quasi-2D charge-transfer salts has been studied in great detail employing measurements of the de Haas-van Alphen (dHvA) and Shubnikov-de Haas (SdH) effect, the angular-dependent magnetoresistance (AMRO) and the cyclotron resonance, see Chapt. 5. For comprehensive reviews on the FS studies, we refer to [208, 209, 205]. In general, the FS topologies determined experimentally are in remarkable agreement with the predictions of the above-mentioned semi-empirical band-structure calculations.

4.3 Electronic Properties

4.3.1 Transport and Optical Properties

Electrical Resistivity

The molecular metals discussed here are fairly good conductors at room temperature, with resistivities that vary over wide ranges depending on the particular compound and on the current direction with respect to the crystal axes. The pronounced anisotropies in the transport properties, encountered in these materials, reflect the strongly directional-dependent electron-transfer integrals. For the $(TM)_2X$ series, one typically finds $\rho_a : \rho_b : \rho_c$ of the order of $1 : 200 : 30.000$, where a denotes the stacking axis. These numbers correspond to a ratio of the overlap integrals t_a, t_b, and t_c of about $10 : 1 : 0.1$ with t_a values in the range $0.1 - 0.24$ eV for TMTTF and around 0.36 eV for the TMTSF compounds [210, 146], cf. Table 4.1. Owing to these strongly anisotropic hopping terms, the electronic structure can be viewed at first glance as one-dimensional. Typical temperature profiles of the resistivity ρ_a for the $(TM)_2X$ compounds are shown in Fig. 4.6. Below a temperature T_ρ, where the resistivity shows a minimum, the resistivity of the sulfur-containing $(TMTTF)_2X$ compounds changes from a metallic-like high-temperature into a thermally-activated low-T behavior.

Such a charge response is well consistent with what one would expect for a one-dimensional Mott insulator. Owing to the small structural dimerization of the chains, such a Mott-insulating state could result either from the half-filled nature of the band (due to the dimerization) or, if the on-site and nearest-neighbor interaction U and V, respectively, are strong enough, from the quarter-filled nature of the band. The underlying mechanism is the so-called Umklapp scattering, where in an interaction process the momentum of the electrons is conserved only modulo a vector of the reciprocal lattice $\boldsymbol{Q} =$

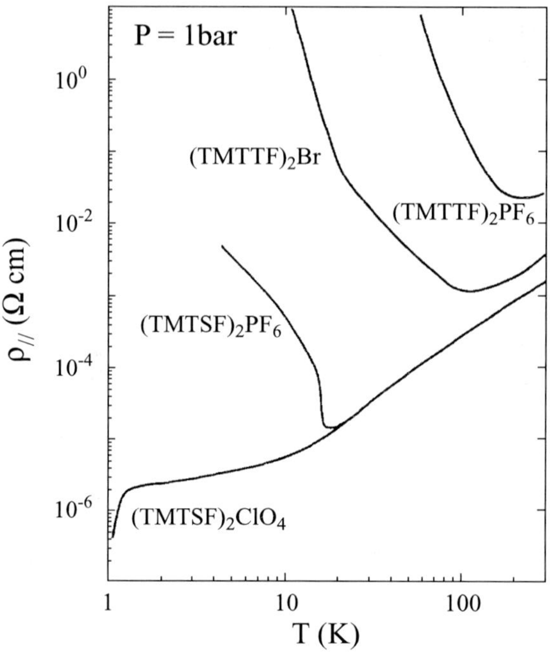

Fig. 4.6. Resistivity along the axis of highest conductivity vs temperature for various (TMTTF)$_2$X and (TMTSF)$_2$X salts at ambient-pressure conditions in a double-logarithmic plot, after [211].

$\boldsymbol{k}_1 + \boldsymbol{k}_2 - \boldsymbol{k}_3 - \boldsymbol{k}_4$.[3] For a one-dimensional system at half filling, this process involves the scattering of two electrons from one side of the Fermi surface — which for a 1D system consists of only two points at $\pm k_F$ — to the opposite side with a momentum transfer of $4k_F = 2\pi/a$, with a the lattice spacing. To produce such an Umklapp process at quarter filling, four particles have to be transferred across the Fermi surface in order to accomplish the proper momentum transfer, i.e. $8k_F = 2\pi/a$. While for half filling, an arbitrarily weak repulsive interaction is enough to turn the system into an insulator, the interactions have to be strong and of finite range for the quarter-filled situation. It is a matter of current debate which of these scenarios applies for the (TM)$_2$X series, see, e.g. [90, 212] for a discussion on this issue.

Upon further cooling through $T_{SP} \simeq 20\,\mathrm{K} < T_\rho$ (not shown in Fig. 4.6), the (TMTTF)$_2$PF$_6$ salt undergoes a phase transition into a spin-Peierls (SP)-distorted nonmagnetic ground state; see e.g. [211] and references cited therein. With the application of moderate pressure, both T_ρ and T_{SP} were found to decrease. By increasing the pressure to $p \geq 10\,\mathrm{kbar}$, the spin-Peierls ground state becomes replaced by an antiferromagnetic Néel state similar to the one found at ambient-pressure conditions for (TMTTF)$_2$Br.

In the selenium-containing (TMTSF)$_2$PF$_6$ salt, the metallic range extends down to lower temperatures until the sudden increase in the resistivity in-

[3] Electrons with wave vectors $\boldsymbol{k}_1$ and $\boldsymbol{k}_2$ are scattered in states with $\boldsymbol{k}_3$ and $\boldsymbol{k}_4$ with $\boldsymbol{Q}$ being a vector of the reciprocal lattice.

dicates the transition into an insulating ground state. The magnetic (SDW) character of the transition was proven by NMR experiments which revealed a strong broadening in the resonance line due to the occurrence of internal magnetic fields below the transition temperature T_{SDW} [213]. Above a critical pressure of about 6 kbar, the SDW state of $(TMTSF)_2PF_6$ becomes unstable giving way to superconductivity at lower temperatures, cf. the phase diagram in Fig. 4.31 in Sect. 4.6.

Interestingly enough, when the spin-Peierls salt $(TMTTF)_2PF_6$ is exposed to sufficiently high pressure in excess of 43.5 kbar, a superconducting state can be stabilized [214] which completes the sequence of ground states indicated in the generic phase diagram proposed by D. Jérome [215].

In their metallic regime, the resistivity of the $(TM)_2X$ salts along the most conducting direction decreases monotonically with a power-law temperature dependence $\rho \propto T^{\alpha}$, where the exponent α, depending on the temperature interval, varies between 1 and 2, see e.g. [216, 146], and the discussion in Sec. 4.3.1. For $(TMTSF)_2PF_6$ for example, $\alpha \approx 1.8$ between 300 and 100 K and approaches approximately 2 at lower temperatures down to the metal-SDW transition [146]. A T^2 dependence in the resistivity has been frequently observed not only in the quasi-1D [216] but also for some $(ET)_2X$ salts, see below.

The resistivity for the various $(ET)_2X$ and related compounds can be roughly classified into two distinct types of temperature dependencies. While some of the materials show a more or less normal metallic-like behavior, i.e. a monotonic decrease of $\rho(T)$ upon cooling, a pronounced $\rho(T)$ maximum at several tens of Kelvin has been found for a number of $(ET)_2X$ compounds, cf. Figs. 4.7 and 4.8. Among them are the κ-$(ET)_2X$ salts with polymer-like anions such as X = $Cu(NCS)_2$ or $Cu[N(CN)_2]Br$ [217, 46, 218, 219], the α-$(ET)_2NH_4Hg(SCN)_4$ [220] as well as the κ- and λ-type BETS salts [121]. The occurrence of the same kind of $\rho(T)$ anomaly in $(DMET)_2AuBr_2$ [221] proves that (i) this feature is not a property specific to ET- or BETS-based salts and (ii) does not rely on the presence of Cu ions. Figure 4.7 shows the in-plane resistivity of κ-$(ET)_2Cu(NCS)_2$ as a function of temperature at various pressures [217]. With decreasing temperatures, $\rho(T)$ first increases to a maximum at around 100 K before a metallic behavior sets in at lower temperatures. Under hydrostatic pressure, the maximum shifts to higher temperatures and becomes progressively suppressed. This is accompanied by a significant reduction of T_c (see also inset of Fig. 4.30). The origin of the anomalous $\rho(T)$ hump has been discussed by many authors and various explanations have been suggested including the formation of small polarons [222], a metal-metal phase transition [223], a valence instability of Cu [224], an order-disorder transition of the ethylene endgroups of the ET molecules [225, 156, 165] as well as a crossover from localized small-polaron to coherent large-polaron behavior [226]. Alternatively, the anomalous resistivity behavior has been linked to the strongly correlated nature of the electrons [227, 228]. Within a dynami-

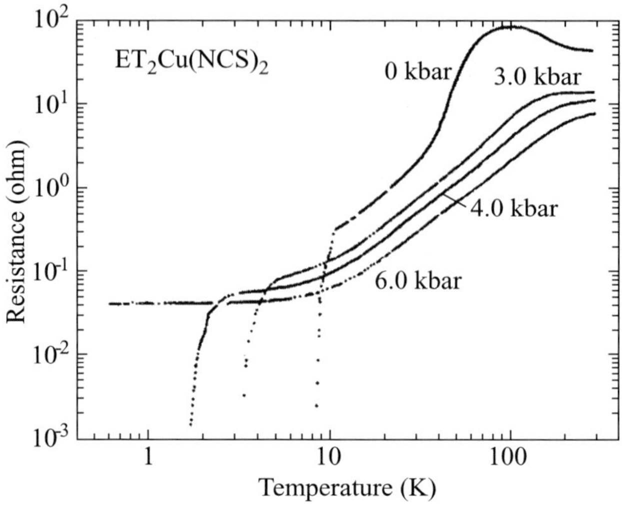

Fig. 4.7. Temperature dependence of the resistivity measured along the in-plane b-axis of κ-$(ET)_2Cu(NCS)_2$ at various pressures. After [217].

cal mean-field (DMFT) approach, J. Merino et al. [227] predicted a smooth crossover from coherent Fermi-liquid excitations with $\rho \propto T^2$ at low temperatures to incoherent (*bad metal*) excitations at higher temperatures. Using such DMFT calculations for a simple Hubbard model, P. Limelette et al. [228] recently attempted to provide even a quantitative account for the ρ(T) behavior of pressurized X $=$ $Cu[N(CN)_2]Cl$ over an extended temperature range covering the anomalous resistivity maximum. In this context, it is interesting to note that for the κ-$(ET)_2Cu[N(CN)_2]Br$ and related systems, the resistivity maximum has been found to be sample dependent: using a different synthetic route, superconducting crystals have been prepared that lack the anomalous resistivity hump [229, 230, 231]. These striking sample-to-sample variations have motivated a comparative investigation of differently prepared κ-$(ET)_2Cu[N(CN)_2]Br$ crystals in [160]. In Fig. 4.8(a), the interlayer resistivity is shown for a collection of such κ-$(ET)_2Cu[N(CN)_2]Br$ single crystals with different residual resistivity ratios $RRR = \rho(300\,\mathrm{K})/\rho(0\,\mathrm{K})$, where $\rho(0)$ has been determined from an extrapolation of the normal-state resistivity just above T_c to $T = 0$ [160]. The salient results of these and related studies [232] are (i) in the absence of significant differences in the crystals' structural parameter and chemical composition, as proved by high-resolution X-ray and electron-probe microanalysis, the sample-dependent resistivity profiles indicate a significant influence of real structure phenomena, i.e. disorder and/or defects which affect the inelastic scattering in these systems. (ii) No correlations have been found between the strength of these scattering contributions and other characteristic properties such as the glass transition at $T_g = 77$ K, the temperature $T^* \sim 40\,\mathrm{K}$, where the temperature dependence of the resistivity changes rather abruptly, or the superconducting transition temperature T_c [232]. As demonstrated in Fig. 4.8(b), where the resistivity near

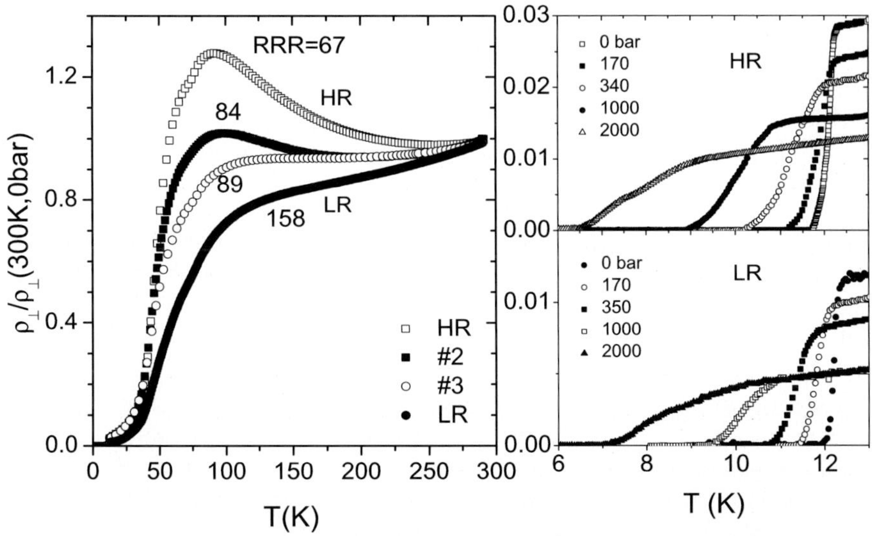

Fig. 4.8. (a) Interlayer resistivity as a function of temperature for various κ-$(ET)_2Cu[N(CN)_2]Br$ single crystals. (b) Low-temperature data at various hydrostatic pressure values for the LR- (lower panel) and HR- (upper panel) crystal of (a). After [160].

T_c is shown for the LR and HR crystal, representing the two extreme cases of the data sets in Fig. 4.8(a), the markedly different inelastic scattering in those two crystals has no significant influence on T_c and its pressure dependence [160].

A closer look at the resistivity of κ-$(ET)_2Cu(NCS)_2$ below the maximum discloses a rapid change in the slope (inflection point) at temperatures T^* around 45 – 50 K [233], cf. Fig. 4.7. A similar behavior is found for κ-$(ET)_2Cu[N(CN)_2]Br$ [234] and pressurized κ-$(ET)_2Cu[N(CN)_2]Cl$ [228]. This effect has been attributed to a crossover from incoherent bad metal excitations at higher temperatures to coherent Fermi liquid excitations at low temperatures [227, 228]. Alternatively, it has been suggested that the temperature T^* marks a kind of density-wave-type *phase transition* [155, 235, 236], see also Sect. 4.3.2 for a detailed discussion. At temperatures $T \leq T_0 < T^*$ below the inflection point, the resistivity turns into an approximate $\rho(T) = \rho_0 + AT^2$ behavior until superconductivity sets in around 10 K. As mentioned above, a resistivity roughly following a T^2 law — even at elevated temperatures — is not an exception in the present molecular conductors, see e.g. [86]. In some high-quality crystals of β- and κ-type $(ET)_2I_3$, it has been observed over an extraordinarily wide temperature range up to temperatures as high as 100 K [237]. It has been argued that the T^2 dependence of the resistivity indicates a dominant role of electron-electron scattering in the present materials [238].

Indeed, such a temperature dependence has been observed at low temperatures in materials with strong electron-electron interactions such as transition metals or heavy-fermion compounds. It has been found that within a given class of materials, the coefficient A scales with the square of the Sommerfeld coefficient of the electronic specific heat, γ, [239, 240], $A \propto (1/E_F^*)^2 \propto m^{*2}$, where E_F^* and m^* are the effective Fermi energy and effective quasiparticle mass, respectively.

The above scaling implies that upon small variations of a control parameter x of the system, such as chemical composition or external pressure, the product $A(x) \cdot T_F^*(x)^2 \propto A(x) \cdot E_F^*(x)^2$ with $T_F^* = E_F^*/k_B$ should stay constant. By identifying the temperature T_0, i.e., the upper limit of the T^2 range in the resistivity, with the effective Fermi energy T_F^*, P. Limelette et al. have verified this invariance for pressurized κ-(ET)$_2$Cu[N(CN)$_2$]Cl [228].

This conclusion has been questioned, however, by the above studies on the differently prepared κ-(ET)$_2$Cu[N(CN)$_2$]Br crystals which disclosed striking sample-dependent coefficients A and upper limits T_0 for the T^2 dependence [160]. The product $A(x) \cdot T_F^*(x)^2$, calculated for the various crystals, was found to differ by a factor up to 3 indicating an origin of the T^2 dependence different from coherent Fermi-liquid excitations.

In line with this interpretation are recent studies of the pressure dependence of superconducting and transport properties of the compound β''-(ET)$_2$SF$_5$CH$_2$CF$_2$SO$_3$ [241, 242]. The slope A of the T^2 dependence, see Fig. 4.9(a), should be proportional to m_c^{*2}, the square of the effective cyclotron mass, that can be extracted from Shubnikov-de Haas experiments. The authors find that m_c^* decreases approximately linearly with increasing pressure[4] from $2\,m_0$ at ambient conditions to about $1.2\,m_0$ at 10 kbar [243]. $T_c(p)$ then can be well described by the modified McMillan equation, see (6.20), by assuming $\lambda = m_c^*/m_b - 1$, where λ is the superconducting coupling parameter, see Chapt. 6, and m_b the band mass. Given an electronic origin of the T^2 dependence (Fermi-liquid description), the ratio A/m_c^{*2} should remain constant for all pressures. However, as seen in Fig. 4.9(b), A reduces with pressure much faster than m_c^{*2}. Indeed, A is roughly proportional to m_c^{*6}, see inset of Fig. 4.9(b), a dependence that is clearly at odds with a Fermi-liquid nature of the T^2 behavior [241, 242].

Alternatively, the T^2 law has been attributed to the scattering of electrons by phonons via electron-libron [244] or a novel electron-phonon scattering mechanism proposed for the high-T_c cuprates [237] invoking electron-electron interactions [245, 246]. For the discussion of the temperature dependence of the resistivity, it is important to bear in mind, however, that due to the large pressure coefficients of the resistivity of about $\partial \ln \rho/\partial p \simeq -20\,\%$/kbar at room temperature together with the extraordinarily strong thermal contraction, it is difficult to make a comparison with theoretical predictions. Since the theory usually describes the temperature dependence at constant volume,

[4] No topological change of the Fermi surface under pressure is observed.

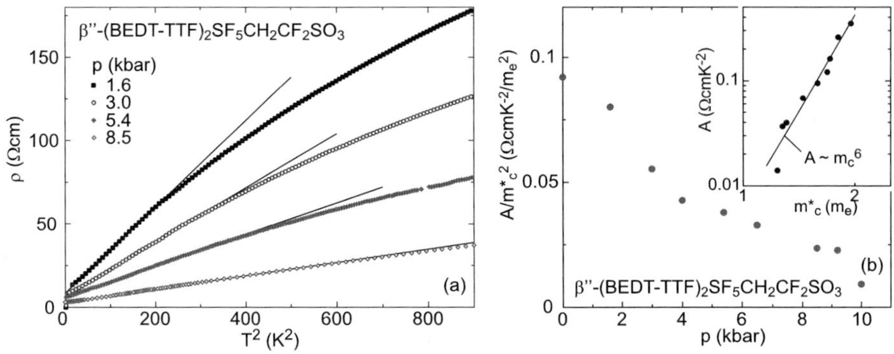

Fig. 4.9. (a) Resistivity of β''-$(ET)_2SF_5CH_2CF_2SO_3$ for different pressures plotted vs. T^2. The straight lines are low-temperature fits according to $\rho = \rho_0 + AT^2$. (b) Pressure dependence of A/m_c^{*2}. The inset shows A as a function of m_c^*. After [241].

a detailed comparison is meaningful only after transforming the constant-pressure into constant-volume profiles by taking into account the thermal expansion of the material.

The Mean Free Path Problem

Similar to the quasi-1D $(TM)_2X$ salts, the room-temperature resistivities of the quasi-2D $(ET)_2X$ materials are generally very high. For κ-$(ET)_2Cu(NCS)_2$ for example, one finds for the intraplane resistivities $\rho_b \approx 6 \cdot 10^4\,\mu\Omega$cm and $\rho_c \approx 3 \cdot 10^4\,\mu\Omega$cm [247], which exceed the values found for Cu by more than a factor of 10^4. In accordance with their quasi-2D electronic structure, a pronounced in-plane vs out-of-plane anisotropy has been observed which amounts to $10^{-3} \sim 10^{-5}$ [86, 209]. Remarkably enough, the low-temperature interlayer resistivity shows unusually high values of up to $1\,\Omega$cm — being a factor of 10^6 larger than that of Cu at room temperature — which nevertheless reveals a metallic temperature dependence. In the frame of a Boltzmann transport theory, this number would correspond to a mean free path ℓ which is much shorter than the interatomic distances. The question has thus arisen as to whether the interlayer transport is coherent or incoherent, i.e. whether there is a coherent motion of band states associated with well-defined wave vectors or if the motion from layer to layer is diffusive and a Fermi velocity perpendicular to the layers cannot be defined [248]. This question has been addressed in recent magnetoresistance studies on the κ-$(ET)_2Cu(NCS)_2$ salt [249]. These experiments are based on the idea that for a Fermi surface which is extended perpendicular to the conducting planes, a magnetic field applied exactly parallel to the layers can cause closed orbits on the sides of the Fermi surface. From a peak in the interlayer resistance, close to the exact $\theta = 90°$ orientation where the field is aligned parallel to the planes, an interlayer transfer integral $t_\perp \approx 0.04\,$meV has been estimated [249]. This has to be compared with $t_{\|} \sim 150\,\mathrm{meV}$ for the intralayer transfer integral [250]. According to this work, the Fermi surface of the κ-$(ET)_2Cu(NCS)_2$ salt is extended along the interlayer direction corresponding to a coherent transport. The same conclusion has been drawn for the β-$(ET)_2IBr_2$ system based on the observation of magnetic quantum oscillations [251, 252]. The authors found nodes in the de Haas-van Alphen (dHvA) signal for magnetic fields perpendicular to the planes. This has been attributed to the existence of a corrugated Fermi surface along the interlayer direction with two slightly different extremal orbits, resulting in beating nodes in the dHvA signal.

However, for the even more anisotropic β''-$(ET)_2SF_5CH_2CF_2SO_3$ salt, such evidence is missing. From the absence of beating nodes in dHvA and Shubnikov-de Haas experiments [253, 254], any possible FS corrugation must be extremely small. Furthermore, additional tests for the existence of a 3D FS failed for this material [255], making it a good candidate for a 2D metal with incoherent interlayer transport envisioned by R.H. McKenzie and P. Moses [248].

Dimensional Crossover and Non-Fermi-Liquid Behavior

For three-dimensional metals, Fermi-liquid theory has provided a most elegant and useful tool to treat the problem of electron-electron interactions, see e.g. [256, 257]. This theory assumes a one-to-one correspondence between the excitations of the interacting electron liquid and those of a noninteracting electron gas, where the effects of interactions are cast into a few phenomenological parameters such as the effective mass, m^*, and the so-called Landau parameters [258]. For some materials, the Fermi-liquid model holds even for very strong interactions when the effective mass exceeds the free-electron mass by a factor of several hundred [259].

In recent years, however, a growing number of systems has been studied intensively which cannot be described by this theory. Prominent examples for such non-Fermi liquids can be found in the underdoped regime of high-temperature superconductors [260] or in strongly correlated intermetallic compounds close to a so-called quantum critical point where the transition into a long-range magnetically-ordered state has been tuned to zero temperature [261]. Another scenario where the Fermi liquid concept fails completely is when the motion of the electrons is restricted to one dimension, where — because of the Coulomb interaction — the motion of an electron is tightly linked to that of the others. As a result, individual excitations are replaced by collective ones in 1D. Because of such collective phenomena, a single fermionic excitation splits into one collective excitation carrying charge and another one carrying spin, both of which propagate at different velocities. This break-up of an electron into two elementary excitations represents one of the fundamental properties of a so-called Luttinger-liquid, the concept which replaces the Fermi liquid in 1D, see, e.g. [262, 263]. [5]

A question which has been of general interest for the quasi-1D molecular conductors is whether these materials represent realizations of such a Luttinger-liquid or whether a Fermi-liquid description is still adequate. Arguments in favor of a Luttinger-liquid behavior have been derived from various observations, not all of which have been generally accepted. Owing to the strongly anisotropic hopping integrals of the $(TM)_2X$ salts of typically $t_a = 0.25\,\text{eV}$, $t_b = 0.025\,\text{eV}$ and $t_c = 0.0015\,\text{eV}$ [175], cf. Sect. 4.1, there are relatively well-separated energy scales so that one may find a temperature range in which the intrachain hopping dominates, i.e., an effectively 1D regime, and a subsequent crossover to higher dimensions. In the absence of electron-electron interactions, this crossover will occur at a temperature where the thermal energy, k_BT is of the same order as the interchain hopping energy $t_\perp$. At temperatures $k_BT > t_\perp$, the warping of the FS caused by interchain hopping is washed out and the system behaves like one with a perfectly flat FS. On the contrary, if the temperature is much below the interchain hopping energy, the warping of the FS affects the motion of the electrons and the system is now two- or three-dimensional. In the presence of electron-electron interactions, the interchain hopping is renormalized so that the crossover temperature becomes reduced compared to that of the free fermions, see, e.g. [212]. Thus interactions tend to make the system more one-dimensional.

A peculiarity of the $(TM)_2X$ systems is the band filling: owing to the $2:1$ stoichiometry, there is half a carrier (hole) per TM molecule. Thus for uniformly spaced molecules, i.e. one formula unit per unit cell, the unit cell contains half a carrier, corresponding to a quarter-filled conduction band. However, as mentioned in Sect. 4.1 a dimerization has been found in the various $(TM)_2X$ salts which is more strongly pronounced in the sulfur series $(TMTTF)_2X$. These commensurate band fillings lead to important modifications of the Luttinger-liquid scenario as they open a new scattering channel via Umklapp processes (see above) for carriers between both sides of the Fermi surface. Instead of the gapless Luttinger-liquid excitations, a gap may open in the charge sector due to these Umklapp-scattering processes and the system becomes a Mott-Hubbard-type insulator. Yet, as a consequence of the one-dimensionality, which requires the separation of spin- and charge-excitations, the spin sector remains gapless. The size of the charge gap depends strongly on the band filling and the strength of the electron-electron

[5] A model for such 1D systems has been proposed by Tomonaga and Luttinger. It is based on a linearized energy spectrum for excitations close to the Fermi level and takes into account the relevant Coulomb repulsions which are responsible for electron scattering with momentum transfer $2k_F$ and 0. This model results in a power-law decay of all correlation functions (spin susceptibility at $2k_F$ and $4k_F$, CDW, superconductivity) with a nonuniversal exponent K_ρ.

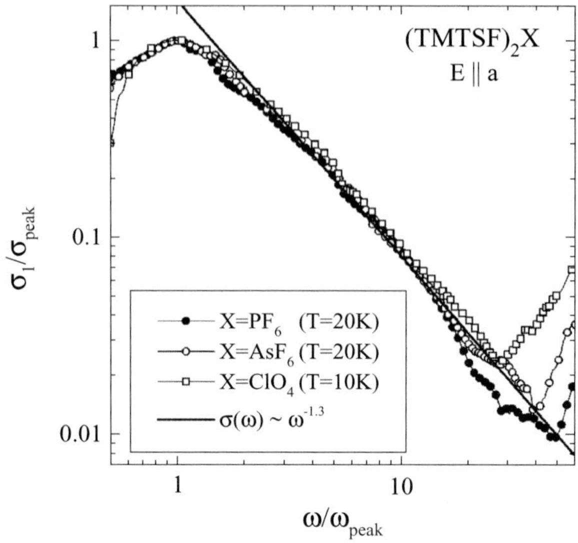

Fig. 4.10. Optical conductivity along the chain axis for various members of the $(TMTSF)_2X$ family. The data have been scaled to the individual gap values. After [265].

interaction, see below. The charge gap has been directly seen in optical conductivity measurements [264, 265] which reveal a reduction of the gap from about 2000 cm^{-1} for $(TMTTF)_2PF_6$ to about 200 cm^{-1} for $(TMTSF)_2PF_6$ with almost all spectral weight residing at frequencies above the gap. For the metallic compounds, a very narrow Drude peak containing only 1% of the spectral weight has been observed which is responsible for the dc transport in this system [266]. The optical conductivity data for frequencies above the gap were found to follow the power-law dependence predicted for a Luttinger liquid very well [265], see Fig. 4.10. The Luttinger interaction parameter K_ρ can be determined from a fit of the conductivity along the chain axis, $\sigma(\omega)$, to an expression

$$\sigma(\omega) \propto \omega^{4n^2 K_\rho - 5}. \tag{4.3}$$

This quantity characterizes the strength of the interactions between the charge carriers with $K_\rho < 1$ for repulsive interactions. The parameter n in (4.3) denotes the order of the commensurability with $n = 1$ for half filling and $n = 2$ for a quarter-filled band. The larger the commensurability, the smaller K_ρ needs to be for the system to become insulating. For half filling, i.e. $n = 1$, an infinitesimally small repulsive interaction $K_\rho < 1$ is sufficient to turn the system into an insulator.

In contrast, for a quarter-filled band ($n = 2$), the critical value is $K_\rho = 1/4$, meaning that very strong non-local interactions are necessary for an insulating state to be stabilized. A consistent fit of both the exponent in $\sigma(\omega)$ and the gap was possible only by assuming a quarter-filled band, i.e. $n = 2$, yielding $K_\rho = 0.23$. This indicates that the electron-electron interactions are quite strong and that a finite range of the interactions at least to nearest neighbors should be taken into account. The optical data are consistent with the interpretation of the insulating state as a quarter-filled Mott insulator, suggesting that the dimerization plays a minor role at least for the $(TMTSF)_2X$ family. As far as dc transport is concerned, one expects for the electronic contribution to the longitudinal resistivity a power-law dependence,

$$\rho_{\|} \propto T^{4n^2 K_\rho - 3}, \tag{4.4}$$

at temperatures in excess to the charge gap $T > \Delta_\rho$ [90]. For $(TMTSF)_2PF_6$ the longitudinal resistivity varies like $T^{0.93}$ for temperatures from 300 to 150 K, once the constant-volume corrections have been taken into account, corresponding to $n^2 K_\rho = 0.98$, i.e. $K_\rho = 0.25$ for $n = 2$. These measurements of K_ρ are consistent with photoemission data [267, 268, 269, 270].

Optical Measurements

Optical measurements provide a most powerful tool for characterizing the physical properties of molecular solids. By employing electromagnetic waves covering wide ranges of energy, various kinds of excitations can be probed. From these measurements, important information can be derived, not only on the basic electronic, magnetic and vibrational properties but also on fundamental interactions [271].

Optical investigations, by means of infrared and Raman measurements, provide important information on electronic parameters such as the plasma frequency, the optical masses and also the bandwidths and collision times of the carriers. In addition, they permit an investigation of vibrational properties and their coupling to the charge carriers. Using polarized light, it is also possible to look for anisotropies in these quantities, as e.g. in the effective masses. The optical properties of quasi-1D and -2D organic conductors have been reviewed by several authors [272, 146, 273, 274, 275, 276], see also [86, 277]. For a detailed discussion on the normal- and superconducting-state optical properties of the quasi-2D $(ET)_2X$ salts, see [278, 276]. A summary of Raman results on these salts is given in [279, 280].

For an ordinary metal, the optical properties consist of contributions from (i) conduction electrons, (ii) phonons and (iii) electronic interband transitions. These contributions can be satisfactorily described by the phenomenological models discussed below. Although low-dimensional molecular conductors, in general, exhibit a more complicated optical response, these models may serve as a first approximation.

(i) **Conduction Electrons**

The response of non-interacting electrons to an alternating electrical field $E(t) = E_0 \exp(-\mathrm{i}\omega t)$ has been described by Drude in his classical approach. It is based on the assumption that the density of states is constant and that the low-frequency conductivity is solely determined by the scattering rate. The frequency-dependent complex conductivity $\tilde{\sigma} = \sigma_1 + \mathrm{i}\sigma_2$ is given by

$$\tilde{\sigma}(\omega) = \frac{ne^2\tau}{m} \cdot \frac{1 - \mathrm{i}\omega\tau}{1 + \omega^2\tau^2}, \tag{4.5}$$

with τ the relaxation or scattering time between two collisions, m the carrier mass, and n the electron density. The sum rule

$$\int_0^\infty \sigma_1(\omega)\mathrm{d}\omega = \frac{\pi ne^2}{2m} = \frac{\omega_p^2}{8} \tag{4.6}$$

relates the area under the real part of the conductivity to the plasma frequency

$$\omega_p = \left(\frac{4\pi ne^2}{m_p}\right)^{1/2}. \tag{4.7}$$

Usually, the effects of many-body interactions, such as electron-phonon and electron-electron interactions are treated by introducing a renormalized mass m^*. Hence, for a known charge carrier concentration, the measurement of ω_p enables the determination of the effective carrier mass and by this the many-body effects. Note that (4.7) which links the effective mass m_p to the plasma frequency ω_p has been derived from Lindhard's dielectric function. In contrast to the cyclotron effective mass, probed, e.g., in a quantum-oscillation experiment (cf. Sect. 4.5) which depends on electronic states at the Fermi surface, the plasma effective mass m_p includes all the occupied states, not only those close to the Fermi level, see e.g. [281] for a discussion on the different effective masses, i.e., band mass (band dispersion of electrons), cyclotron mass (periodic motion of electrons on the Fermi surface along orbits in position and momentum space in the presence of an external magnetic field), and plasma frequencies (associated with collective oscillations of the charged Fermi liquid).

In the dc limit the real part of the conductivity is

$$\sigma_1(\omega = 0) = \sigma_{\mathrm{dc}} = \frac{ne^2\tau}{m}. \tag{4.8}$$

(ii) **Phonons**

The absorption of electromagnetic waves by phonons can be described in a first approach by a Lorentz model for the complex dielectric constant $\tilde{\epsilon} = \epsilon_1 + \mathrm{i}\epsilon_2$

$$\tilde{\epsilon}(\omega) = 1 + \frac{\omega_p^2}{(\omega_0^2 - \omega^2) - \mathrm{i}\omega/\tau}, \tag{4.9}$$

which can be expressed in terms of a real and imaginary part of the optical conductivity:

$$\sigma_1(\omega) = \frac{\omega}{4\pi}\epsilon_2(\omega) = \frac{\omega_p^2}{4\pi} \cdot \frac{\omega^2/\tau}{(\omega_0^2 - \omega^2)^2 + \omega^2/\tau^2} \tag{4.10}$$

and

$$\sigma_2(\omega) = \frac{\omega}{4\pi}[1 - \epsilon_1(\omega)] = -\frac{\omega_p^2}{4\pi} \cdot \frac{\omega(\omega_0^2 - \omega^2)}{(\omega_0^2 - \omega^2)^2 + \omega^2/\tau^2}. \tag{4.11}$$

Here ω_0 denotes the oscillator frequency, $1/\tau$ the broadening of the oscillator due to damping, and ω_p the oscillator strength also referred to as the plasma frequency.

(iii) **Interband Transitions**

Their contributions to the optical conductivity usually manifest themselves in broad bands located in the mid- and near-infrared region, but also all the way up to the visible and above. It is common to model these effects by a sum of Lorentz oscillators

$$\tilde{\sigma}(\omega) = \sum_j \frac{\omega_{pj}^2}{4\pi} \cdot \frac{\omega}{\mathrm{i}(\omega_{0j}^2 - \omega^2) + \omega/\tau_j}. \tag{4.12}$$

Data Analysis

Figure 4.11 gives a schematic view of the above three contributions to the reflectivity and optical conductivity. The quantity, which is usually measured for conducting materials, is the optical reflectivity $R(\omega)$. If this quantity has been determined over a sufficiently wide spectral range, the complex electrodynamic response can be obtained.

The Kramers-Kronig relation gives the phase shift $\phi_r(\omega)$ of the reflected signal

$$\phi_r(\omega) = \frac{\omega}{\pi}\int_0^\infty \frac{\ln R(\omega') - \ln R(\omega)}{\omega^2 - \omega'^2}\mathrm{d}\omega', \tag{4.13}$$

from which the complex conductivity $\tilde{\sigma} = \sigma_1 + \mathrm{i}\sigma_2$ and dielectric constant $\tilde{\epsilon} = \epsilon_1 + \mathrm{i}\epsilon_2$ can be derived:

$$\sigma_1 = \frac{\omega}{4\pi}\epsilon_2(\omega) = \frac{\omega}{4\pi} \cdot \frac{4\sqrt{R(\omega)}[1 - R(\omega)]\sin\phi_r}{[1 + R(\omega) - 2\sqrt{R(\omega)}\cos\phi_r]^2}, \tag{4.14}$$

and

$$\sigma_2 = \frac{\omega}{4\pi}[1 - \epsilon_1(\omega)] = \frac{\omega}{4\pi}\left(1 - \frac{[1 - R(\omega)]^2 - 4R(\omega)\sin^2\phi_r}{[1 + R(\omega) - 2\sqrt{R(\omega)}\cos\phi_r]^2}\right). \tag{4.15}$$

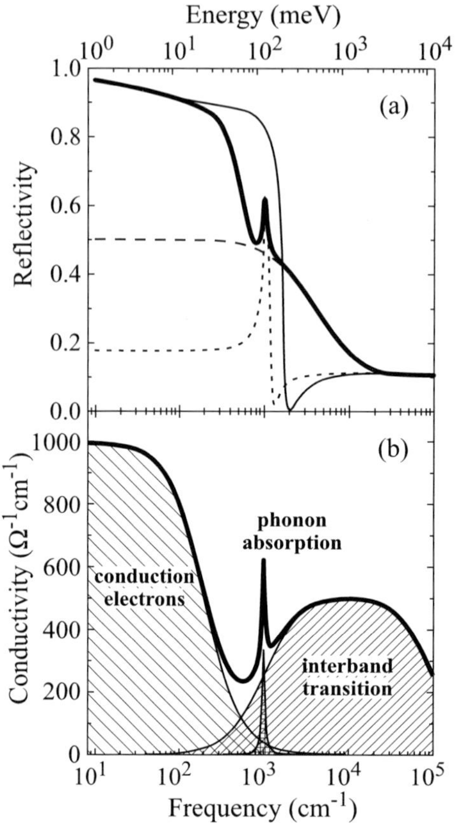

Fig. 4.11. Schematic representation of the frequency-dependent (a) reflectivity and (b) conductivity of a typical molecular conductor with contributions from (i) conduction electrons (thin solid line), (ii) interband transitions (dashed line) and (iii) a single phonon at 1000 cm^{-1} (dotted line). The total conductivity and reflectivity are shown by thick solid lines. After [276].

$(TM)_2X$ and $(ET)_2X$ salts

First extensive optical studies of the electronic properties of $(TM)_2X$ by C.S. Jacobsen et al. [282, 283, 284] provided information on the energy of charge-transfer processes and on the electron-phonon coupling: the large absorption features observed in the optical conductivity of the TMTTF salts have been assigned to intra- and intermolecular vibrations [285]. These studies have been supplemented by a series of more detailed investigations covering also the low-frequency range, see [264, 265, 275] and references therein. In accordance with the expectations for a strongly anisotropic metal with an open Fermi surface and strong electron-electron interactions, the optical response of the $(TMTTF)_2X$ and $(TMTSF)_2X$ salts was found to deviate markedly from that of an ordinary metal. The main feature is a gap which decreases from about 2000 cm^{-1} for $(TMTTF)_2PF_6$ to 200 cm^{-1} for $(TMTSF)_2PF_6$ with 99% of the spectral weight residing in the high-energy spectrum above the gap structure. For the metallic compounds, there is in addition a very narrow Drude peak — quite different from an ordinary Fermi liquid. As discussed earlier in this section, the data above the gap follow the expectation for a Luttinger liquid indicating that these compounds are well described by the Luttinger-liquid theory for not too low temperatures of the order of 200 K.

Fig. 4.12 shows reflectivity and conductivity data of κ-$(ET)_2Cu(NCS)_2$ at room temperature and $T = 12\,K$. Similar to what has been observed for many other molecular conductors, the room-temperature spectra reveal only a very small, if any, Drude contribution, although the materials are relatively good conductors within the planes. A narrow zero-frequency contribution, however, shows up below 50 K, cf. Fig. 4.12. The width of the low-temperature conductivity peak is only about $40\,cm^{-1}$ for κ-$(ET)_2Cu(NCS)_2$ and the related superconducting compounds of the κ-$(ET)_2X$ family [287]. In these metallic/superconducting compounds, the low-frequency spectral weight grows, to some extent, on the expense of the mid-infrared spectral weight upon cooling. This contrasts with the behavior of κ-$(ET)_2Cu[N(CN)_2]Cl$, where spectral weight shifts to higher frequencies and an optical gap opens around $900\,cm^{-1}$ upon cooling below 50 K [288], see Fig. 4.13. The opening of a gap in the charge sector of this material has been assigned to the Mott metal-insulator transition driven by strong Coulomb interactions. Applying dynamical mean field theory, J. Merino et al. [227] have calculated the optical conductivity for an on-site Coulomb repulsion $U \approx W = 4t$. They found a narrow Drude peak at low temperatures $T/t = 0.1$ which is already significantly suppressed for $T/t = 0.15$. In addition, their calculations revealed a low-temperature feature at a frequency $\omega = U/2$ due to interband transitions from a coherent

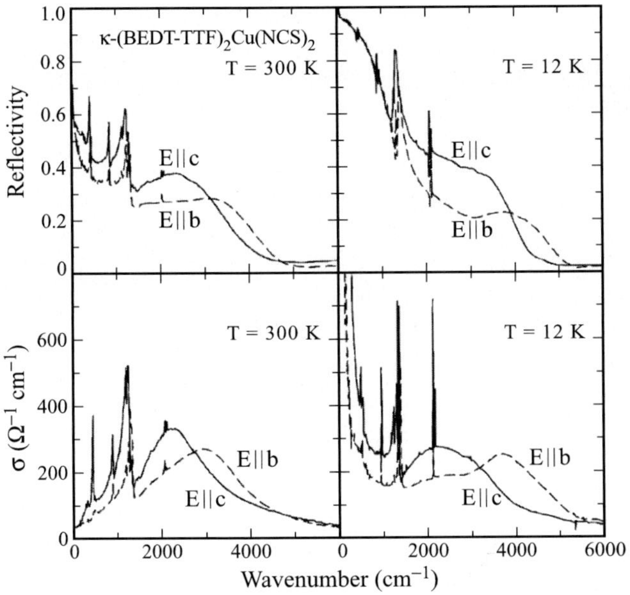

Fig. 4.12. Frequency dependence of the optical reflectivity and conductivity of κ-$(ET)_2Cu(NCS)_2$ at $T = 300\,K$ and $12\,K$ for $E \| b$ and $E \| c$. After [286].

quasiparticle peak at the Fermi energy to the upper Hubbard band and from the lower Hubbard band into the quasiparticle band.

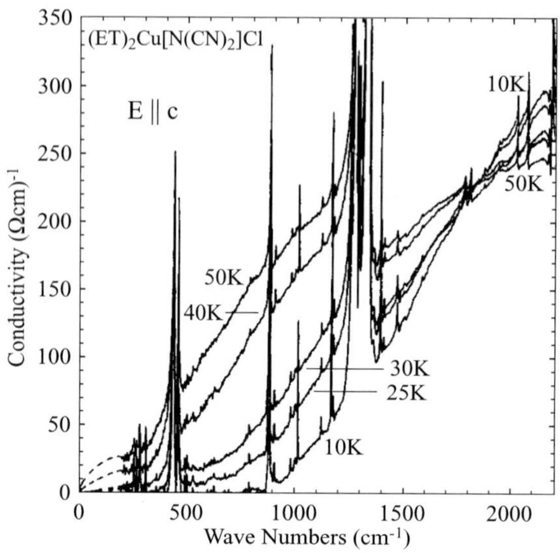

Fig. 4.13. Optical conductivity of κ-$(ET)_2Cu[N(CN)_2]Cl$ at various temperatures as a function of the wave number for infrared light polarized parallel to the c-axis. After [288].

A feature shared by many other quasi-2D molecular conductors is the existence of a broad mid-infrared band with a maximum centered around 2000 to 3000 cm^{-1}. The dependence of its intensity on the arrangement of the donor molecules, i.e. the packing pattern, led to the assignment of this feature to interband transitions across the dimerization gap of the donor molecules [289]. As Fig. 4.12 exemplifies, this mid-infrared band is particularly strongly pronounced for the κ-phase salts, consistent with the above assignment [290, 287]. Alternatively, an interpretation in terms of photon-assisted hopping of small polarons has been proposed [226].

Besides the broad electronic contribution, the infrared spectra of molecular conductors reveal narrow bands at frequencies of the molecular vibrations, cf. Fig. 4.12. These features, which are asymmetric in shape, are caused by electronic excitations which couple to intramolecular vibrations via the so-called electron-molecular vibration (EMV) coupling. The characteristic energies of those modes which couple the strongest are typically between 400 and 1600 cm^{-1}. In addition, a coupling to the low-energy lattice phonons can be observed. Due to the large molecules involved, their frequencies lie below approximately 200 cm^{-1}, see the following paragraph for a discussion of the electron-phonon coupling.

Electron-Phonon Coupling

Infrared reflectivity measurements can also be used to obtain information on the electron-phonon interaction. In molecular crystals, the coupling between the conduction electrons and the phonons is twofold. One kind of interaction, the so-called electron-molecular-vibration (EMV) coupling, involves the *intra*molecular vibrations which are characteristic of the molecular structure. This has to be distinguished from the electron-*inter*molecular-vibration coupling which refers to the interaction of the charge carriers with motions of almost rigid molecules around their equilibrium positions and orientations (translational and librational modes).

Electron-Molecular-Vibration Coupling

It is well known [291] that electrons in the HOMOs of the TTF molecule and its derivatives couple strongly to the totally symmetric (A_g) molecular vibrations via the modulation of the HOMO energy, E_{HOMO}, by the atomic displacements. The linear EMV coupling constant g_i for mode i is defined as:

$$g_i = \frac{1}{h\nu_i} \cdot \frac{\partial E_{\mathrm{HOMO}}}{\partial Q_i}, \tag{4.16}$$

where Q_i is an intramolecular normal coordinate and ν_i the mode frequency. The effective electron-intramolecular-phonon coupling constant λ_i can then be calculated using

$$\lambda_i = 2g_i^2 h\nu_i N(E_F), \tag{4.17}$$

where $N(E_F)$ is the density of states at the Fermi level.

The sharp features superimposed on the electronic background in Figs. 4.12 and 4.13 can be attributed to molecular vibrations. An assignment of these features to the various vibrational modes ([290, 89, 292, 293, 103, 278] and references therein) is feasible by comparing spectra of different isotopically labelled salts with calculations based on a valence-force-field model by M.E. Kozlov et al. [294, 295]. For the ET molecule, the modes with the strongest coupling constants are those involving the central carbon and sulfur atoms [294, 295, 296] at which the HOMOs have the largest amplitudes [82] (Fig. 2.1). The frequencies of these C=C stretching and ring-breathing modes are $\nu_2 = 1465\,\mathrm{cm}^{-1}$ ($g_2 = 0.165$), $\nu_3 = 1427\,\mathrm{cm}^{-1}$ ($g_3 = 0.746$) and $\nu_9 = 508\,\mathrm{cm}^{-1}$ ($g_9 = 0.476$), where the calculated coupling constants are given in the brackets. Despite these sizeable coupling constants, several studies — especially those of the mass isotope shifts on T_c for the ET salts [297] — indicate that the EMV coupling seems to play only a minor role in mediating the attractive electron-electron interaction for the superconductivity, cf. Sect. 6.6.

Electron-Intermolecular-Vibration Coupling

While much experimental data are available on the EMV coupling, relatively little is known about the coupling of the charge carriers to the low-lying intermolecular phonons. This interaction is provided by the modulation of the effective charge-transfer integrals $t_{\rm eff}$ between neighboring molecules during their translational or librational motions. Within the Eliashberg theory, the dimensionless electron-intermolecular-phonon coupling constant λ is given by

$$\lambda = 2\int \frac{\alpha^2(\omega)}{\omega} F(\omega) \mathrm{d}\omega, \tag{4.18}$$

where $\alpha(\omega)$ is the electron-phonon-coupling constant, ω the phonon frequency and $F(\omega)$ the phonon density of states. The Eliashberg function $\alpha^2(\omega)F(\omega)$ can, in principle, be derived from tunneling characteristics of strong-coupling superconductors or via point-contact measurements. The latter experiments have been carried out by A. Nowack et al. on the β-type $(ET)_2X$ salts with $X = I_3$ and AuI_2 [298, 299] yielding $\lambda \simeq 1$. Some of the frequencies of the intermolecular modes have been determined by employing Raman and far-infrared measurements [300, 301, 302, 303]. More recent studies including inelastic-neutron [304] and Raman-scattering [88, 305, 306, 303] have focused on an investigation of the role of intermolecular phonons for superconductivity. These experiments yielded quite sizeable superconductivity-induced phonon-renormalization effects [6] which clearly indicate a significant coupling of the superconducting charge carriers to the intermolecular phonons and suggest an important role of these modes in the pairing interaction.

A strong coupling of the charge carriers to the intermolecular phonons can be inferred also from temperature-dependent Raman scattering studies of the phonon dynamics in κ-$(ET)_2Cu(NCS)_2$ and κ-$(ET)_2Cu[N(CN)_2]Br$ [292, 308, 88, 305, 306, 303]. The observed anomalous temperature dependence of the low–frequency phonons around and below T_c were found to be consistent with an isotropic gap $2\,\Delta$ close to 2.8 meV [306]. From the reported frequency shifts, the electron–phonon coupling constants λ_i have been calculated yielding a total coupling constant of $\lambda_{tot} = 0.97 \pm 0.11$ [306].

Probing the Charge on the Molecule

As has been shown by various authors, the positions of intramolecular modes reveal systematic shifts upon changing the oxidation state of the molecule [309, 310, 311]. It was found that the fully symmetric Raman-active modes ν_2, ν_3, and ν_6 reveal a significant and linear shift to lower frequencies as the charge per ET molecule changes from 0 to $+e$ or even $+2e$. Hence, measurements on the intramolecular vibrations can be used to study the phenomenon

[6] In general, the effect of superconductivity on the phonon frequencies is as small as $2\Delta/W$, i.e. of the order of 10^{-2} or less, see e.g. [307], and Sect.6.7

of charge ordering, i.e. periodic deviations from a homogeneously charged donor molecule, see Sec. 4.4.3. Information on the charge-ordering pattern is obtained by employing an analysis of the symmetry of the vibrational bands [311].

4.3.2 Thermal and Magnetic Properties

For both classes of quasi-1D and -2D charge-transfer salts, a variety of different ground states has been found depending on parameters such as chemical composition, external pressure, or magnetic field. A striking feature common to both families is a long-range magnetically ordered state which shares a common phase boundary with the superconducting state.

Quasi-One-Dimensional $(TM)_2X$

We start the discussion with the most anisotropic sulfur-based compounds $(TMTTF)_2X$ with X = PF_6 or Br. As shown in Fig. 4.6, the charge dynamics of these salts are characterized by a metallic response above a temperature $T_\rho \sim 100 - 250\,K$ followed by an activated behavior for $T < T_\rho$. This indicates an opening of a charge gap Δ_ρ which has been assigned to the joint role of Coulomb correlations and low dimensionality. According to transport measurements, the charge gap, which is related to the minimum in the resistivity at T_ρ by $\Delta_\rho \approx T_\rho \pi$, decreases from about 700 K for X = PF_6 to 300 K for the Br salt. This is in line with the results from photoemission, optical and dielectric studies, yielding $\Delta_\rho \approx 500 - 1500\,K$ (PF_6) and $300 - 500\,K$ (Br) [270, 312]. Remarkably enough, the spin degrees of freedom as probed, e.g., by magnetic susceptibility or ESR measurements, remain unaffected by the drastic changes in the transport properties. In addition, these magnetic studies reveal a phase transition from a paramagnetic high-temperature into a non-magnetic low-temperature state. For the X = PF_6 salt, for example, the combination of magnetic susceptibility [313], NMR [314] and X-ray [315] investigations showed that the transition into a non-magnetic state at $T_{SP} = 19\,K$ is accompanied by a periodic lattice modulation, i.e., a tetramerized state. This so-called spin-Peierls transition — the magnetic analogue to the Peierls instability in quasi-1D conductors — is a consequence of the coupling of the 1D spin system to the 3D phonons: a one-dimensional arrangement of localized spins with uniform antiferromagnetic exchange coupling J can lower its energy by adopting a modulated structure, see Sect. 4.4.2. As a result, the exchange coupling constant between neighboring spins alternates between J_1 and J_2, $J_{1,2} = J[1 \pm \delta]$ and δ the alternation parameter. This leads to the formation of spin singlet $S = 0$ pairs and a singlet-triplet gap Δ_σ opens in the magnetic excitation spectrum. Figure 4.14(a) shows the spin susceptibility χ_s (crosses) for $(TMTTF)_2PF_6$ as derived from the integrated ESR absorption [316]. The experiments revealed an isotropic spin susceptibility with a broad maximum around 340 K followed by a rapid drop below

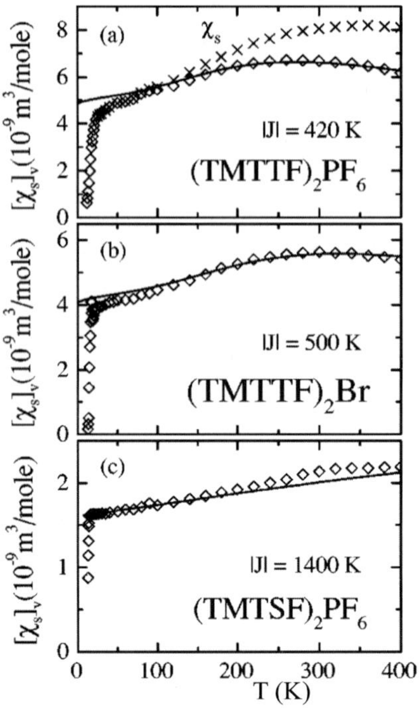

Fig. 4.14. Temperature dependence of the spin susceptibility in b' direction as obtained by measurements of the ESR intensity (crosses) and converted to constant volume (diamonds) of (a) $(TMTTF)_2PF_6$, (b) $(TMTTF)_2Br$, and (c) $(TMTSF)_2PF_6$. The lines correspond to fits using different models as described in the text. After [316].

$T_{SP} = 19\,\mathrm{K}$.

To compare the experimental results with theoretical predictions which usually refer to constant-volume processes, the data have been corrected for the thermal contraction of the materials. To this end, the susceptibility data taken at constant pressure $(\chi_s)_p$ were transformed into susceptibilities at constant volume $(\chi_s)_V$ (diamonds) by applying the same temperature-dependent scaling factor $(\chi_s)_V/(\chi_s)_p$ derived for $(TMTSF)_2PF_6$ from NMR and X-ray investigations [211]. This procedure leads to an overall reduction of the susceptibility at high temperature by about 25% compared to the $(\chi_s)_p$ data, cf. Fig. 4.14. The high-temperature maximum in $(\chi_s)_V$ resembles the susceptibility of a homogeneous antiferromagnetic S $= 1/2$ Heisenberg chain with nearest-neighbor exchange coupling J described by the Hamiltonian

$$\mathcal{H} = J \sum S_i \cdot S_{i+1}. \tag{4.19}$$

The thermodynamic and magnetic properties of (4.19) for finite systems were first studied by J.C. Bonner and M.E. Fisher [317]. Using a numerical expression for the Bonner-Fisher results and employing a g factor of 2.002, M. Dumm et al. were able to fit the high-temperature data around the maximum reasonably well [316], yielding an antiferromagnetic exchange coupling constant $|J| = 420\,\mathrm{K}$, cf. solid line in Fig. 4.14(a). From a similar procedure applied to the constant-volume susceptibilities of the related

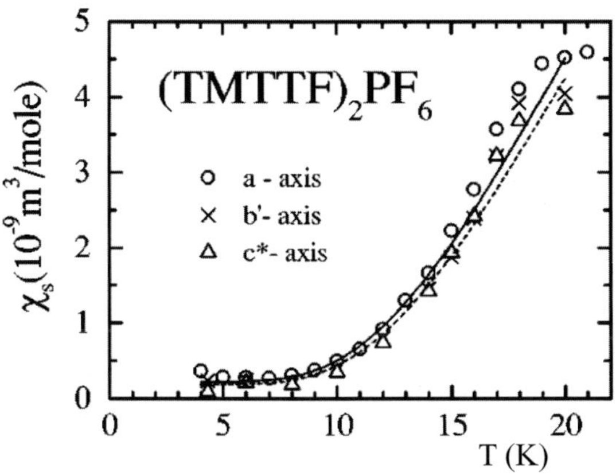

Fig. 4.15. Low-temperature part of the spin susceptibility as obtained from measurements of the ESR intensity for $(TMTTF)_2PF_6$ along the three crystallographic axes. The solid line corresponds to a fit according to (4.20). After [316].

salts $(TMTTF)_2ClO_4$ and $(TMTTF)_2Br$, cf. Fig. 4.14(b), coupling constants of $|J| = 430\,\text{K}$ (ClO_4) and 500 K (Br) have been inferred [316]. According to these results, the spin degrees of freedom of the $(TMTTF)_2X$ salts are well described by uniform antiferromagnetic Heisenberg chains with localized $S = 1/2$ spins residing on a pair of TMTTF molecules consistent with the weak structural dimerization of these compounds, cf. Fig. 4.14.

Details of the magnetic susceptibilities near the spin-Peierls transition are shown in Fig. 4.15. The figure shows the isotropic reduction of $(\chi_s)_V$ below $T_{SP} = 19\,\text{K}$. At low temperatures, i.e., for $T \leq 12\,\text{K}$, the data can be described by an expression given by L. Bulaevskii [318] for an alternating antiferromagnetic $S = 1/2$ Heisenberg chain

$$\chi_s(T) = \frac{\alpha}{T} \exp\left(-\frac{J_1\beta}{T}\right), \tag{4.20}$$

where α and β are constants tabulated in [318]. Using the above exchange constant $|J| = 420\,\text{K}$ results in an alternation parameter $\gamma = [1-\delta]/[1+\delta] = 0.91$, corresponding to $\delta = 0.0476$. According to the mean-field theory by E. Pytte [319], δ is related to the magnetic excitation gap by

$$\delta(T) = \frac{\Delta_s(T)}{1.637J}, \tag{4.21}$$

yielding $\Delta_\sigma = 32.3\,\text{K}$ for $(TMTTF)_2PF_6$ in the $T = 0$ limit.

The lattice response at the spin-Peierls transition has been measured by high-resolution thermal expansion measurements. Figure 4.16(a) shows the expansion coefficient along the b-axis, i.e. perpendicular to the stacking axis, for $(TMTTF)_2PF_6$ [320]. At $T_{SP} = 18\,\text{K}$ (see inset of Fig. 4.16, left panel), a pronounced second-order phase transition anomaly has been observed.

The above experimental results demonstrate particularly clearly that for the present quasi-1D systems, the spins and charges behave independently.

Indeed, the decoupling of spin-and charge-degrees of freedom is one of the hallmarks of a so-called Tomonaga-Luttinger liquid — the concept which describes interacting Fermions in one dimension, cf. Sec. 4.3.1.

On going to the selenium analogue $(TMTSF)_2PF_6$, the metallic character prevails and the system stays conducting down to low temperatures until at around 12 K the resistivity suddenly increases. The 1D-character of the material, i.e., a Fermi surface consisting of a pair of open sheets exhibiting good nesting properties, cf. Fig. 4.4, suggests that this transition marks a $2k_F$ density-wave instability. In fact, the anisotropy in the magnetic susceptibility below the transition temperature, with b' the easy and c^* the hard axis [321], the observation of an antiferromagnetic resonance [322] and the detection of local fields by NMR and μSR [323, 324] clearly proved the magnetic character of the ground state. Proton NMR experiments suggested the formation of a sinusoidal SDW with an incommensurate wave vector $\boldsymbol{q}_{SDW} = (0.5, 0.2-0.24, 0.06)$ and an amplitude of $0.08 - 0.085\,\mu_B$ [325, 326]. By means of detailed X-ray investigations, J.P. Pouget et al. [327] were able to detect, below T_{SDW}, satellite reflections with $\boldsymbol{q}_1 = (0.5, 0.25, 0.25)$ and $\boldsymbol{q}_2 = 2\boldsymbol{q}_1$, where $\boldsymbol{q}_1$ is in accordance with the wave vector seen in the above NMR measurements. These results indicate a peculiar mixed $2k_F$ CDW-SDW ground state to be realized in this material. A pronounced coupling to the lattice degree of freedoms has also been inferred from the sizeable anomaly in the coefficient of thermal expansion at T_{SDW} displayed in Fig. 4.16, right panel.

Remarkably enough, the high-temperature magnetic susceptibility of the metallic $(TMTSF)_2PF_6$ salt, after correction for the thermal contraction, shows a similar temperature dependence as the non-metallic sulfur compounds, although the maximum in $(\chi_s)_V$ around 350 K appears less clearly pronounced, cf. Fig. 4.14(c) [316]. In an attempt to determine the exchange coupling constant that couples the itinerant spins in this compound, M. Dumm et al. applied a result derived by W.A. Seitz and D.J. Klein [328] for a linear half-filled Hubbard model in the limit of strong Coulomb repulsion U, i.e., $t_a/U = 0.2$. As a result of their fitting procedure, which, however, fails to describe the maximum in $(\chi_s)_V$ around 350 K, the authors propose $|J| \approx 1400$ K for this compound.

On going from $(TMTTF)_2PF_6$ to $(TMTSF)_2ClO_4$, the metallic character becomes reinforced as indicated by the substantial increase in the room-temperature conductivity from $\sigma_\parallel \approx 30\,\Omega^{-1}\text{cm}^{-1}$ ($(TMTTF)_2PF_6$) to $\sigma_\parallel \approx 500\,\Omega^{-1}\text{cm}^{-1}$ ($(TMTSF)_2ClO_4$) and the system undergoes a superconducting transition at 1.2 K [36]. Despite these marked differences in their metallic properties, the spin susceptibilities behave rather similarly with a monotonic drop from 300 to 50 K by about 40%.

The various above-mentioned symmetry-broken ground states indicate that at low temperatures, three-dimensional couplings are active. The cross-over temperatures, where the systems change from predominantly one- over

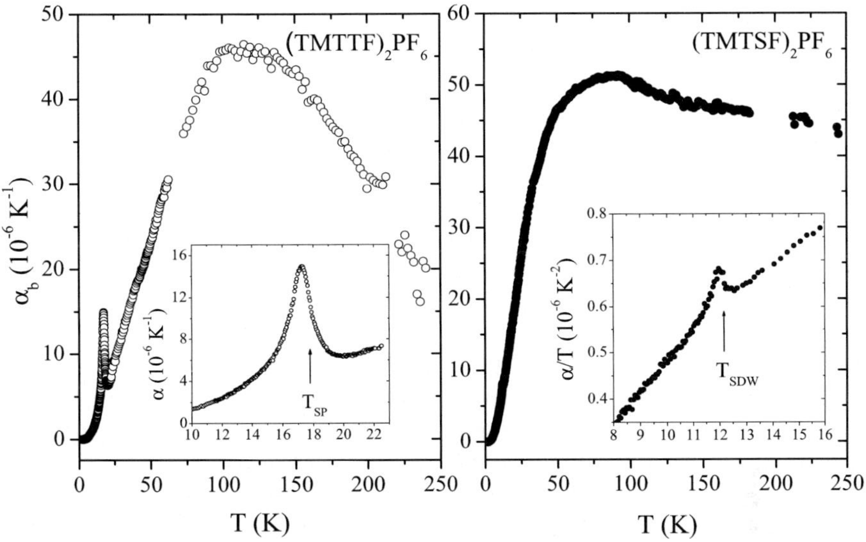

Fig. 4.16. Coefficient of thermal expansion along the b-axis for $(TMTTF)_2PF_6$ (left) and $(TMTSF)_2PF_6$ (right). The insets show an expansion of the data around the spin-Peierls and SDW phase transitions, respectively. After [320].

two- to eventually three-dimensional behavior are of the order of t_i/k_B and t_l/k_B, where t_i and t_l denote the transfer integrals for the intermediate- and low-conductivity directions, respectively [329]. Thus, at sufficiently high temperatures $T > t_i/k_B$, short-range 1D fluctuations predominate and 1D theory should be a good starting point for understanding the electronic properties of these materials.

Quasi-Two-Dimensional (BEDT-TTF)$_2$X

For the quasi-2D $(ET)_2X$ compounds, the Fermi-surface topologies indicate that nesting is less favorable and, indeed, in most of the materials the Fermi surface remains intact and the systems stay conducting down to low temperatures. However, as has been pointed out by C. Hotta et al. [193], see also [194], nesting might also be of importance for these materials. According to their work, the key parameter which determines the shape of the Fermi surface, and by this, the possibility of a nesting-induced SDW instability, is the degree of splitting of the anti-bonding HOMO bands, cf. Fig. 4.1. Nevertheless, as for the $(TM)_2X$ salts, superconductivity is close to an antiferromagnetically ordered state, see the pressure-temperature phase diagram for the κ-phase $(ET)_2X$ salts in Sect. 4.6. In this respect, the salt with the complex anion X = Cu[N(CN)$_2$]Cl is of particular interest as it is an antiferromagnetic insulator which, by the application of a small hydrostatic pressure of only 300 bar, can be transformed into a superconductor with the highest T_c of 12.8 K among this family [105]. For an understanding of the superconducting state and its interplay with magnetism for this class of materials, an identification of the nature of the magnetic state is of crucial importance. There has been a debate on whether the magnetic order results from the nesting of the open parts of the Fermi surface, i.e., an itinerant SDW-type of magnetism, or whether it is of local character driven by strong electron-electron correlations. First magnetic measurements aiming at a determination of the spin structure revealed indications for an antiferromagnetic transition at $T_N = 45$ K and a weak ferromagnetic behavior below 22 K [330]. This interpretation has been questioned by the results of subsequent susceptibility measurements by K. Miyagawa et al. [168], yielding a Néel-temperature of $T_N = 27$ K. These authors assigned the weak ferromagnetic moment below 23 K to the spontaneous canting within the ET layers of localized spins having an easy axis perpendicular to the layers. Further insight into the low-temperature magnetic structure has been gained by detailed investigations of the ^{13}C- and ^{1}H-NMR [331, 332]. From ^{13}C-NMR spectra on a powdered sample, an only weakly temperature-dependent line shape has been observed at high temperatures $T \geq 60$ K. Upon approaching the Néel-temperature from above, however, the spectra broaden significantly which has been attributed to short-range antiferromagnetic ordering within the ET layers [333].

Supplementary information has been received from ^{1}H-NMR on single-crystalline material, see Fig. 4.17. It has been argued in [333] that because

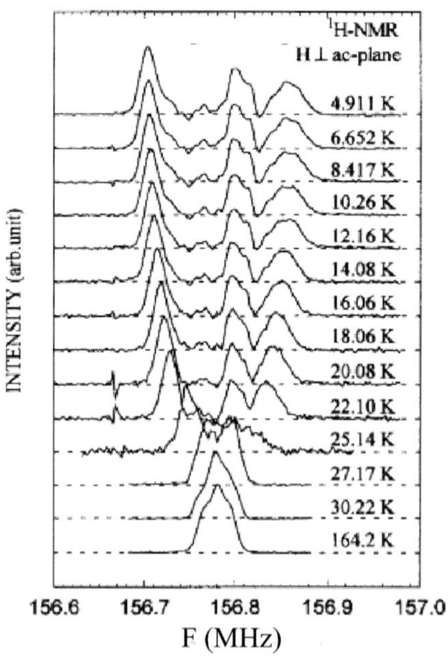

Fig. 4.17. ^{1}H-NMR spectra of single crystalline κ-(ET)$_2$Cu[N(CN)$_2$]Cl at a field of 3.7 T perpendicular to the conducting *ac*-plane at different temperatures. After [168].

of the large hyperfine fields acting on the ^{13}C nuclei as a consequence of the magnetic ordering, the spectra spread over a wide frequency range and cannot be detected by the applied radio frequency pulse. Therefore, the ^{1}H-nuclear probe is superior in this case because its hyperfine coupling to the conduction electrons is much smaller. Below T_N, the broadened line splits into three distinct lines, indicating the presence of inhomogeneous local fields at the nuclear site, consistent with the antiferromagnetic nature of the transition. Although the site assignment of the split lines was not possible due to the complexity that comes from the presence of eight inequivalent ^{1}H sites, two fundamental conclusions have been drawn from these observations. First, the discrete number of lines indicates that the ordering wave vector is commensurate with the lattice. Second, from the size of the line splitting, an ordered moment of $(0.4 - 1.0)\,\mu_B$ has been estimated [168]. From these measurements, it has been determined that the easy magnetization direction of the ordered spins is perpendicular to the conducting layers and that a small canting of the spins below 23 K causes a weak ferromagnetic moment parallel to the layers. More recent ^{13}C-NMR experiments confirmed the commensurate structure and yielded a moment of $0.45\,\mu_B/(\mathrm{ET})_2$ [334].[7] The commensurate magnetic structure along with a relatively large moment are at variance to what is expected for a conventional SDW scenario, which suggests that the magnetic ordering is of local character, i.e., driven by strong electron-electron correlations.

It is instructive to compare the magnetic susceptibility of the insulating antiferromagnet κ-$(\mathrm{ET})_2\mathrm{X}$, X = Cu[N(CN)$_2$]Cl with that of the related metallic κ-phase salts with X = Cu[N(CN)$_2$]Br and Cu(NCS)$_2$, see Fig. 4.18(a). While at high temperatures, the susceptibilities of the various salts are only weakly temperature dependent, a rather steep decrease is found below about 50 – 60 K. Upon further cooling, the drop in $\chi(T)$ for the X = Cu[N(CN)$_2$]Br and Cu(NCS)$_2$ salts indicates the onset of superconductivity while the sharp increase below about 30 K for the X = Cu[N(CN)$_2$]Cl reflects the magnetic transition discussed above. These data can be compared with the results of the spin-lattice relaxation rate T_1^{-1} divided by temperature as derived from ^{13}C-NMR measurements [332, 331, 333, 336, 163], see Fig. 4.18(b). The data in Fig. 4.18(b) were found to differ markedly from what is expected for a simple metal. First, $(T_1T)^{-1}$ for the κ-phase compounds reveals an overall enhancement by a factor 5 – 10 compared to the usual Korringa-type behavior of uncorrelated electrons, as realized to a good approximation in the α-$(\mathrm{ET})_2\mathrm{NH_4Hg(SCN)_4}$ salt [337]. Second, upon cooling, $(T_1T)^{-1}$ for the metallic salts with X = Cu(NCS)$_2$ and Cu[N(CN)$_2$]Br gradually increases down to a temperature $T^* \simeq 50$ K, below which a steep decrease sets in. Apparently, this anomalous feature has its correspondence

[7] The magnetic moment of deuterated κ-$(\mathrm{D_8\text{-}ET})_2$Cu[N(CN)$_2$]Br showing an antiferromagnetic insulating ground state (see Sect. 4.6) similar to κ-$(\mathrm{ET})_2$Cu[N(CN)$_2$]Cl has been found to be $0.26 - 0.3\,\mu_B/(\mathrm{ET})_2$ [334, 335].

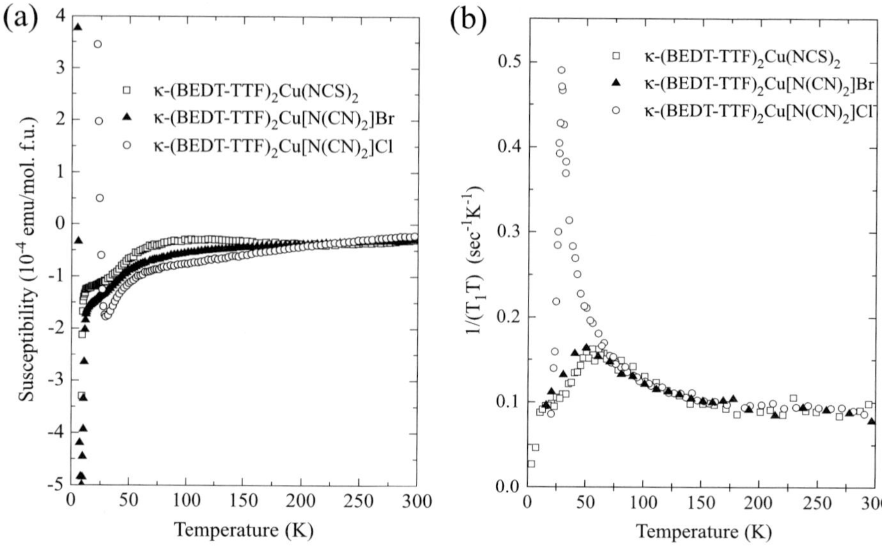

Fig. 4.18. Temperature dependence of (a) the static spin susceptibility of κ-(ET)$_2$X salts (a diamagnetic core contribution of -4.7×10^{-4}emu/mol has been subtracted) and (b) the ^{13}C-nuclear spin-lattice-relaxation rate divided by temperature for various κ-(ET)$_2$X salts. After [332].

in the drop of the static susceptibility, see Fig. 4.18(a). For the non-metallic κ-(ET)$_2$Cu[N(CN)$_2$]Cl, $(T_1T)^{-1}$ continues to increase upon cooling with a divergent peak around 27 K indicative of a magnetic ordering. The temperature dependence of $(T_1T)^{-1}$ above 27 K has been assigned to critical magnetic fluctuations [332, 331].

The unusual temperature dependence of $(T_1T)^{-1}$ and its significant enhancement around 50 K have been assigned to the effect of strong antiferromagnetic spin fluctuations with a finite wave vector $\boldsymbol{Q}$ [332, 333, 336, 163]. The latter might be related to the ordering wave vector that characterizes the antiferromagnetic state in κ-(ET)$_2$Cu[N(CN)$_2$]Cl. It has been emphasized by A. Kawamoto et al. [332] and K. Kanoda [333] that, despite the rapid drop below $T^* \simeq 50$ K, the overall enhancement of $(T_1T)^{-1}$ persists down to low temperature until the onset of superconductivity reflecting the highly correlated nature of the metallic state in these compounds. NMR investigations at various pressures revealed that with increasing the pressure, the maximum at T^* shifts to higher temperatures while its size becomes progressively reduced [336]. At pressures above 4 kbar the peak is replaced by a normal Korringa-type behavior, i.e., $(T_1T)^{-1}$ = const. In [333, 338] the abrupt reduction in $(T_1T)^{-1}$ below $T^* \simeq 50$ K has been linked phenomenologically to a pseudo-gap behavior as has been discussed also for the normal state of the high-T_c cuprates, see, e.g. [339] for a review. Actually, the formation of a pseudo-gap for the κ-phase $(ET)_2X$ salts was first proposed by V. Kataev et al. [340] based on their ESR results yielding a reduction of the spin susceptibility around 50 K. The same conclusion has been drawn from the above static magnetic susceptibility measurements [332] and investigations of the Knight-shift, K_s [336].

Cooling through $T^* \simeq 50\,\mathrm{K}$ does not only cause anomalies in the above magnetic properties but also leads to clear signatures in transport, acoustic, optical and thermodynamic quantities. As mentioned above, for both superconducting compounds a distinct peak shows up in the temperature derivative of the electrical resistivity $\partial\rho/\partial T$ [233, 234, 160] indicating a change in the charge dynamics. A pronounced softening of the longitudinal ultrasound velocity for κ-$(ET)_2Cu[N(CN)_2]Br$ and κ-$(ET)_2Cu(NCS)_2$ with distinct minima at $T^* \simeq 38\,\mathrm{K}$ and 46 K, respectively, has been attributed to a coupling between acoustic phonons and antiferromagnetic fluctuations [341, 342]. An even more strongly pronounced dip was found in the ultrasound velocity for pressurized κ-$(ET)_2Cu[N(CN)_2]Cl$ [343]. This feature reaches its maximum amplitude near 34 K and $p \simeq 210\,\mathrm{bar}$, which lies close to the critical end-point of the first-order metal-insulator transition in the temperature-pressure phase diagram, see Fig. 4.34. Most remarkably, the ultrasound velocity anomaly remains visible also for pressures much below 210 bar, i.e. in the insulating regime. It was argued in [343] that the growth of the ultrasound velocity anomaly seen upon approaching the critical point reflects a diverging compressibility of the electronic degrees of freedom, consistent with the predictions by dynamical mean-field theory (DMFT) [344]. More recently, S. Hassan et al. [345] have calculated the lattice response and the accompanied ultrasound velocity anomaly at the Mott transition by considering the compressible Hubbard model and treating the electron correlations within DMFT. Their calculations revealed, indeed, a pronounced ultrasound velocity anomaly which is largest along the crossover line continuing smoothly the first-order phase-transition above the critical endpoint, cf. phase diagram in Sec. 4.6. No explanation, however, could be given for the strong bending of this line and its continuation into the insulating range observed in the experiments by D. Fournier et al. [343].

Alternatively, an interaction between the phonon system and magnetism has been suggested by Y. Lin et al. based on their Raman scattering experiments [346, 280]. Theoretical studies have attempted to explain both the acoustic and Raman experiments by a correlation-induced crossover from a coherent Fermi liquid at low temperatures to an incoherent bad metal at high temperatures [347, 227]. According to this work, pronounced phonon anomalies as well as anomalous transport and thermodynamic properties are expected to occur at the crossover temperature T^*.

More insight into the nature of the anomaly at T^* and its interrelation with superconductivity can be obtained by directly probing the lattice re-

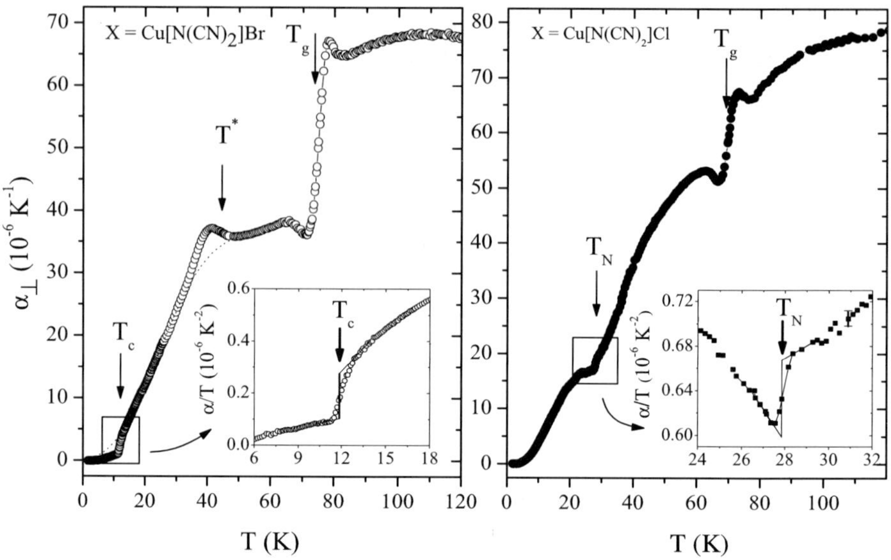

Fig. 4.19. Linear thermal expansion coefficient perpendicular to the planes of highest conductivity, $\alpha_\perp$, as a function of temperature for superconducting κ-$(ET)_2Cu[N(CN)_2]Br$ (left panel) and insulating κ-$(ET)_2Cu[N(CN)_2]Cl$ (right panel). The inset shows details of $\alpha_\perp$ for both salts as $\alpha_\perp/T$ vs T at the superconducting (X=$Cu[N(CN)_2]Br$) and antiferromagnetic (X=$Cu[N(CN)_2]Cl$) phase transition. Arrows indicate different kinds of anomalies as explained in the text. After [155]

sponse using high-resolution thermal expansion measurements [156, 157, 155]. Figure 4.19 compares the linear coefficient of thermal expansion, $\alpha(T) = \partial \ln l(T)/\partial T$, where $l(T)$ is the sample length, measured perpendicular to the planes for the superconducting κ-$(ET)_2Cu[N(CN)_2]Br$ (left panel) with that of the non-metallic κ-$(ET)_2Cu[N(CN)_2]Cl$ salt (right panel). For both compounds, various anomalies have been observed as indicated by the arrows [155, 348]. These are (i) large step-like anomalies at $T_g = 70 \sim 80\,\mathrm{K}$ which are due to a kinetic, glass-like transition associated with the ethylene endgroups, cf. Sect. 3.3 and (ii) a distinct peak in $\alpha(T)$ at T^*. The latter feature, also observed for the superconductor κ-$(ET)_2Cu(NCS)_2$ (not shown) is absent in the non-metallic κ-$(ET)_2Cu[N(CN)_2]Cl$ salt, cf. right panel of Fig. 4.19 [155]. As demonstrated in the insets of Fig. 4.19, the transitions into the superconducting ($T_c = 11.8\,\mathrm{K}$) and antiferromagnetic ($T_N = 27.8\,\mathrm{K}$) ground states for the X=$Cu[N(CN)_2]Br$ and X=$Cu[N(CN)_2]Cl$ salts, respectively, are accompanied by distinct second-order phase transition anomalies in the coefficient of thermal expansion.

Figure 4.20 shows the anomalous contribution, $\delta\alpha_i(T) = \alpha_i(T) - \alpha_{ib}(T)$, to the uniaxial thermal expansion coefficients along the principal axes, $\alpha_i(T)$, at T_c and T^* for the superconducting salts X=Cu[N(CN)$_2$]Br (left panel) and X=Cu(NCS)$_2$ (right panel) obtained after subtracting a smooth background α_{ib} (dotted line in the left panel of Fig. 4.19) [235]. Judging from

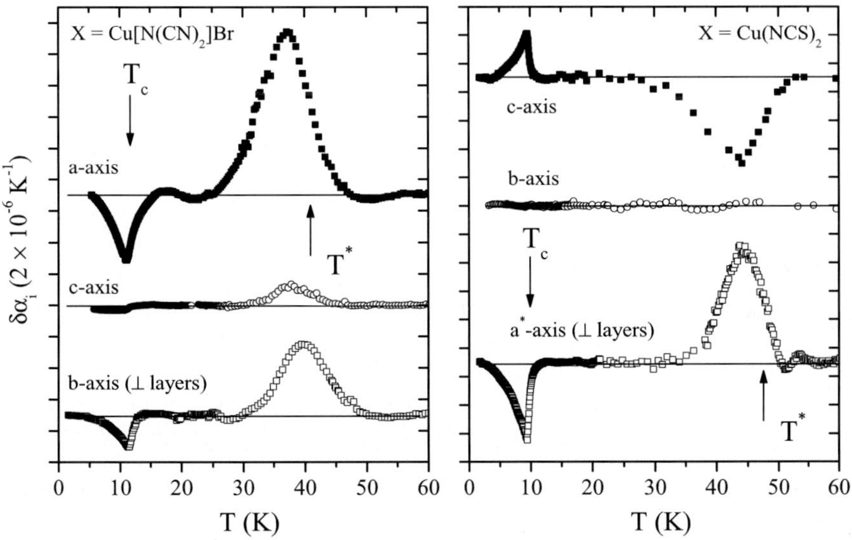

Fig. 4.20. Anomalous contributions, $\delta\alpha_i(T) = \alpha_i(T) - \alpha_{ib}(T)$ to the uniaxial thermal expansion coefficients $\alpha_i(T)$, obtained by subtracting a smooth background contribution $\alpha_{ib}(T)$, along the three principal axes for the superconducting salts κ-(ET)$_2$Cu[N(CN)$_2$]Br (left panel) and κ-(ET)$_2$Cu(NCS)$_2$ (right panel). After [235].

the shape of the anomalies at T^*, i.e. their sharpness and magnitude, it has been suggested that this feature is reminiscent of a second-order phase transition [235]. Moreover, Fig. 4.20 uncovers an intimate interrelation between the phase-transition anomalies at T_c and T^*: while both features are correlated in size, i.e. a large (small) anomaly at T_c complies with a large (small) one at T^*, they are anticorrelated in sign. A positive peak at T_c goes along with a negative anomaly at T^* and vice versa. According to the Ehrenfest relation

$$\left(\frac{\partial T^\star}{\partial p_i}\right)_{p_i\to 0} = V_{\text{mol}} \cdot T^\star \cdot \frac{\Delta\alpha_i}{\Delta C}, \tag{4.22}$$

which relates the uniaxial-pressure dependence of a second-order phase-transition temperature $T^\star$ to the discontinuities in α_i, $\Delta\alpha_i$, and that of the specific heat, ΔC, the opposite signs of the discontinuities $\Delta\alpha_i$ at T_c and T^* imply that the uniaxial-pressure coefficients of these two critical temperatures are strictly anticorrelated. In [155, 235], it has been argued that the transition at T^* is of electronic origin and most likely related to the Fermi-surface topology. Based on the above uniaxial-pressure results, it has been proposed that T_c and T^* mark competing instabilities on disjunct parts of the Fermi surface [155, 235]: while the instability at T^* most likely involves only the minor quasi-1D fractions (see Fig. 4.4 and [336]), the major quasi-2D parts are subject to the superconducting instability at lower temperatures. As a result, these studies hint at the opening of a real gap associated with T^* on the small 1D-parts of the FS as opposed to a pseudo-gap on the major quasi-2D fractions. The condensation of parts of the FS into a density-wave below T^* would imply the onset of anisotropies in magnetic and transport properties. In fact, this has been found in recent orientational-dependent studies on both compounds [236]: cooling through T^* is accompanied by the onset of a small but distinct anisotropy in the magnetic susceptibility. As can be seen in Fig. 4.21, T^* affects also the charge degrees of freedom where below T^* the b-axis transport becomes more resistive compared to that along the c-axis [236]. These authors proposed that below T^* a static or fluctuating charge-density-wave (CDW) on minor parts of the FS coexists with the metallic phase on the remaining quasi-2D fractions. In [349], it has been argued that such an interpretation could not explain the smooth evolution with field of the magnetic breakdown oscillations observed in κ-$(ET)_2Cu(NCS)_2$ [350]. Instead, based on their findings of a spatially inhomogeneous metallic state, they suggested that two distinct types of domain exist below the glassy transition at $T_g \simeq 70\,K$. These two domains are characterized by slightly different Fermi-surface topologies. While one type of domain may have a Fermi

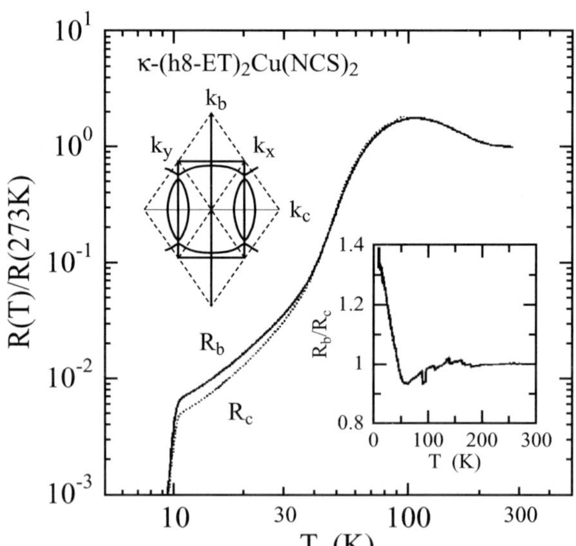

Fig. 4.21. Temperature dependence of the normalized resistance along the in-plane b- and c-axis of κ-$(ET)_2Cu(NCS)_2$. The top left inset shows the FS and the Brillouin zone. After [236].

surface prone to nest (probably at $T \simeq T^*$), the remaining Fermi surface remains unnested down to low temperature [349].

It is fair to say that an understanding of the origin of the T^* anomaly has not been achieved yet. However, the position of the T^* line in the generalized temperature-pressure phase diagram, cf. Sect. 4.6, indicates that it is intimately related to the critical point ($T_{\text{crit}}, p_{\text{crit}}$) which marks the endpoint of the first-order metal-insulator transition: $T^*(p)$ either meets with or evolves into the first-order metal-insulator transition.

4.4 Instabilities Involving Spin- and Charge-Degrees of Freedom

Fermi surface instabilities leading to CDW-, SDW-, and spin-Peierls transitions, as well as charge ordering have been introduced in the preceding sections. Since all these instabilities involving spin- and charge-degrees of freedom are intertwined with each other, we will recapitulate here the conceptual backgrounds and phenomena from a more general point of view.

4.4.1 Charge- and Spin-Density-Waves

As introduced in Sect. 4.2, the density-wave condensates such as CDW and SDW are commonly induced when a nesting condition $E(\boldsymbol{k}) = E(\boldsymbol{k}+\boldsymbol{Q})$ is satisfied sufficiently well for the underlying Fermi surface (FS). For simplicity, consider a quasi-1D tight-binding band with a small but finite interchain transfer integral t_b,

$$E(k_x, k_y) = \hbar v_F(|k_x| - k_F) - 2t_b \cos(k_y b). \tag{4.23}$$

The FS is shown in Fig. 4.22(a), where a pair of open sheets perfectly nest onto each other by the nesting vector $\boldsymbol{Q} = (2k_F, \pi/b, 0)$. This nesting makes the electronic response function

$$\chi(\boldsymbol{q}) = \sum_{\boldsymbol{k}} \frac{f[E(\boldsymbol{k}+\boldsymbol{q})] - f[E(\boldsymbol{k})]}{E(\boldsymbol{k}+\boldsymbol{q}) - E(\boldsymbol{k})} \tag{4.24}$$

divergent at $\boldsymbol{q} = \boldsymbol{Q}$, turning a metal into an insulator with an energy gap $E_G = 2\Delta_{DW}$ opening at E_F, see Fig. 4.22(b). Two different metal-insulator (MI) transitions to either a CDW or SDW ground state are well known: first, a CDW, which is a hybridized compressional wave of both the electron density and the lattice distortion induced by the electron-phonon interaction, see Fig. 4.22(c). In this CDW condensate, which was predicted in 1955 by R. Peierls [32],[8] a spontaneous breakdown of translational symmetry induces a

[8] The terminology Peierls-instability or -transition is often used for a CDW in a quasi-1D system.

lattice modulation with a periodicity corresponding to the nesting vector $\boldsymbol{Q}$. Second, a SDW condensate, Fig. 4.22(d), which is a compressional wave made of the electron spin-densities with up and down spins staying away from each other due to the Coulomb repulsion. The SDW transition which is induced by electron-electron interaction is not necessarily accompanied by a periodic lattice distortion.[9] A. Overhauser [355, 356, 357] predicted a SDW ground state in 1960 for isotropic metals like alkaline metals.[10]

We introduce the electron number- and spin-densities; $\rho(\boldsymbol{r}) = \rho_\sigma(\boldsymbol{r}) + \rho_{-\sigma}(\boldsymbol{r})$ and $\sigma(\boldsymbol{r}) = \rho_\sigma(\boldsymbol{r}) - \rho_{-\sigma}(\boldsymbol{r})$, respectively, where $\sigma =\uparrow, \downarrow$. Here $\rho_\sigma(\boldsymbol{r})$ and $\rho_{-\sigma}(\boldsymbol{r})$ are spin-dependent number-densities expressed as

$$\rho_\sigma(\boldsymbol{r}) = \frac{1}{2}\rho_0[1 - A\sin(\boldsymbol{Q}\cdot\boldsymbol{r} - \theta)] \tag{4.25}$$

and

$$\rho_{-\sigma}(\boldsymbol{r}) = \frac{1}{2}\rho_0[1 + A\sin(\boldsymbol{Q}\cdot\boldsymbol{r} + \theta)], \tag{4.26}$$

where ρ_0 is a constant background density. The CDW condensate with $\sigma = 0$, Fig. 4.22(c), is obtained at $\theta = \pi/2$:

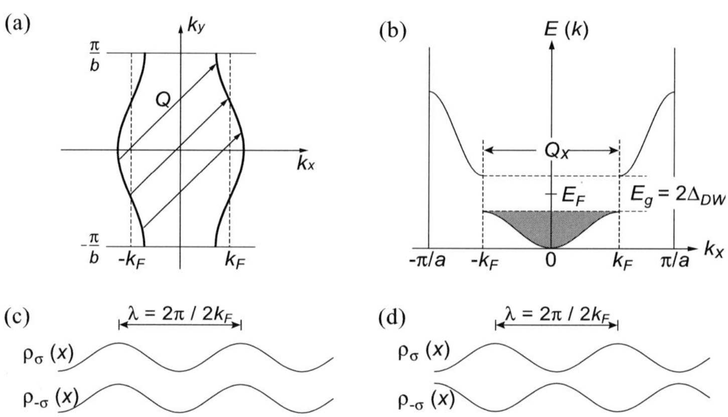

Fig. 4.22. (a) Quasi-1D FS in the k_x-k_y space for the tight-binding band (4.23). Arrows show the nesting vector $\boldsymbol{Q} = (2k_F, \pi/b, 0)$. (b) The energy band in a DW state. Spin-dependent number-density-waves: (c) in-phase for CDW and (d) anti-phase for SDW, both having the wave length $\lambda = 2\pi/2k_F$.

---

[9] In Cr metal, a second harmonic CDW ($4k_F$) accompanies the incommensurate SDW through the electron-phonon interaction [351, 352, 353, 354].

[10] Since the first observations of a CDW in TTF-TCNQ and of a SDW in Cr, a number of experimental and theoretical studies have been devoted to this subject so far yielding evidence for a CDW in transition-metal chalcogenides such as $NbSe_3$ and the blue bronzes like $K_{0.3}MoO_3$ [358, 359, 360, 361] and a SDW in Cr alloys and compounds [353, 354]. Following the investigations on the prototype TTF-TCNQ, many low-dimensional molecular metals have served as model systems for studying both types of condensates [359, 362, 361, 363, 86].

$$\rho(\boldsymbol{r}) = \rho_0[1 - A\cos(\boldsymbol{Q} \cdot \boldsymbol{r})], \tag{4.27}$$

while an SDW with $\rho = \rho_0$, Fig. 4.22(d), at $\theta = 0$:

$$\sigma(\boldsymbol{r}) = \rho_0 A \sin(\boldsymbol{Q} \cdot \boldsymbol{r}). \tag{4.28}$$

Both condensates with electron-hole pairing with momentum $2\hbar k_F$ due to nesting can be treated within a mean-field theoretical framework. From a mathematical point of view this is equivalent to the BCS weak-coupling theory [364, 365] for superconductivity due to electron-electron pairing ($\boldsymbol{k}\uparrow, -\boldsymbol{k}\downarrow$) with zero momentum. As a consequence, one finds a second-order phase transition at T_{DW} from a high-temperature metallic state to a low-temperature DW ground state with an order parameter Δ_{DW}:

$$2\Delta_{DW} = 3.52 k_B T_{DW} \tag{4.29}$$

and

$$k_B T_{DW} = 1.14 E_F \exp\left(-\frac{1}{\lambda_{DW}}\right). \tag{4.30}$$

Here λ_{DW} is a dimensionless coupling constant, $\lambda_{CDW} = gN(E_F)$ and $\lambda_{SDW} = UN(E_F)$, where g and U are the electron-phonon coupling constant and the electron-electron Coulomb repulsion, respectively. The condensation energy is given by $-(1/2)N(E_F)\Delta_{DW}^2$ and the amplitude A in (4.25) and (4.26) is expressed as $A = \Delta_{DW}/\hbar v_F k_F \lambda_{DW}$.

In the following, we will discuss TTF-TCNQ and $(TMTSF)_2PF_6$ as canonical systems for a CDW and SDW instability, respectively.

CDW in TTF-TCNQ

As noted in Chap. 1, the first molecule-based metal was the quasi-1D system TTF-TCNQ [30, 31, 210, 90]. The crystal has a monoclinic structure with the space group $P2_1/c$ in which both donor (TTF) and acceptor (TCNQ) molecules uniformly stack along the b-axis in a segregated way forming a column or chain (see Fig. 3.2 in Sect. 3.2). The average valence state $TTF^{+\delta}$-$TCNQ^{-\delta}$ with $\delta = 0.56 - 0.67$ was determined from X-ray photoelectron spectroscopy [366]. This indicates that both the hole and electron bands are responsible for the metallic conduction. The conductivity σ_b along the most conducting b-axis increases with decreasing temperature, followed by an MI transition around 50 – 60 K [31, 367, 368, 369, 370, 371].[11] The conductivity is highly anisotropic: σ_b is about two orders of magnitude higher than the

[11] The divergently increasing conductivity [31, 367] in approaching the transition temperature led to a great deal of turmoil in the field placing hopes on possible superconducting fluctuations associated with the Peierls instability. The conductivity, however, was proved to be caused at least partly by the inhomogeneous current paths depending on the sample and the electrical contacts [367].

interchain conductivities σ_a and σ_c, consistent with the results of optical reflectance measurements [372].

X-ray [373, 374, 375] and neutron-scattering [376, 377] measurements provided structural evidence for the Peierls instability. Figure 4.23(a) shows X-ray diffuse scattering intensities around the Bragg reflection (021) as a function of the wave number Q along the b^*-axis [374, 375]. Two distinct peaks are seen at $Q = 0.295\,b^*$ and $0.41\,b^*$, the latter of which is equivalent to $0.59\,b^*$ ($= b^* - 0.41\,b^*$). These signatures of diffuse scattering were identified as occurring at $2k_F$ and $4k_F$, since $2k_F = (\delta/2)\,b^* \simeq 0.3\,b^*$.[12] Both $2k_F$ and $4k_F$ peaks grow with decreasing temperature, indicating the existence of strong fluctuations preceding the Peierls instability in the metallic state. Figure 4.23(b) shows the dispersion of the transverse acoustic phonon propagating along the b-axis and polarized along c [378, 379]. A dip appears around $0.3\,b^*$ corresponding to the $2k_F$ diffuse scattering in Fig. 4.23(a) and becomes deeper at lower temperatures. This is a manifestation of the so-called Kohn anomaly indicating a phonon softening, i.e., a phonon mode whose frequency approaches zero toward the structural phase transition where the lattice loses the restoring force against the relevant distortion (schematically illustrated by the dashed curve).

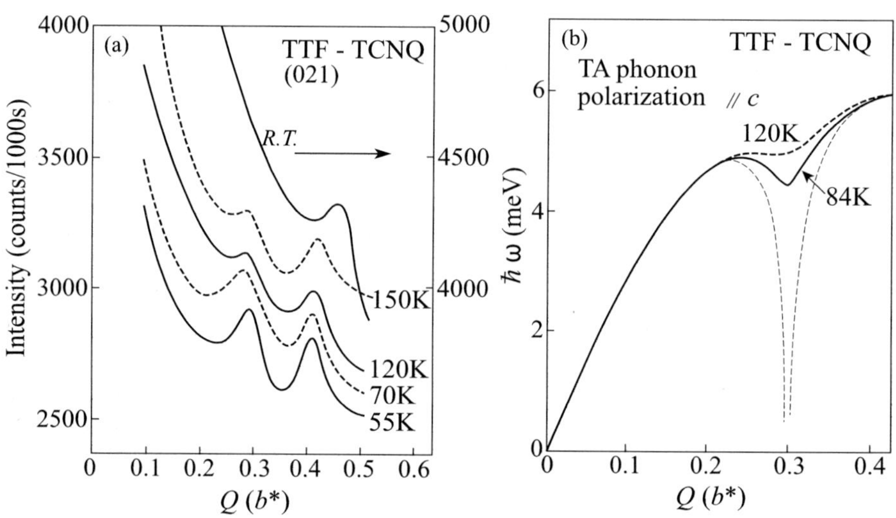

Fig. 4.23. (a) X-ray diffuse scattering intensities as a function of $Q(b^*)$ near the (021) Bragg reflection in TTF-TCNQ, after [374, 375]. (b) The phonon dispersion of the transverse acoustic mode propagating along the b^*-axis with the polarization along c, after [378, 379]. The dashed curve schematically represents the phonon quenching at $Q = 0.295\,b^*$ expected for the lattice displacement.

[12] As described in Chap. 2, the experimental observations of a superstructure turned out to be a powerful tool in determining precisely the charge transfer δ.

A comparison of X-ray and neutron data suggested that the $2k_F$ and $4k_F$ modulations occur on the TCNQ and TTF chains, respectively. These precursory fluctuations in the metallic state result in a formation of 3D ordering of $2k_F$- and $4k_F$-CDWs below 53 K and 49 K, respectively, and eventually a superlattice with $\boldsymbol{Q} = (0.25\,a^*, 0.295\,b^*, c^*)$ is formed below 38 K. These superlattice formations were visualized in a real-space picture and its Fourier transform by scanning tunneling spectroscopy [380]. Since a purely 1D system (without any interchain couplings) cannot undergo an ordering transition at finite temperatures, these successive transitions were discussed theoretically in terms of phase couplings between the chains [381, 382, 383].

The application of pressure changes this incommensurate superlattice into a commensurate one with $Q = (1/3)\,b^*$ [384], where the conductivity σ_b sharply decreases [385]. This can be ascribed to the so-called phase locking due to the commensurability with the lattice [386, 387], which can impede the sliding motion of the fluctuating CDW carrying an electric current under the influence of an electric field [362, 361].[13]

SDW in $(TMTSF)_2PF_6$

As described above, this quarter-filled quasi-1D metal exhibits a MI transition around 12 K ($= T_{SDW}$) into a SDW ground state [390, 391]. The single-particle gap $2\Delta_{SDW}$ of 30 – 40 K estimated from the dc conductivity agrees with equation (4.29). Other parameters are obtained within a weak-coupling framework (4.30); $\lambda_{SDW} = 0.26$, the coherence length $\xi_0 = \hbar v_F/\pi\Delta_{SDW} = 320$ Å, and $U = 2.0$ eV [363]. However, the experimentally observed SDW moment of $\mu = 0.08\,\mu_B$ is much larger than the theoretical value $(4\Delta_{SDW}/U)\,\mu_B = 0.01\,\mu_B$. Moreover, the order parameter evaluated from NMR and muon spin rotation measurements [323] shows a sharp upturn just below T_{SDW}, which deviates significantly from the BCS-type temperature dependence, $[1 - (T/T_{SDW})]^{-1/2}$. For the mean-field theories based on the quasi-1D Hubbard model, refer to [392, 393].

For this SDW condensate, a number of studies have been carried out, including the nonlinear transport [394, 395] and high-frequency conductivity [396], to explore the dynamics of the SDW. Figure 4.24 shows the fractional change of the nonlinear conductivity as a function of pulsed electric fields applied along the chain axis [394]. The data indicate an Ohmic behavior at low

[13] Since the energy of an incommensurate DW condensate is independent of the phase [388], a translational motion of sliding DWs is possible when an electric field is applied [386]. Theoretically this sliding motion is a gapless Goldstone mode, i.e., a recovery of the broken translational symmetry, so that no energy is necessary to excite the collective motions of the amplitude and phase. In a real crystal, however, DWs are pinned at defects in a crystal via the electrostatic interaction. In the absence of defects, they would carry a super-current. This was proposed in 1954 by H. Fröhlich [389] as a mechanism of superconductivity before the BCS theory [364] came out in 1957.

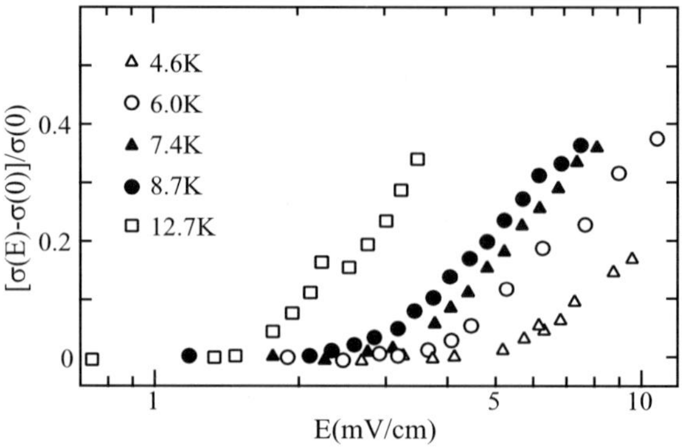

Fig. 4.24. Non-linear conductivity $[\sigma(E) - \sigma(E \to 0)]/\sigma(E \to 0)$ vs the logarithmic electric field E in $(TMTSF)_2PF_6$ measured with clamped contacts at various temperatures. After [394].

fields, followed by a field-dependent conductivity above the threshold field E_T that increases with decreasing temperatures. This nonlinear transport in the SDW state is quite similar — not only qualitatively but even quantitatively — to those in CDW condensates [362, 363, 361] with E_T being of the same order of magnitude. This can be interpreted as depinning of the CDW condensate. However, this similarity has remained enigmatic for reasons described below.

In general, at low electric fields both condensates remain pinned at defects in the crystal. At sufficiently high fields they can overcome the pinning potential, becoming depinned, and start to carry a collective current. The difference between the two types of DWs is that the pinning interaction (usually nonmagnetic) for $\rho_\sigma(\boldsymbol{r})$ and $\rho_{-\sigma}(\boldsymbol{r})$ works collaboratively for the CDW, but not for the SDW, i.e., the SDW pinning can be expected to be an effect of higher-order perturbation and thus should be much smaller than the CDW pinning. For details refer to [397] and references therein. Moreover, the high-frequency conductivity spectrum indicates a prominent oscillator strength at several GHz due to pinned SDWs [361, 363, 396]. This suggests that the pinning frequency is also quite similar to those observed in CDW condensates. This again is puzzling because the oscillator strength of a SDW with an effective mass, expected to be the band mass, should be much smaller than that of a CDW with much larger mass of about 100 or more times the band mass as a consequence of the concomitant lattice displacement. The optical conductivity issue was readdressed theoretically with a random phase approximation for the extended Hubbard model by taking into account the collective mode relevant charge fluctuations [398].

For the problem of the mixed CDW-SDW in $(TMTSF)_2PF_6$ as mentioned above, the $2k_F$ diffuse scattering indicates fluctuation effects of the lattice distortion similar to TTF-TCNQ below 200 K which surprisingly disappear near T_{SDW} [327]. This behavior is in contrast to that in $(TMTTF)_2PF_6$ in which the fluctuations diverge at the spin-Peierls (SP) transition temperature $T_{SP} = 15$ K. From these facts it was taken that the mixed state is of purely electronic origin without any accompanying lattice distortions. The coexistence was discussed in connection with the charge-ordering phenomenon, which will be addressed later. Another puzzling observation was that the CDW fades out at lower temperatures below 3 – 4 K [399], just where successive SDW subphases appear which are associated with anomalies in NMR [400, 401], specific heat [402], dielectric properties [403], magnetic susceptibility [404] and magnetotransport [405, 406].

4.4.2 Spin-Peierls State

The spin-Peierls ground state was first discovered in 1:1 TTF salts with bisthiolene metal complexes as an acceptor [407]. These materials, which are Mott-Hubbard insulators at half filling with one electron per TTF-site, have served as a model system for a 1D magnetic Heisenberg chain. Instead of the Hamiltonian (4.19) for the homogeneous chain, an inhomogeneous case of alternating exchange interactions J_1 and J_2 has been considered [319]:

$$\mathcal{H} = \sum_{i=0}^{N/2} (J_1 \boldsymbol{S}_{2i} \cdot \boldsymbol{S}_{2i-1} + J_2 \boldsymbol{S}_{2i} \cdot \boldsymbol{S}_{2i+1}) . \tag{4.31}$$

Here the alternation parameter, $\gamma_{ex} = J_2/J_1$ is introduced. A particularly interesting situation arises for uniform antiferromagnetic $S = 1/2$ Heisenberg

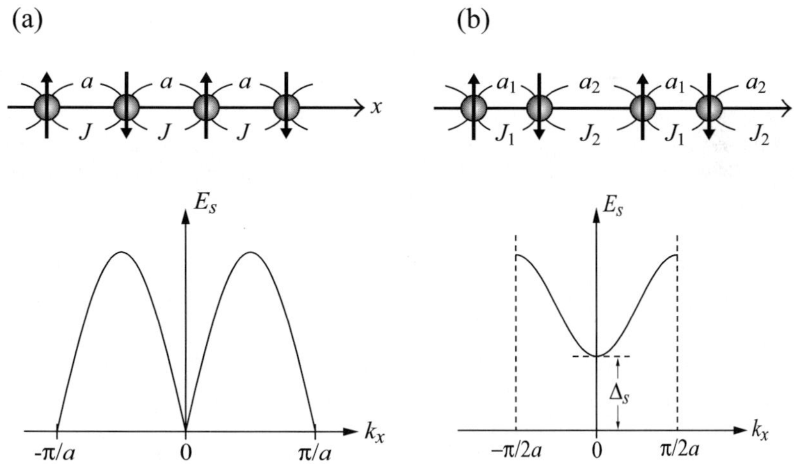

Fig. 4.25. Schematic illustration of the dispersion of low-lying magnetic excitations. (a) Homogeneous case ($\gamma_{ex} = 1$, i.e. uniform exchange coupling constant J) with a 1D periodicity of a. Note that for quantum spins $S = 1/2$, the dispersion relation shown corresponds to the lower boundary of the two-spinon continuum. (b) Inhomogeneous case ($\gamma_{ex} < 1$) with alternating J_1 and J_2 and dimerization $a_1 + a_2 = 2a$. Here the singlet ground state is separated by an energy gap Δ_s from the lowest-lying band of triplet excited states.

chains ($\gamma_{ex} = 1$), where the excitations are so-called spinons with $S = 1/2$ (in contrast to spin-1 magnons) and thus are fermions. These spinons are created as pairs giving rise to the two-spinon continuum. The lower bound of this continuum has the dispersion $E_s \sim (J/2\pi)|\sin(k_x a)|$, shown in Fig. 4.25, which is gapless. In the limit of large U/t in the Hubbard model ($V_{ij} = 0$ in (4.2), $J_1 = J_2 = 2t^2/U$ [408], and the gapless nature of the excitation spectrum gives a non-zero susceptibility χ_s at $T = 0$. In contrast, for the alternating case $\gamma_{ex} < 1$ with a doubling of the unit cell ($a_1 + a_2 = 2a$), a gap Δ_s opens up in the spin excitation spectrum and the ground state becomes nonmagnetic. This is the case in the spin-Peierls (SP) state.[14] The driving force of the SP transition is the magneto-elastic coupling between the one-dimensional spin system and the three-dimensional phonons. Upon cooling through T_{SP}, there is an elastic distortion which transforms the uniform chain at $T > T_{SP}$ to a dimerized chain with two, unequal alternating exchange coupling constants for $T < T_{SP}$. This T-dependent dimerization is accompanied by the opening of a spin gap between a nonmagnetic singlet ($S = 0$) ground state and the lowest-lying band of triplet ($S = 1$) excitations. The size of the gap is related to the degree of dimerization. Thus in a SP system, the cost of elastic energy due to the dimerization is overcompensated by the magnetic energy gain due to the opening of the singlet-triplet excitation gap [319, 411].

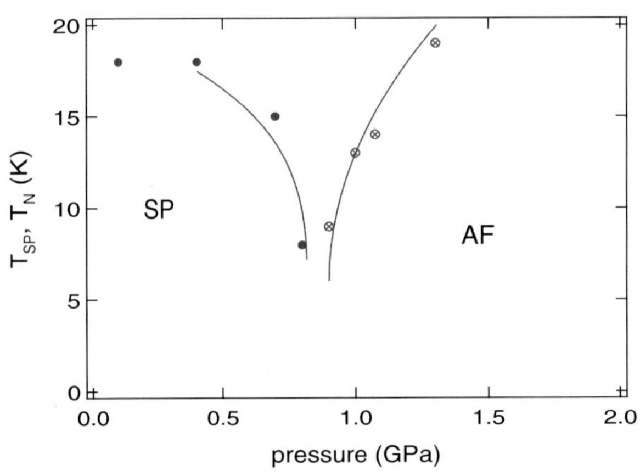

Fig. 4.26. Temperature vs pressure phase diagram determined by ^{1}H-NMR measurements on $(TMTTF)_2PF_6$. After [412].

In contrast to the above half-filled case, $(TMTTF)_2PF_6$ at quarter filling shows a characteristic interplay between AF and SP states. Figure 4.26 demonstrates the pressure dependence of T_{SP} and T_N determined by ^{1}H-NMR measurements [412]. These ground states appear in the Mott-Hubbard

14 A SP ground state was also observed in inorganic $CuGeO_3$ with Cu^{2+} ($S = 1/2$) [409]. Here, direct measurements of the spin excitations on sizeable single crystals have been made by inelastic neutron scattering. A review is available given in [410].

insulator below $T_\rho > 100$ K (Sect. 4.3.1, Fig. 4.6). The low-temperature SP state at ambient pressure forms out of the high-temperature charge-ordered state [413], as will be described below. Of particular importance is that both phase boundaries are sharply suppressed to less than 9 K around the critical pressure 0.9 GPa. This is reminiscent of a quantum critical point [414], i.e. a phase transition at $T = 0$, which can divide the dimerized SP and uniform AF states in a 1D system.

Another typical example for a SP state was found in MEM-$(TCNQ)_2$, where MEM is *N-methyl,N-ethyl-morpholinium*, which is a Mott-Hubbard insulator below 335 K [415]. The system exhibits a temperature-dependent magnetic susceptibility of 1D Bonner-Fisher type [317] with an AF exchange interaction, followed by a sharp decrease below 19 K indicating a spin-singlet state. The latter was identified as a SP ground state associated with dimerization within the TCNQ chains [416].

4.4.3 Charge-Ordering Phenomena

In addition to FS instabilities leading to DW condensates, electron correlation effects can be the source for another type of MI transition in low-dimensional systems. In particular, the long-range inter-site Coulomb repulsion can play a decisive role in the charge-ordering (CO) phenomena. This has been the subject of theoretical investigations in quarter-filled 1D metals, see [176] for a recent review. The model which has been used is an extended Hubbard Hamiltonian,

$$\mathcal{H} = \sum_i \sum_\sigma \left(t[1 + (-1)^i \Delta_d] c^\dagger_{i,\sigma} c_{i+1,\sigma} + h.c. \right) + \sum_i U n_{i\uparrow} n_{i\downarrow} + \sum_i V n_i n_{i+1}, \tag{4.32}$$

where Δ_d measures the degree of dimerization. Equation (4.2) corresponds to the homogeneous case, $\Delta_d = 0$, without dimerization. Although the mean-field calculations [192, 417, 418] ignore any quantum fluctuations of charges and spins which are expressed by space-time correlations as $< \hat{\rho}(\boldsymbol{r}_2, t_2)\hat{\rho}(\boldsymbol{r}_1, t_1) >$ and $< (\hat{\sigma}(\boldsymbol{r}_2, t_2)\hat{\sigma}(\boldsymbol{r}_1, t_1) >$, respectively, they may provide a physical insight into the ground states, as schematically illustrated in Fig. 4.27.

In case (a) with $\Delta_d = 0$ and $V < V_c$ $(= 0.34t)$ [419], where the on-site Coulomb repulsion U dominates the uniform-transfer band, a $2k_F$-SDW with the charge-spin sequence of $(0.5_\downarrow, 0.5_\downarrow, 0.5_\uparrow, 0.5_\uparrow)$ can be stabilized as illustrated in Fig. 4.22(d) [420]. At large V $(> V_c)$, however, electrons tend to stay away from each other. As a consequence they form a charge-density pattern $(\delta, 1-\delta, \delta, 1-\delta)$ which corresponds to a $4k_F$-CDW with a 2-fold periodicity compared to the lattice. In this case, a $2k_F$-SDW and a $4k_F$-CDW can coexist either coherently with the sequence $(0, 1_\downarrow, 0, 1_\uparrow)$ for $\Delta_d = 0$ (case (b)), or incoherently. The latter state with a charge-spin modulation of $(\delta_\downarrow, 1-\delta_\downarrow, \delta_\uparrow, 1-\delta\uparrow)$ occurs for finite dimerization $\Delta_d \neq 0$ (case (c))

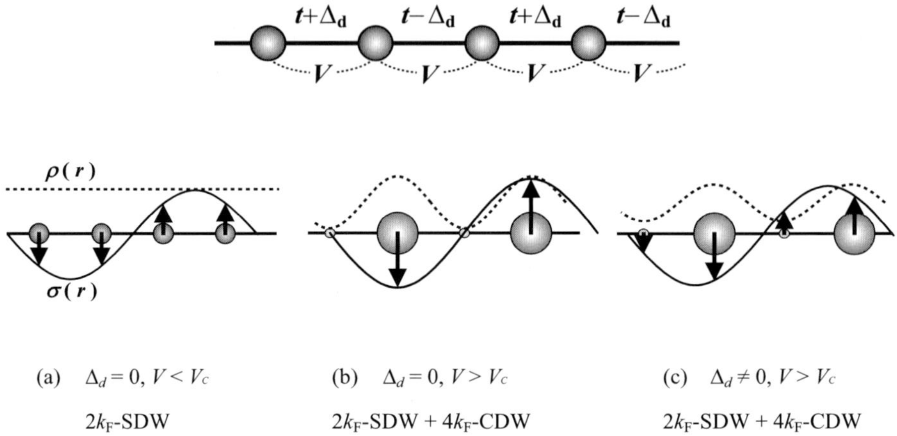

Fig. 4.27. Schematic illustrations for a quarter-filled 1D chain with alternating transfer integrals (top), and the charge- and spin-arrangements expected from a mean-field approximation. Dotted and solid lines represent the charge- and spin-density patterns, respectively.

[192, 421] and is called a charge disproportionation (CD). By further taking into account the next-nearest neighbor Coulomb repulsion, a $2k_F$-CDW can coexist with both $2k_F$- and $4k_F$-SDWs as $(1-\delta_\downarrow, \delta_\downarrow, \delta_\uparrow, 1-\delta_\uparrow)$ [417, 418] which was proposed for the mixed state in $(TMTSF)_2PF_6$ [327] mentioned above.

So far a few prototype materials showing the above-mentioned CO phenomena at quarter filling have been found. The first example is the quasi-1D metal $(DI\text{-}DCNQI)_2Ag$ with $\Delta_d = 0$ [422] (R1 = R2 = Iodine, see Fig. 7.1 for the crystal structure). For CO in $(DCNQI)_2Cu$, see Sect. 7.1.2. This material is a semiconductor, even at room temperature, and undergoes an AF ordering at 5.5 K [423]. The ^{13}C-NMR measurements on the Ag-salt, in which the cyano-groups are enriched with ^{13}C and where strong hyperfine coupling to π-electrons can be expected, showed that below 220 K the single line becomes split into broad and sharp peaks with comparable intensities [424]. The splitting develops with decreasing temperature, followed by a broad single peak due to $2k_F$ AF fluctuations. These findings were ascribed to the formation of a $4k_F$-CD ($4k_F$-CDW) with a density ratio of 3:1, the latter being consistent with the $4k_F$ diffuse scattering observed in X-ray diffraction [425]. Therefore, case (b) in Fig. 4.27 may explain this CD and the low-temperature AF order.

The second example is the $(TMTTF)_2X$ salt with the centrosymmetric anion X = AsF_6 and finite dimerization ($\Delta_d = 0.01 - 0.005$) [426]. This material undergoes a MI transition with a sharp increase of the resistivity below $T_{CO} = 100$ K, where the ^{13}C-NMR line (here enriched at the central double bond) becomes split into two subpeaks [413]. The CD, with a ratio of 3:1 evaluated from T_1^{-1} measurements, agrees with the theoretical prediction (case (c) in Fig. 4.27) for the CD and spin-density modulation [192]. Below 10 K at ambient pressure, this CD state allows a coexistence with a SP state [413, 427]. Under pressure, the CD is suppressed and a pure SP phase is induced [428]. Moreover, the CD at ambient pressure is associ-

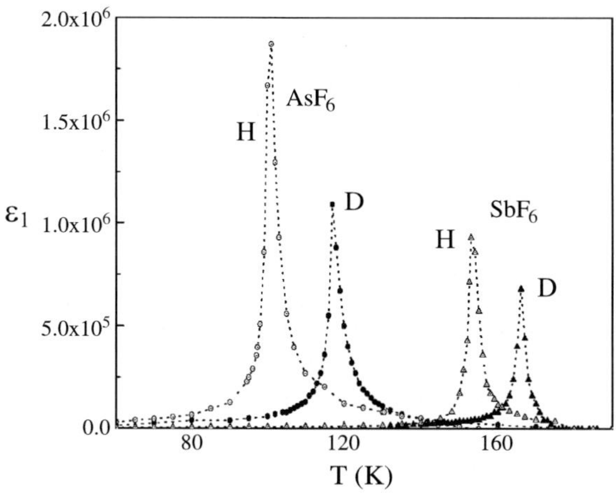

Fig. 4.28. Temperature dependence of the real part of the dielectric constant at 100 kHz for deuterated and non-deuterated $(TMTTF)_2AsF_6$ and $(TMTTF)_2SbF_6$. After [429].

ated with a ferroelectric ordering as observed in the low-frequency dielectric constant [430, 431, 429]. Figure 4.28 shows the real part ε_1 of the complex dielectric constant as a function of temperature for pristine (TMTTF-H_{12})$_2$X and deuterated (TMTTF-D_{12})$_2$X (X = AsF_6, SbF_6). The data indicate a Curie law, $\varepsilon_1 \propto (T - T_{CO})^{-1}$ below and above T_{CO} and a Debye relaxation time diverging at T_{CO} as usual in ferroelectric materials. This ordering was interpreted as follows. At high temperatures, the unit cell with inversion symmetry consists of two equivalent TMTTF molecules. The symmetry breaking within the unit cell in the CD state may induce permanent electric dipoles and, if the interchain interactions are ferroelectric and the dipole arrangements on neighboring chains are phase-coherent, a 3D ferroelectric ordering may occur. It is noticeable that the charge ordering and associated ferroelectric transitions observed in X = AsF_6 and SbF_6 exhibit a large isotope shift of $T_{CO}(= T_C)$: upon deuteration T_{CO} shifts to higher temperatures by 16 K and 13 K, respectively. From ESR-, NMR- and X-ray measurements [432] it has been inferred that the deuteration enhances the ratio of the intra- to inter-dimer distance by about 0.2 % and that this ratio, which is a measure of the degree of uniform stacking, increases with decreasing temperature. Therefore the deuterated salt may arrive at some critical ratio, triggering the CO transition at higher temperatures. Similar behaviors are found in the non-centrosymmetric ReO_4 salt with T_{CO} = 230 K and an anion-ordering temperature of 156 K [433].

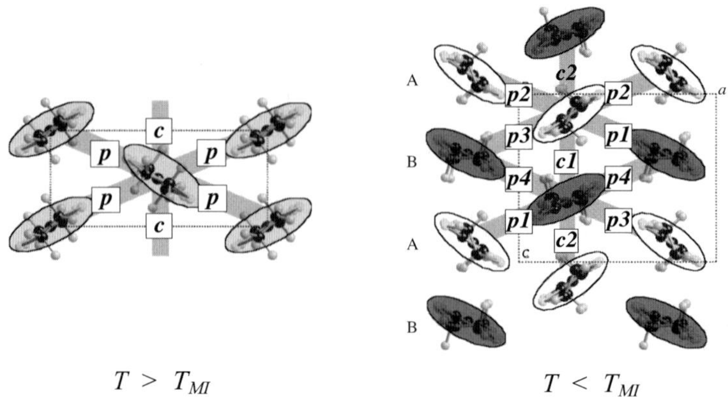

Fig. 4.29. Charge-density patterns in the *ac*-plane of θ-$(ET)_2RbZn(SCN)_4$: (left) metallic state above $T_{MI} = 200\,K$ and (right) charge-ordered insulating state below T_{MI}. After [84].

For prototype quasi-2D systems, we mention θ-$(ET)_2RbZn(SCN)_4$ yielding a uniform stacking with no dimerization. This quarter-filled quasi-2D compound exhibits a MI transition at $T_{MI} = 200\,K$ accompanied by a structural phase transition, i.e. a doubling of the ET-stacking *c*-axis [133]. From ^{13}C-NMR [434, 435], Raman and infrared spectroscopies [436, 311], and X-ray diffraction measurements [84], the low-temperature phase called θ_d-phase was identified to be in the CD state with horizontal stripes in the 2D basal plane, consistent with mean-field theory [437]. Figure 4.29 illustrates the patterns with the transfer integrals for both phases θ (left) and θ_d (right) [84]. These charge distributions were determined from X-ray structural analysis by taking advantage of the fact that the intramolecular bonding lengths of the ET molecule sensitively depend on the valence state as described in Sect. 2.2, Fig. 2.3. While all the ET sites in Fig. 4.29 (left panel) are covered with grey ellipsoids indicating an average valence of 0.5 in the high-temperature θ-phase, the low-temperature θ_d-phase (right panel of Fig. 4.29) is characterized by a horizontal stripe pattern of alternating open and shaded ellipsoids corresponding to quasi-neutral ($\delta = 0 - 0.2$) and quasi-ionic ($\delta = 0.8 - 1.0$) ET states, respectively. In this CO state, a zig-zag chain of localized spins running along the *a*-axis is expected. These spin chains may be responsible for the Bonner-Fisher-type [317] broad peak in the magnetic susceptibility [133] suggesting an AF exchange coupling along the p4 transfer route (Fig. 4.29 (right panel)). In fact, the magnetic properties can be understood within a 1D Heisenberg model for zig-zag spin chains along the stripes [176]. At temperatures below 25 K, the system turns into a SP state just as predicted [437].

For CD phenomena in general, polarized Raman and infrared spectroscopies are complementary to conventional NMR and X-ray tools. These optical measurements evaluate both the degree of CD from the ring-stretching mode of the central carbon double bond with a weak EMV coupling (4.16) and the symmetry of CD from another stretching mode with a strong EMV effect [311]. This technique [438] was also applied to other CO phenomena as e.g. in α-(ET)$_2$I$_3$ with a horizontal stripe pattern [189, 439].

4.5 Many-Body Effects

The simple tight-binding band-structure calculations, discussed in Sec. 4.1, are based on the assumption of a periodic potential set up by the ions or molecules which are rigidly fixed in a perfect periodic arrangement. This approach neglects all the interactions an electron may have with its surrounding. However, since the ions or molecules can be charged, an electron passing through the crystal will tend to distort the lattice on its way due to the Coulomb force that acts between these charged objects. This leads to a strain field which accompanies the electrons' motion. Likewise, electrons interact with each other, i.e., they repel one another, due to the Coulomb repulsive force between them. In usual three-dimensional metallic compounds with a relatively high density of conduction electrons, the Coulomb potential is screened by these electrons and becomes effectively short ranged. The screening is characterized by the Thomas-Fermi screening length q_{TF}^{-1} defined by

$$q_{TF}^2 = 4\pi e^2 N(E_F), \tag{4.33}$$

where $N(E_F)$ is the density of states at the Fermi energy. For a normal metal such as copper with a charge carrier density of about $n = 8.5 \cdot 10^{22}\,\mathrm{cm}^{-3}$, q_{TF}^{-1} is usually less than 0.5 Å. An efficient screening, i.e., a strong suppression of Coulomb interactions, is the justification for the above band-structure calculations to provide a reasonable description of the electronic properties of simple metals. The effects of many-body interactions such as electron-phonon and electron-electron interactions, neglected in the simple tight-binding scheme, can be accounted for by considering the electrons as quasiparticles, i.e., non-interacting objects with spin $S = 1/2$, which obey Fermi-Dirac statistics, but whose properties are renormalized. This quasiparticle concept, proposed by L.D. Landau in the 1950s, enabled the treating of the electrons as independent particles to be continued, even in the presence of considerably strong interactions. It forms the foundation of the Landau Fermi-liquid theory which has become what is called the *standard model* of metals. The quantity which reflects these renormalization effects in a most direct way is the effective mass m^*. As the electron moves through the crystal, it produces other quasiparticle excitations which move along with it, thereby changing its mass. For the ET-based materials, typical values of the cyclotron effective mass, obtained from fitting the temperature dependence of the amplitude of magnetic quantum oscillations to the Lifshitz-Kosevich formula (cf. Sec. 5.2.1), are in the range

Table 4.3. Fermi surface parameters determined from quantum-oscillation studies on various quasi-2D $(ET)_2X$ and $(BETS)_2X$ salts including the cyclotron effective masses in units of the free electron mass m_0, the oscillation frequency as well as the oscillation orbit.

	m_c^* (m_0)	frequency (Tesla)	orbit	Ref.
α-$(ET)_2KHg(SCN)_4$	2	570	α	[443]
β-$(ET)_2I_3$	4.7	3730		[444]
κ-$(ET)_2I_3$	1.9 ± 0.1	570	α	[445]
κ-$(ET)_2I_3$	3.9 ± 0.1	3880	β	[445, 446]
κ-$(ET)_2Cu(NCS)_2$	3.5 ± 0.1	598	α	[96, 447]
κ-$(ET)_2Cu(NCS)_2$	6.9 ± 0.8	3800	β	[447]
κ-$(ET)_2Cu[N(CN)_2]Br$	5.4 ± 0.1	3798	β	[448]
κ-$(ET)_2Cu[N(CN)_2]Br$	6.7 ± 0.2	3810	β	[449]
κ-$(ET)_2Cu[N(CN)_2]Cl$ (at 2 kbar)	1.7	577	α	[450, 451]
κ-$(ET)_2Cu[N(CN)_2]Cl$ (at 2 kbar)	3.5	3880	β	[450, 451]
β''-$(ET)_2SF_5CH_2CF_2SO_3$	2	199		[253, 254]
λ-$(BETS)_2GaCl_4$	3.6	650	α	[452]
λ-$(BETS)_2GaCl_4$	6.3	4030	β	[452]

$m_c/m_0 \sim 1-7$ (where m_0 is the free-electron mass), suggesting a significant role of many-body effects.

In fact, for the present materials, there are several reasons why screening is less effective and electron-electron interaction effects may become important. First, the low dimensionality, i.e., the confinement of the electron motion to one or two dimensions, strongly hampers the electron's ability to screen the Coulomb potential. Similarly, the rather low density of conduction electrons or holes of $n_h \sim 10^{21}\text{cm}^{-3}$, which is about two orders of magnitude lower than in ordinary metals, causes a less effective screening. Moreover, in [440, 441], it has been pointed out, that in materials with a large ionic dielectric constant ϵ_0, the screening length can substantially increase as $q_{TF}^{-2} \propto \epsilon_0$. For the insulating α-phase of $(ET)_2I_3$, a background dielectric constant of $\epsilon_0 \simeq 15-20$ has been reported [442].

The knowledge of the effective cyclotron mass is, however, not sufficient to determine the size of the many-body effects unless the band mass, m_b, is known precisely [453]. Estimates for the latter quantity, which accounts for the motion of the electron in the periodic potential of the underlying crystal lattice, can vary by as much as a factor of 2 – 3 depending on the method used for calculating the band structure: For the κ-$(ET)_2Cu(NCS)_2$ system, for example, semi-empirical band-structure calculation based on the effective-dimer model revealed $m_b^\alpha = 0.64\, m_0$ and $m_b^\beta = 1.27\, m_0$ for the α- and β-orbits [250], cf. Sect. 5.2.1. Employing the generally accepted experimental cyclotron masses of $m_c^\alpha = 3.5\, m_0$ and $m_c^\beta = 6.5\, m_0$ [96, 447, 250], these numbers predict a mass renormalization greater than a factor 4. On the other hand, considerably enhanced band masses of $m_b^\alpha = (1.72 \pm 0.1)\, m_0$ and $m_b^\beta = (3.05 \pm 0.1)\, m_0$ have been derived from first-principle self-consistent local-density calculations [188], which would correspond to a many-body mass-enhancement factor of about 2. The cyclotron effective masses, determined from quantum-oscillation experiments for various quasi-2D salts, are collected in Table 4.3.

The mass renormalization leads to a compression of the spacing between the Fermion energy levels, thereby increasing the number of quasiparticles that are excited at a given temperature, corresponding to an increase of the density of states by a factor m^*/m_b. This can be probed directly by measurements of the specific heat, for example. In magnetic fields sufficiently high to suppress superconductivity, the specific heat at low temperatures ($T \leq 3 - 4\,$K) for the various $(ET)_2X$ salts were found to follow a $C(T) = \gamma T + \beta T^3$ dependence, where γT is the electronic contribution (γ is the Sommerfeld coefficient) and $\beta T^3 \propto (T/\theta_D)^3$ the specific heat of the phonons at low temperatures (θ_D denotes the Debye temperature).

For a quasi-2D metal consisting of stacks of metallic layers with interlayer spacing s, the Sommerfeld coefficient $\gamma^{2D} = C/T$ is related to the thermodynamic effective mass m_{th}^* by [461]

Table 4.4. Sommerfeld coefficient γ and corresponding thermodynamic effective mass m_{th}^* in units of the free-electron mass m_0 calculated by using (4.34) for various quasi-2D $(ET)_2X$ salts and λ-$(BETS)_2GaCl_4$. The Debye temperature is calculated via $\Theta_D = (\frac{12}{5}\pi^4 Rn/\beta)^{1/3}$ with R the gas constant and n the number of atoms per formula unit. β is the cubic term in $C(T) = \gamma T + \beta T^3$. The molar volume V_{mol} is also given. For the values of the interlayer distance s, see Table 3.1.

	$\gamma\ (\frac{\mathrm{mJ}}{\mathrm{molK}^2})$	Θ_D (K)	Ref.	V_{mol} (cm^3)	m_{th}^*/m_0
α-$(ET)_2NH_4Hg(SCN)_4$	25 – 26	221	[112]	601	5.3 ± 0.4
β-$(ET)_2I_3$	24 ± 3	197 ± 5	[454]	515	4.4 ± 0.6
κ-$(ET)_2I_3$	18.9 ± 1.5	218 ± 7	[455]	508	3.7 ± 0.3
κ-$(ET)_2Cu(NCS)_2$	23 ± 1	205 – 230	[456, 457]	490.5	4.7 ± 0.2
κ-$(ET)_2Cu[N(CN)_2]Br$	25 ± 2	200 ± 10	[458]	499	4.7 ± 0.4
β''-$(ET)_2SF_5CH_2CF_2SO_3$	18.7 ± 1	221 ± 7	[459]	539	3.8 ± 0.2
λ-$(BETS)_2GaCl_4$	14.4 ± 3.4	197 ± 0.4	[460]	533.5	3.1 ± 0.7

$$\gamma^{2D} = \frac{\pi k_B^2 V_{\mathrm{mol}}}{3} \cdot \frac{m_{th}^*}{\hbar^2} \cdot \frac{1}{s}, \tag{4.34}$$

where V_{mol} is the molar volume. For the interlayer spacings s ranging from about 15 to 18 Å for κ-(ET)$_2$X (see Table 3.1), and using the Sommerfeld coefficients and molar volumes given in Table 4.4, Eq. (4.34) yields values for the thermodynamic effective mass of these κ-phase salts ranging from $m_{th}^* = 3.7\,m_0$ for κ-(ET)$_2$I$_3$ to $m_{th}^* = 4.7\,m_0$ for κ-(ET)$_2$Cu(NCS)$_2$. Slightly lower values for m_{th}^* can be obtained from the relation for an electron gas in three dimensions

$$\gamma^{3D} = \frac{\pi^{2/3} k_B^2 V_{\mathrm{mol}}}{3^{2/3}} \cdot \frac{m_{th}^*}{\hbar^2} \cdot n_{\mathrm{3D}}^{1/3}, \tag{4.35}$$

where n_{3D} is the carrier concentration. Since the thermodynamic mass measures the integral density of states at the Fermi level, these numbers agree reasonably well with the cyclotron masses listed in Table 4.3. As mentioned in Sect. 4.3.1, there is also a significant coupling of the π-electrons to the phonons. In the present molecular materials, this interaction involves the coupling both to the high-frequency intramolecular vibrations (EMV) as well as to the low-frequency intermolecular modes. From Raman scattering studies of the phonon dynamics around T_c of κ-(ET)$_2$X salts with X = Cu[N(CN)$_2$]Br and Cu(NCS)$_2$, a total electron-phonon coupling constant of $\lambda_{tot}^{el-ph} = 0.97 \pm 0.11$ has been estimated [306].

Thus, the experimental observation of a significant mass renormalization clearly indicates that both electron-electron and electron-phonon interactions are of importance in these quasi-2D molecular conductors.

Pressure Dependence of the Effective Mass

The pressure dependence of the cyclotron effective masses has been examined for various κ-(ET)$_2$X salts with X = Cu(NCS)$_2$ [250, 462], Cu[N(CN)$_2$]Br [463, 449] and Cu[N(CN)$_2$]Cl [450, 451]. The results follow the same general trend, a decrease of the effective masses with increasing pressure. Qualitatively, this is in accordance with the expectation: with increasing pressure, the donor molecules are pushed closer together so that the overlap integrals increase. This results in an increase of the bandwidths and a reduction of the effective masses. By means of pressure-dependent Shubnikov-de Haas experiments on κ-(ET)$_2$Cu(NCS)$_2$, J. Caulfield et al. [250] inferred a striking interrelation between changes in the effective masses and the suppression of superconductivity. Figure 4.30 shows the hydrostatic-pressure dependence of the effective masses for κ-(ET)$_2$Cu(NCS)$_2$ [250]. As the pressure increases, m_c^α rapidly decreases. Above some critical pressure of about $4-5$ kbar, the rate of suppression of m_c^α becomes much weaker. Since this is about the same pressure above which superconductivity becomes completely suppressed (see inset of Fig. 4.30), an intimate interrelation between mass enhancement and superconductivity has been suggested [250]. (For a discussion of the pressure

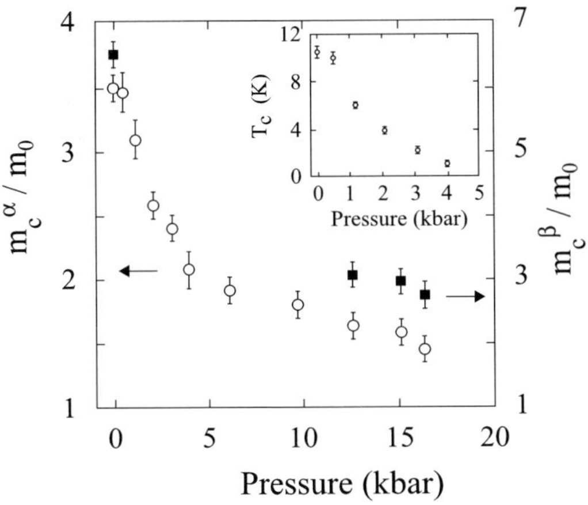

Fig. 4.30. Cyclotron effective masses normalized to the free-electron mass of the α- (open circles, left scale) and β- (full squares and right scale) orbits as a function of hydrostatic pressure for κ-$(ET)_2Cu(NCS)_2$. The inset shows T_c against pressure. After [250].

dependence of the Dingle temperature, see Sect. 5.2.1.) This is consistent with recent results of reflectivity measurements which showed that the pressure dependence of the 'optical masses', thought to be strongly influenced by interband processes and thus closely related to the bare band masses, do not show such a crossover behavior [464]. Consequently, the pressure-induced reduction of the effective cyclotron masses has to be associated with a decrease in the strength of the electron-electron and/or electron-phonon interactions.

4.6 Phase Diagrams

As discussed in the previous sections, the ground-state properties of the organic charge-transfer salts can be tuned by variations of the chemical composition or external parameters such as pressure or magnetic field. It has been suggested that for the κ-$(ET)_2X$ compounds — in analogy to the $(TM)_2X$ family — changes in the anion X correspond to discrete pressure shifts on the pressure scale.

$(TM)_2X$ Salts

Figure 4.31 comprises in a pressure-temperature plane results of various experiments on the quasi-1D charge-transfer salts $(TMTTF)_2X$ and their selenium analogues $(TMTSF)_2X$. The arrows labelled $a - d$ indicate the position of the various salts at ambient pressure. The generic character of the phase diagram — first proposed by D. Jérome et al. [215] — has been demonstrated by pressure studies on the $(TMTTF)_2PF_6$ salt [465, 466, 214], for which the ambient-pressure ground state is a dimerized spin-Peierls state, cf. Fig. 4.26. With increasing pressure, the system was found to pass through the whole sequence of ground states, as shown in Fig. 4.31, and eventually becoming superconducting at high pressures above 43.5 kbar [466, 214].

On the low-pressure (left) side of the phase diagram in Fig. 4.31, the molecular stacks can be considered as only weakly-coupled chains, i.e., the system is close to be truly 1D. In fact upon cooling, the $(TMTTF)_2PF_6$ compound behaves very much like a canonical 1D conductor where spin- and charge-degrees of freedom are decoupled: below $T_\rho = 250\,K$ the resistivity increases by several orders of magnitude due to charge localization while the spin susceptibility remains unaffected [210, 146, 467]. The phase below T_ρ has been assigned to a Mott insulating state. Upon further cooling to $T_{SP} = 19\,K$, a spin gap opens and the system enters a distorted spin-Peierls ground state. More recently, another phase transition (not included in Fig. 4.31) has been discovered in this salt at $T_{CO} = 70\,K$, accompanied by a huge anomaly in the dielectric constant [430] and identified as a charge-ordering transition dividing the TMTTF molecules in two non-equivalent species with unequal electron densities [413].

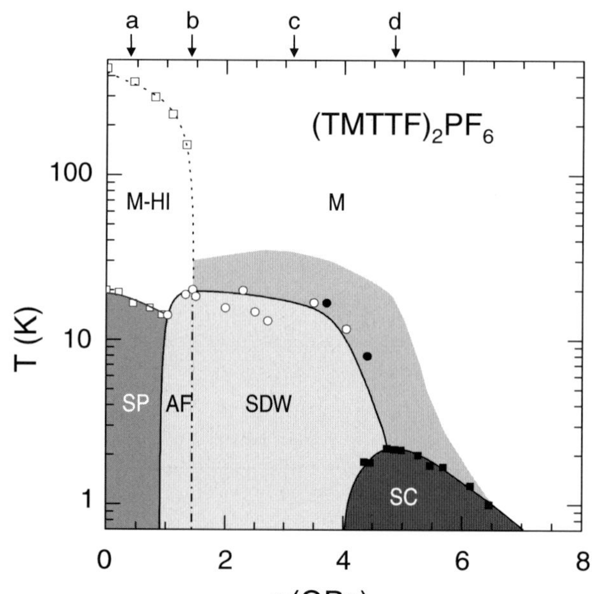

Fig. 4.31. Temperature-hydrostatic-pressure phase diagram for $(TMTTF)_2PF_6$. The following abbreviations have been used: Mott-Hubbard insulating state (M-H), metallic (M) and superconducting (SC) state, spin-Peierls (SP), commensurate (AF) and incommensurate (SDW) antiferromagnetic spin-density-wave state. Arrows indicate positions of salts with various anions X at ambient pressure: (a) $(TMTTF)_2BF_4$, (b) $(TMTTF)_2Br$, (c) $(TMTSF)_2PF_6$ and (d) $(TMTSF)_2ClO_4$. After [214].

In $(TMTTF)_2Br$, a long-range magnetic order is established below the spin-density-wave transition at $T_{SDW} \approx 13\,K$. Upon moving toward the right side of the phase diagram, which can be achieved either by varying the anion or by the application of hydrostatic pressure, inter-stack interactions become more important. In this region of the phase diagram the electron-phonon interaction is less dominant and electron-electron interactions along with the good nesting properties of the Fermi surface (cf. Fig. 4.4) lead to a spin-density-wave ground state as observed, e.g. in the Bechgaard salt $(TMTSF)_2PF_6$ at ambient pressure. After suppression of the SP phase in $(TMTTF)_2PF_6$ with increasing pressure, a commensurate antiferromagnetic state is adopted before an incommensurate SDW phase is stabilized. With increasing pressure, T_{SDW} becomes progressively reduced until, above some critical pressure, the systems remain metallic and superconductivity replaces the SDW ground state. The effect of pressure is to increase the π-orbital overlap also in the transverse direction, i.e. perpendicular to the stacking axis. As a result, the good nesting properties are destroyed and the systems become more 3D in character. According to NMR experiments [146, 211] and recent transport measurements under hydrostatic pressure [466, 214], strong SDW correlations are still active in the metallic state even when the SDW instability is replaced by superconductivity in $(TMTSF)_2ClO_4$. The range of strong SDW correlations for pressurized $(TMTTF)_2PF_6$ derived from these experiments is indicated in Fig. 4.31 by the shaded region above the SDW and SC phase boundaries.

κ-(BEDT-TTF)$_2$X Salts

Figure 4.32 shows a conceptual phase diagram proposed by K. Kanoda for the strongly dimerized $(ET)_2X$ salts [338, 337]. It is based on the strongly correlated nature of the π-electrons and assumes that the ground state is controlled by a single parameter U_{eff}/W, i.e. the strength of the effective on-site (dimer) Coulomb repulsion U_{eff} relative to the width of the conduction band W [333, 338]. The dotted lines indicate the positions of the various compounds at ambient pressure. According to this phase diagram, the antiferromagnetic insulator X = $Cu[N(CN)_2]Cl$ and the correlated metal κ-$(H_8\text{-ET})_2Cu[N(CN)_2]Br$ lie on opposite sites of a bandwidth-controlled Mott transition. The deuterated κ-$(D_8\text{-ET})_2Cu[N(CN)_2]Br$ salt is situated right at the antiferromagnetic insulator (AFI) to superconductor (SC) phase boundary [335]. It has been proposed that a *partial* substitution of the 2×4 H-atoms in the ethylene endgroups of the ET molecule by D-atoms permits the κ-$(ET)_2Cu[N(CN)_2]Br$ system to be finely-tuned across the AFI/SC border [472, 473, 474, 475]. The close proximity of an antiferromagnetic insulating to a superconducting phase has been considered — in analogy to the high-T_c cuprates — as an indication that both phenomena are closely connected to each other, see e.g. [104].

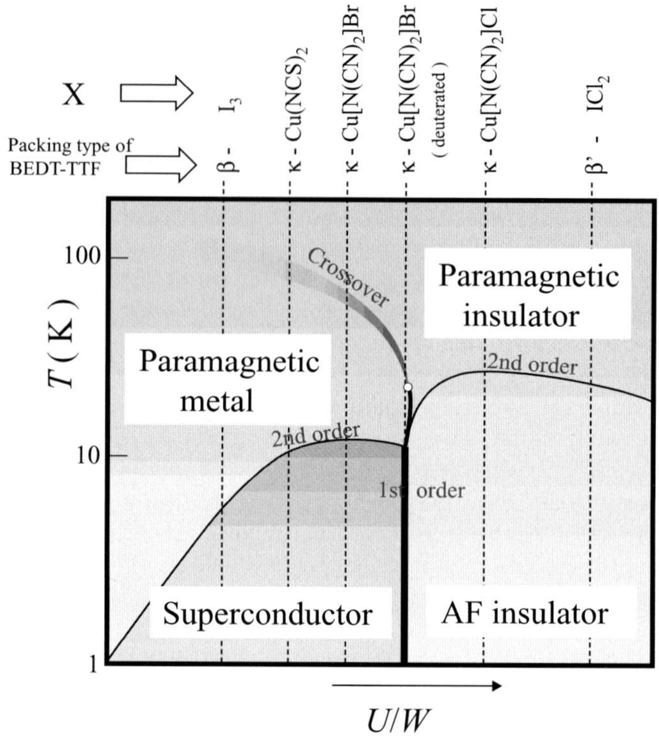

Fig. 4.32. Conceptual phase diagram for $(ET)_2X$ as proposed by Kanoda [338]. Note that hydrostatic pressure decreases the ratio U_{eff}/W, i.e. the low-pressure side is on the right end of the phase diagram. The positions of the various salts with their packing type and anion X at ambient pressure are indicated in the upper part of the diagram. After [337].

Figure 4.33 summarizes experimental data of a thermal expansion study on the various κ-$(ET)_2X$ compounds in a pressure-temperature phase diagram [155]. The positions of the various salts at ambient pressure are indicated by the arrows.[15] Particular attention was paid to the pronounced anomalies at elevated temperatures. At $T_g \sim 70 - 80\,\mathrm{K}$ (dotted line), a glass-like transition has been identified. It marks the boundary between an ethylene-liquid at $T > T_g$ and a glassy state at $T < T_g$. At temperatures above T_g, a certain fraction of the ethylene endgroups is thermally excited, i.e. there is a fluctuation between two possible conformations, with a preferential

[15] Note that hydrostatic pressure has been used as an abscissa for the purpose of compatibility with the conceptual phase diagram in Fig. 4.32. It has been pointed out, however, that the uniaxial-pressure dependencies for the various phase boundaries are strongly anisotropic [348, 155, 235] with a non-uniform behavior for the uniaxial-pressure coefficients of both the instability at T^* and that at T_c, cf. Fig. 4.20.

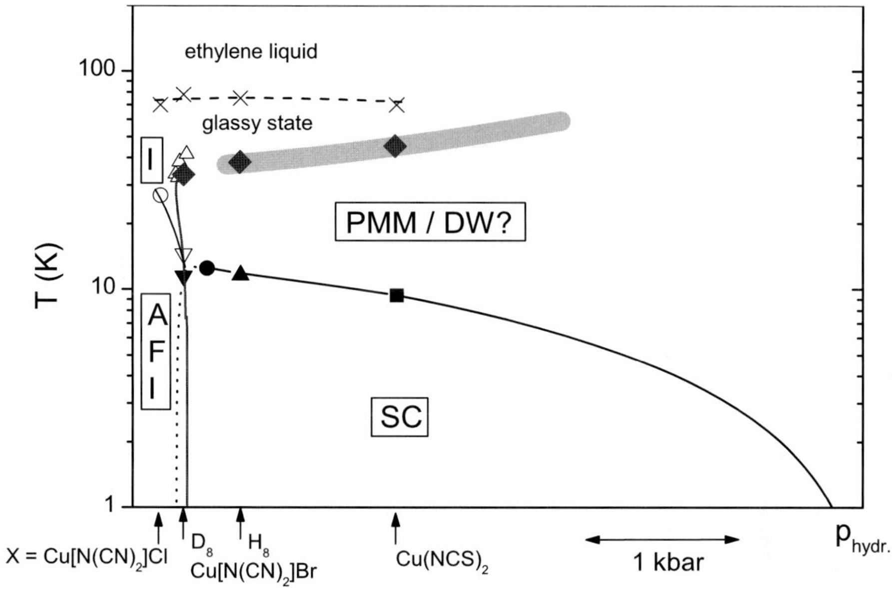

Fig. 4.33. Temperature-pressure phase diagram for κ-$(ET)_2X$. Arrows indicate the positions of various compounds at ambient pressure. The transitions into the superconducting and antiferromagnetic states are represented by closed and open symbols, respectively. Crosses denote the glass-like freezing of ethylene disorder (see Sect. 3.3) and dark-grey diamonds a phase-transition-like anomaly at T^* as derived from thermal expansion measurements, see Figs. 4.19 and 4.20 and [468]. The position of this feature coincides with anomalies in various thermal, magnetic, elastic and optical properties (thick shaded line), see Sect. 4.3.2. The solid lines representing the phase boundaries between the paramagnetic metallic PMM and the superconducting SC or (antiferromagnetic) insulating (AFI) states refer to the results of hydrostatic-pressure studies of T_c and T_N [469, 470, 219]. Also shown is the MI phase transition line (dark grey) and literature results for the critical endpoint $(T_{\rm crit}, p_{\rm crit})$ of the first-order MI transition (open up triangles), see text below and [471]. After [155, 468].

occupancy of one of these orientations depending on the anion. Upon cooling through T_g, however, these orientational degrees of freedom freeze in within a narrow temperature interval which is accompanied by a certain amount of disorder in the ethylene endgroups. The glass-like transition, which is structural in nature, has been shown to cause time dependencies in electronic properties and may have severe implications on the ground-state properties of the κ-$(ET)_2Cu[N(CN)_2]Br$ salt depending on the cooling rate employed at T_g (see Sect. 3.3).

At intermediate temperatures T^*, pronounced anomalies in the coefficient of thermal expansion have been found reminiscent of broadened second-order phase transition. These anomalies coincide with features in various thermal

[233, 476, 477, 236], magnetic [340, 336, 331, 478, 163], acoustic [341, 343] and optical properties [479], see Chapt. 4.3.2. Various explanations have been proposed as to the nature of the anomaly at T^*, including the formation of a pseudo-gap in the density of states [340, 336, 331, 478], a crossover from a coherent Fermi liquid at low temperatures into a regime with incoherent excitations at high temperatures [228, 227], a density-wave-type instability [155, 236, 235], as well as an incipient divergence of the electronic compressibility caused by the proximity to the Mott transition [343]. Recent thermal expansion studies of a deuterated crystal, situated close to the critical pressure marking the end point of the first-order metal-insulator transition in the generalized phase diagram (see below), reveal a distinct anomaly at around 30 K which is attributed to the anomalous lattice response upon approaching the first-order MI transition [468]. This feature is shown to be intimately related to the T^* anomaly at around 40 K for the hydrogenated salt, see also Fig. 4.33.

The Insulator-to-Metal/Superconductor Transition

Of particular interest, for an understanding of the unusual "normal" state properties and the conditions leading to superconductivity in the κ-phase $(ET)_2X$ salts, is the nature of the insulator-to-metal/superconductor transition. The latter can be studied either by hydrostatic or chemical pressure experiments. By chemical means, substitutions on the anion site X allows only discrete shifts on the pressure scale of the order of a few hundred bars. A *partial* substitution of the 2×4 H-atoms by deuterium in κ-$(ET)_2Cu[N(CN)_2]Br$, however, was found to substantially reduce the pressure steps enabling a fine-tuning across the insulator-to-metal/superconductor transition [472, 473, 474, 475]. From the separation of ^{13}C-NMR spectra at low temperatures in the fully deuterated compound, the existence of inhomogeneous electronic states has been suggested originating from a phase separation [335]. According to this study the fully deuterated system is located right at the metal-insulator (Mott) transition, which is of first order, between a metallic state (bearing indications for a pseudo-gap in the density of states) and a commensurate antiferromagnet. The electronic phase separation has been investigated in more detail by T. Sasaki et al. using scanning microregion infrared spectroscopy [480, 471]. In real-space imaging, metallic and insulating domains of size $50 - 100\,\mu m$ are found in partially deuterated samples of κ-$(ET)_2Cu[N(CN)_2]Br$ below the critical end temperature ($30 - 40$ K) of the first-order Mott transition (see below). It has been pointed out in [480, 471] that the macroscopic electronic phase separation near the metal-insulator transition, observed for the present molecular materials, is different from the nano-scale inhomogeneities reported for inorganic correlated electron systems. The latter include, among others, the spatial variations in the superconducting gap in $Bi_2Sr_2CaCu_2O_{8+\delta}$ [481], the stripe formation in the normal state in some other copper oxides [482], or the phase separation seen in the doped manganese oxides [483].

The transition from the insulating state to the metallic/superconducting state can be studied in a particularly clean fashion by applying He-gas pressure to the κ-(ET)$_2$Cu[N(CN)$_2$]Cl salt. In this way the pressure-temperature phase diagram close to the metal-insulator transition of the system κ-(ET)$_2$Cu[N(CN)$_2$]Cl has been explored in great detail by many authors [485, 219, 486, 487, 488, 484, 228, 343, 489, 490] employing a variety of experimental techniques such as resistivity, ac-susceptibility, ^{1}H-NMR and ultrasound velocity. By simultaneous measurements of the ^{1}H-NMR and ac-susceptibility, S. Lefebvre et al. have shown that superconducting and antiferromagnetic phases overlap through a first-order boundary that separates two regions of an inhomogeneous phase coexistence (shaded region in Fig. 4.34) [484], in accordance with previous resistivity studies by H. Ito et al. [487]. It has been argued that this boundary curve merges at a point (282 bar, 13.1 K) with the first-order line (open circles in Fig. 4.34) of the metal-insulator transition at intermediate temperatures. The phase diagram in Fig. 4.34 implies that the first-order metal-insulator (MI) phase boundary ends at a critical point ($p_{\text{crit}}, T_{\text{crit}}$). An extension of this phase diagram has been presented by the same group [228] where, in the normal state, four different transport regimes as a function of pressure and temperature, corresponding to *insulating*, *semiconducting*, *bad metal*, and *strongly correlated Fermi-liquid* behav-

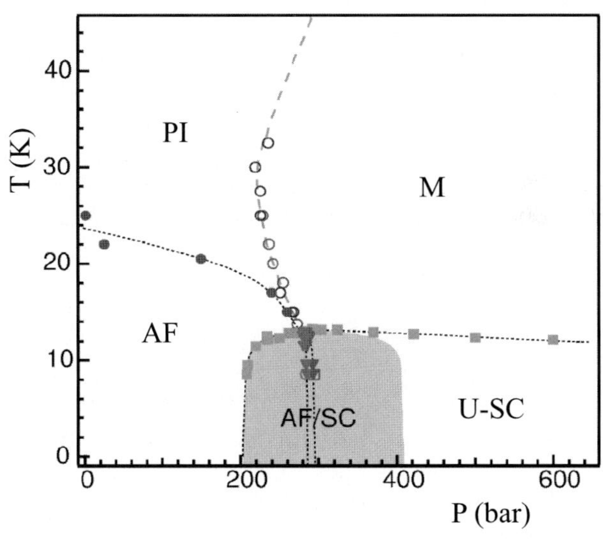

Fig. 4.34. Pressure-temperature phase diagram for κ-(ET)$_2$Cu[N(CN)$_2$]Cl. The antiferromagnetic (AF) transition temperature $T_N(p)$ (closed circles) was determined from the NMR relaxation rate while $T_c(p)$ (closed squares) and $T_{MI}(p)$ (open circles) were obtained from ac-susceptibility measurements. U-SC denotes unconventional superconductivity; the AF/SC boundary line separates two regions of inhomogeneous phase coexistence (shaded area). After [484].

iors, have been identified. Hysteresis experiments have confirmed the first-order nature of the MI transition. The complex physics in this region of the phase diagram has been attributed to strong spatial inhomogeneities [228].

The critical behavior around (p_{crit}, T_{crit}) has been recently explored in detail by F. Kagawa et al. [489, 490] by employing conductivity studies both as a function of temperature and pressure. In this study, a huge discontinuous resistance jump and sharp criticality of the end point ($p_{crit} = 230 - 250$ bar, $T_{crit} = 38 - 40$ K) have been observed as signatures of the first-order Mott transition. The critical exponents, derived from these measurements, however, were found to be inconsistent with known universality classes and it has been proposed that this might reflect an unconventional critical behavior for the Mott transition in two dimensions [490]; see also [491] for a recent review comparing the MI transition in κ-(ET)$_2$X and (DCNQI)$_2$M (cf. Sect. 7.1). A feature not discussed in these studies, and also not included in the phase diagram in Fig. 4.34, is the existence of the $T^*(p)$ line, cf. Fig. 4.33. According to the results of ultrasonic measurements by D. Fournier et al. [343], the $T^*(p)$ line intersects the first-order metal-insulator transition right at (p_{crit}, T_{crit}), cf. Fig. 4.35, with a distinctly different slope. In addition, another anomaly was observed by these authors which was seen to continue the T^* line to the low-pressure side, i.e. into the insulating regime. These observations are difficult to reconcile with the notion that the T^* line marks a crossover line which continues smoothly the first-order M-I line for $T > T_{crit.}$.

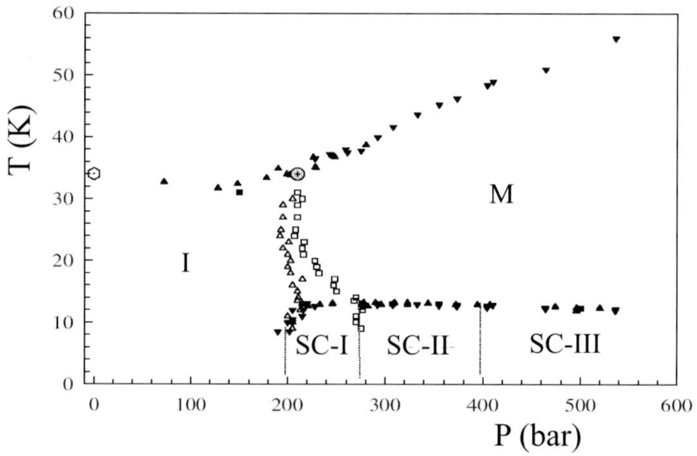

Fig. 4.35. Pressure-temperature phase diagram for κ-(ET)$_2$Cu[N(CN)$_2$]Cl as determined from ultrasonic investigations, after [343]. Different symbols denote different anomalies collected on three different crystals. The upper curve indicates the T^* anomaly which, at the critical point (p_{crit}, T_{crit}) = (210 bar, 34 K) (gray circle), merges with the first-order metal-insulator transition (open triangles and open squares). SC-I to -III indicate different superconducting regions.

5 Magnetic-Field Effects

In general, the Fermi surface (FS) and its related properties in metals have been traditionally studied in detail by investigating magnetic field effects on the conduction electrons and their dynamics [76, 77, 492, 493]. This is possible because the electron's charge and spin can interact with the magnetic field via the Lorentz force-induced orbital motion and the Zeeman interaction, respectively. Such studies, often called *fermiology* (or *fermiography*), have greatly increased our knowledge of Fermi surfaces and related properties in metals and semiconductors. The roots can be traced back to 1930 when the so-called quantum oscillations in the magnetoresistance and magnetic susceptibility were discovered in the semimetal Bi by L.W. Shubnikov, W.J. de Haas and P.M. van Alphen in Leiden [494, 495]. Accordingly, these oscillatory phenomena in the magnetoresistance and magnetization have been labeled, respectively, Shubnikov-de Haas (SdH) and de Haas-van Alphen (dHvA) effects.[1]

Thanks to advanced technologies for magnets, refrigeration, crystal growth and purification, as well as computational methods for band-structure calculations, fermiology has continued to play a central role not only in the understanding of physical properties in metals, but also in exploring the novel phenomena that an ensemble of electrons display when exposed to high magnetic fields and low temperatures. One of the most prominent discoveries is the quantum Hall effect [502] in semiconductor heterostructures with low carrier-number densities and extremely high mobilities of the 2D electrons.[2] These artificial electron systems provided an ideal and unique opportunity to study the physics in the extreme 2D limit [503, 504, 505, 506, 507]. In contrast, the present low-dimensional molecular metals (LDMM) can be viewed as really *quasi*-1D or *quasi*-2D systems where the interchain or interplane couplings (3D effects) are so important that they give rise to cooperative

[1] Several books have been published on this subject by, e.g., I.M. Lifshitz, M.Y. Azbel and M.I. Kaganov [496], D. Shoenberg [497] and A.B. Pippard [498], and edited by J.F. Cochran and R.R. Haering [499] and M. Springford [500]. For the experimental techniques, refer to a review paper by J.R. Anderson and D.R. Stone [501].

[2] The mobility exceeds $10^6\,\mathrm{cm}^2/\mathrm{Vs}$ corresponding to a mean free path ℓ in the order of $10^{-2} - 10^{-3}$ cm, which is $10^2 - 10^3$ longer than in the present molecular metals.

phenomena such as superconductivity, metal-insulator (MI) transitions as well as density-waves or charge ordering. The 3D coupling results in the existence of a corrugated open FS in the quasi-1D systems like the TM salts, and the coexistence of an open FS with a corrugated cylindrical FS in quasi-2D systems like the ET and BETS salts.

This chapter deals with the intriguing phenomena displayed by electrons orbiting on those FSs and magnetic field-induced electronic transitions and/or collective states. Recent reviews on this topic are given in [208, 86, 209, 205, 508].

5.1 Electron Motion in Magnetic Fields

5.1.1 Semi-Classical Description in k-Space

An electron with velocity $\boldsymbol{v}$ moving in a magnetic field $\boldsymbol{B}$ follows the semi-classical equation of motion due to the Lorentz force

$$\hbar \frac{\mathrm{d}\boldsymbol{k}}{\mathrm{d}t} = -e\boldsymbol{v} \times \boldsymbol{B}, \tag{5.1}$$

where $\boldsymbol{k}$ is the wave vector and $-e$ the electron charge. In this section we focus on electrons at the FS which is defined as the equi-energy contour with the Fermi energy E_F. The velocity is given by the gradient of the energy dispersion $E(\boldsymbol{k})$ with respect to $\boldsymbol{k}$,

$$\boldsymbol{v} = \frac{1}{\hbar} \nabla_{\boldsymbol{k}} E(\boldsymbol{k}), \tag{5.2}$$

which is always normal to the FS. The projection of the real space orbit in a plane perpendicular to the field, $\boldsymbol{r}_\perp = \boldsymbol{r} - \hat{\boldsymbol{B}}(\hat{\boldsymbol{B}} \cdot \boldsymbol{r})$, can be obtained by taking the vector product of both sides of (5.1) with a unit vector $\hat{\boldsymbol{B}}$ parallel to the field B. This gives

$$\hat{\boldsymbol{B}} \times \hbar \dot{\boldsymbol{k}} = -eB\left(\dot{\boldsymbol{r}} - \hat{\boldsymbol{B}}(\hat{\boldsymbol{B}} \cdot \dot{\boldsymbol{r}})\right) = -eB\dot{\boldsymbol{r}}_\perp . \tag{5.3}$$

By integrating (5.3) with respect to time, the following relation is obtained:

$$\boldsymbol{r}_\perp(t) - \boldsymbol{r}_\perp(0) = \lambda_B^2 [\boldsymbol{k}(t) - \boldsymbol{k}(0)] \times \hat{\boldsymbol{B}}. \tag{5.4}$$

Here

$$\lambda_B = (\hbar/eB)^{1/2} \tag{5.5}$$

is the so-called magnetic length.[3] Since $(\mathrm{d}\boldsymbol{k}/\mathrm{d}t)\cdot\boldsymbol{v} = (1/\hbar)\mathrm{d}E(\boldsymbol{k})/\mathrm{d}t \equiv 0$, the electron energy is always conserved in a magnetic field and consequently the electrons move along the k-space orbit on the intersection normal to $\boldsymbol{B}$. The electron trajectory can be mapped onto the r-space by rotating the k-space trajectory by $\pi/2$ around $\hat{\boldsymbol{B}}$ and then scaling with λ_B^2.

[3] At $B = 1$ (10) T, $\lambda_B = 256$ (81) Å.

5.1.2 Angular-Dependent Magnetoresistance Oscillations

Here we describe the phenomena called angular-dependent magnetoresistance oscillations (AMRO), which were observed in the magnetoresistance as a function of field orientation with respect to the crystallographic axes.

Quasi-1D Case with a Pair of Corrugated Open FSs

Starting with a quasi-1D tight-binding band with $t_a \gg t_b, t_c$ as expressed in (4.1), we consider the situation when the field $\boldsymbol{B} = (0, B\sin\theta, B\cos\theta)$ is rotated in the bc-plane about the chain a-axis. Here the angle θ is measured from the c-axis. From (5.1) we obtain

$$\frac{\mathrm{d}k_b}{\mathrm{d}t} = \frac{e}{\hbar} v_a B\cos\theta = \frac{1}{\lambda_B^2} v_a \cos\theta, \tag{5.6}$$

and

$$\frac{\mathrm{d}k_c}{\mathrm{d}t} = -\frac{e}{\hbar} v_a B\sin\theta = -\frac{1}{\lambda_B^2} v_a \sin\theta, \tag{5.7}$$

where v_a is the Fermi velocity along a. The angular frequency, $\omega_{b(c)}$, for electrons traversing the Brillouin zone $(2\pi/b(c))$ along $k_{b(c)}$ is given by

$$\omega_b = \frac{b}{\lambda_B^2} v_a \cos\theta, \tag{5.8}$$

and

$$\omega_c = \frac{c}{\lambda_B^2} v_a \sin\theta. \tag{5.9}$$

In general, these two frequencies[4], which are cyclotron-resonance frequencies for open orbits, are incommensurate to each other. However, they become commensurate, $\omega_b/\omega_c = q/p$ (p and q are arbitrary integers), when $\boldsymbol{B}$ is applied just parallel to the lattice vector $\boldsymbol{R}_{pq} = p\boldsymbol{b} + q\boldsymbol{c}$,

$$\tan\theta = \frac{p}{q} \cdot \frac{b}{c}. \tag{5.10}$$

At these so-called *magic angles*, satisfying the commensurability or resonance condition, electrons can perform periodic and coherent motions in the first Brillouin zone if we take a reduced zone scheme.[5] This commensurability concept, which originally has been proposed by A.G. Lebed [512] for the

[4] For the unit cycle, $\int_0^{2\pi/\omega_{b(c)}} \mathrm{d}t |\mathrm{d}k_{b(c)}/\mathrm{d}t| = 2\pi/b(c)$. Exactly the same relations are obtained for the r-space motion by using (5.4). The magnitudes of $\omega_{b(c)}$ at, e.g., $B = 10\,\mathrm{T}$ are expected to be in the microwave range of 100 GHz since $b(c)v_a \sim 1\,\mathrm{cm^2/s}$.

[5] The commensurability resonance effect was revealed in periodically modulated heterostructures of AlGaAs/GaAs [509, 510, 511].

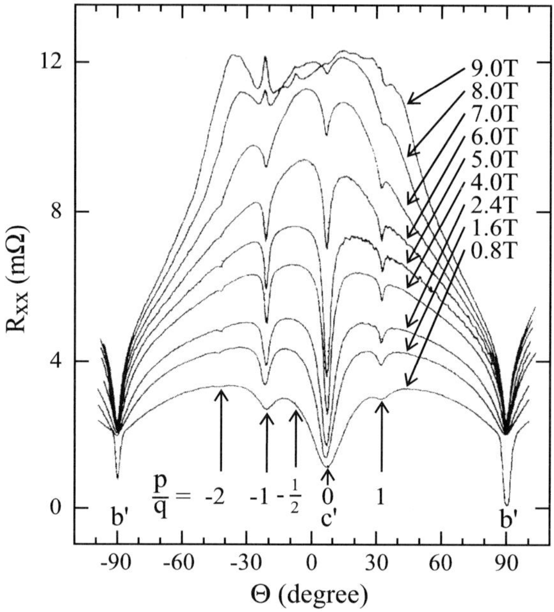

Fig. 5.1. Angular-dependent magnetoresistance R_{xx} of $(TMTSF)_2PF_6$ at 0.5 K under 10 kbar. The sample was mounted in the conventional six-probe geometry with the chain (a) axis (∥ current) oriented normal to the applied field. The magnetic field was rotated in the bc-plane and the angle θ was measured from the c-axis. At the lowest field of 0.8 T, partial superconductivity was observed. After [516].

field-induced SDW phenomena (Sect. 5.3.1), is crucially important also for AMRO observed in $(TMTSF)_2ClO_4$ with a doubling of the periodicity along the b-axis [513, 514, 515].[6] The same phenomena were confirmed more clearly in the pressurized PF_6 salt without such a structural complication [516]. As described in Sect. 4.3, these materials have a characteristic hierarchy in the magnitude of the transfer integrals $t_a : t_b : t_c = 1 : 0.1 : (0.006 - 0.003)$.

Figure 5.1 shows the angular dependent magnetoresistance R_{xx} of pressurized $(TMTSF)_2PF_6$. On top of a smooth background, sharp minima in R_{xx} are seen at $\theta = \pm 90°$ (the b axis or $p/q = \infty$), 33° ($p/q = +1$) and $-22°$ ($p/q = -1$) for all magnetic fields shown. These angles could be identified exactly with the magic angles (5.10) by taking into account the triclinic structure. In the normal metallic state below 5 T, a series of magnetoresistance dips at commensurate magic angles sharpens upon increasing the magnetic field. In addition to the salient dips, some fine features are seen at $-42°$ and $-8°$ that can be assigned to $p/q = -2$ and $-1/2$. For fields above 5 T, the angular dependence is changed by field-induced spin-density-waves (FISDW), see Sect. 5.3.1.

The semiclassical calculations by T. Osada et al. [517] for the conductivity tensor, using a Boltzmann equation based on a constant relaxation-time approximation, indicated that the electrons can drift along the direction of $\boldsymbol{B}$ at

[6] The sample needs to be cooled down slowly through $T_{AO} = 24$ K to achieve a well-developed orientational ordering of the non-centrosymmetric ClO_4^- anions, see Sect. 3.3.

the magic angles and consequently the angular dependence of the interchain conductivity σ_{yy} and σ_{zz} exhibits resonance peaks at the magic angles, see (5.18). It should be noted here that neither magnetoresistance nor AMRO can be expected in $\sigma_{xx} = \rho_{xx}^{-1}$ that were measured in experiments with an external dc current parallel to the chain axis [513, 514, 515, 516].[7] The observation of similar AMRO when $\boldsymbol{B}$ is rotated in the *ac*-plane is also consistent with the commensurability effect [519]. Furthermore, a kink structure was found in the angular-dependent magnetoresistance when $\boldsymbol{B}$ was rotated in the *ab*-plane [520]. This effect, which also had been observed in another quasi-1D salt $(DMET)_2I_3$ [521], has been explained in a way that the kink appears just at the critical angle for a small closed orbit to be formed. This mechanism, due to the peculiar curvature of the corrugated sheets, is physically similar to that in the quasi-2D FS of a corrugated cylinder and will be discussed in the next section.

To note, more detailed measurements have been made for these geometrical effects of electron trajectories on open FSs in $(TMTSF)_2X$ salts [522, 523, 524, 525, 526, 527], in κ-$(BEDT\text{-}TTF)_2Cu(NCS)_2$ [528, 529], and in $(DMET\text{-}TSF)_2AuCl_2$ [530, 531]. An interference effect due to Bragg reflection [532, 533] and a dimensional crossover of electron wave functions [534] have been discussed. Recently T. Osada and his collaborators have indicated novel aspects concerning quasi-1D AMRO. One is an effect of strong electric fields applied parallel to the interchain axis which induces a splitting of the magic-angle resonance into double resonances observed in α-$(ET)_2KHg(SCN)_4$ [535]. The other readdresses the incoherent-transport problem [536, 537, 538, 248, 539] in which AMROs originate from the change of the electron-tunneling probability between neighboring chains or planes [540, 541].

Quasi-2D Case with a Corrugated Cylindrical FS

Figure 5.2(a) shows the angular dependence of the interplane resistance $R_\perp$ of β-$(ET)_2IBr_2$ as a function of θ between the z-axis and the magnetic field of 15 T. Upon tilting $\boldsymbol{B}$ from the z-axis, several oscillations clearly appear and the amplitudes tend to fade out upon approaching 90°, followed by an isolated, sharp single peak just in the vicinity of $\theta = 90°$. These novel phenomena were observed in the present salt by M.V. Kartsovnik et al. [542] and also in the θ-$(ET)_2I_3$ by K. Kajita et al. [543]. The oscillations have the characteristic features that the positions of local maxima in $R_\perp$, which are independent of both magnetic field and temperature, are periodic in $\tan\theta$ (see the inset of Fig. 5.2(a)). For these oscillatory phenomena in the magnetoresistance that cannot be ascribed to a conventional SdH effect, K. Yamaji

[7] A possible reason may be found in the experimental fact that the distribution of in-plane currents is essentially inhomogeneous in highly anisotropic conductors, often leading to an uncontrollable mixing of $\rho_\parallel$ and $\rho_\perp$ [205, 518].

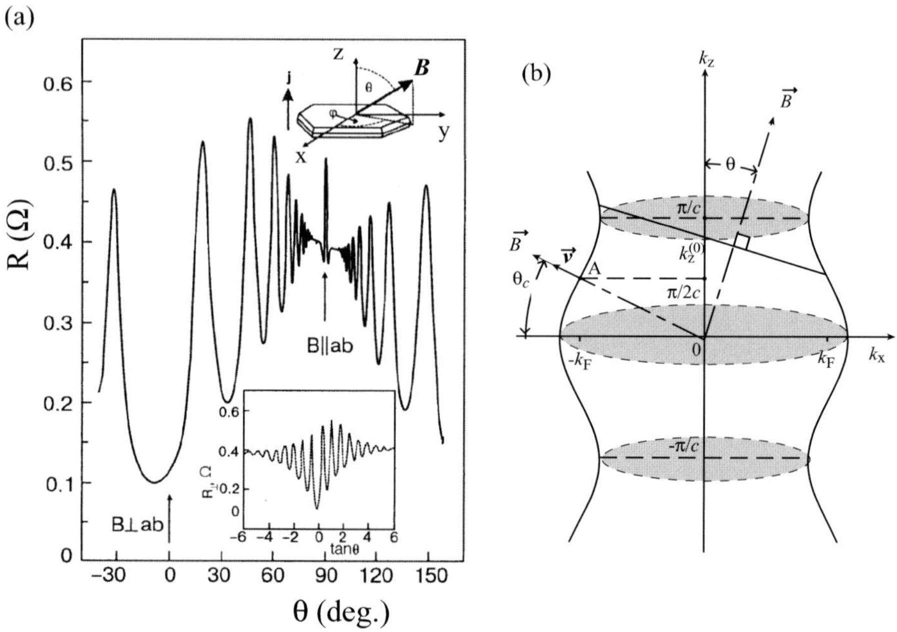

Fig. 5.2. (a) Orientation dependence of the interplane resistance in β-(ET)$_2$IBr$_2$ at $B = 15\,\text{T}$ and $T = 1.4\,\text{K}$. The angle θ is measured from the z-axis as shown in the upper inset. The lower inset shows the resistance as a function of $\tan\theta$. After [508]. (b) Schematic illustration for the corrugated cylinder projected along the k_y-direction. The point denoted by A is the reflection point at $k_z = \pi/2c$.

suggested that, at certain *magic angles* when $\boldsymbol{B}$ is rotated, all the crosssectional areas (perpendicular to $\boldsymbol{B}$) of a corrugated cylinder become equal to each other [544]. Figure 5.2(b) shows the projection of the FS to the $k_x k_z$-plane with the first Brillouin zone $\pi/c \geq k_z \geq -\pi/c$. The quasi-2D electronic dispersion is written as

$$E(\boldsymbol{k}) = \frac{\hbar^2}{2m_\parallel}(k_x^2 + k_y^2) - 2t_z \cos(k_z c). \tag{5.11}$$

Here t_z is the interplane transfer integral assumed to be much smaller than $E_F = \hbar^2 k_F^2/2m_\parallel$. When $\boldsymbol{B}$ is just parallel to the z-axis, there are two extremal orbitals on the belly at $k_z = 0$ and the neck at $k_z = \pm\pi/c$. The area S_k encircled by electrons in the plane perpendicular to $\boldsymbol{B}$ is approximated as

$$S_k \cos\theta = \pi k_F^2 + \frac{4\pi m_\parallel t_z}{\hbar^2} \cdot \cos(k_z^{(0)} c) \cdot J_0(k_F c \tan\theta). \tag{5.12}$$

Here $k_z^{(0)} (= k_z + k_x \tan\theta)$ denotes the intersecting point of the k_z-axis with the orbital plane, and J_0 the *zero*-th order Bessel function. Since $J_0(z) \approx$

$(2/\pi z)^{1/2}\cos(z-\pi/4)$ for $z \geq 1$, the second term in (5.12) vanishes when θ satisfies the condition

$$k_F c \tan\theta = \pi(n - \frac{1}{4}), \tag{5.13}$$

where n is an integer. All the orbits have then identical $S_k = \pi k_F^2/\cos\theta$ and thereby, the electronic energy levels become completely discrete Landau levels. In this situation, electrons cannot perform any drift motions along the z-direction, resulting in a large (theoretically infinite) resistance that appears periodically as a function of $\tan\theta$. While these phenomena were explained by theoretical calculations [545] for $\sigma_{zz}(\theta)$ within a relaxation-time approximation for the Boltzmann equation, a theory [546] within a fully quantum mechanical framework justified Yamaji's semiclassical treatment. Moreover, taking into account the anisotropy in the basal plane, the fine structure of the FS can be determined in principle [547]. Thus, these new oscillatory phenomena have provided a new tool as a caliper for corrugated cylindrical FSs in general.

The isolated single peak near $\theta = 90°$ in Fig. 5.2(a) was discussed in terms of a non-saturating magnetoresistance due to the open orbits [548]. A similar peak was reproduced in β_H-$(ET)_2I_3$ (under 5 kbar) and α-$(ET)_2NH_4Hg(SCN)_4$ and attributed to the appearance of closed orbits that are formed around the convex part of the corrugated cylinder when $\boldsymbol{B}$ is close to the basal plane within θ_c (Fig. 5.2(b)) [549]. It has been noted that this geometrical effect can also apply for the kink anomaly in quasi-1D systems [521, 520] as mentioned in the preceding section. Here we take a cosine band $-2t_\parallel \cos(k_F a)$ instead of the parabolic band in (5.11). The z- and in-plane-components of the Fermi velocity $\boldsymbol{v_F}$ (5.2) are $v_{F,z} = (2t_z c/\hbar)\sin(k_{F,z}c)$ and $v_{F,\parallel} = (2t_\parallel a/\hbar)\sin(k_{F,\parallel}a)$. Since the former oscillates during the motion on the closed orbit, the drift velocity is averaged out, resulting in $\sigma_{zz} = 0$. The geometrical condition for the critical angle θ_c for the closed orbit to appear is satisfied at $k_z = \pi/2c$ where $\boldsymbol{B}$ is parallel to $\boldsymbol{v_F}$ (Fig. 5.2(b));

$$\tan\theta_c = \frac{v_{F,z}(k_z = \pi/2c)}{v_{F,\parallel}} = \frac{t_z}{t_a}\cdot\frac{c}{a}\cdot\frac{1}{\sin(k_F a)}. \tag{5.14}$$

Using the parameters $c/a = 2$, $k_F = \pi/\sqrt{2}a$, and $t_z/t_a = 1/140$, then $\theta_c \sim 1°$ which is in good agreement with the data [549]. This successful explanation has provided another tool to estimate the ratio of the interlayer to intralayer transfer integrals.

It is noted that these magnetoresistance phenomena, characteristic of a corrugated cylindrical FS, have been observed also in other quasi-2D materials like $Tl_2Ba_2CuO_{6+\delta}$ [550], Sr_2RuO_4 [551, 552, 553, 554], intercalated graphite [555, 556], and artificially modulated superlattices [557].

5.1.3 Magneto-Optic Resonances

In addition to the dc (or low frequency) magnetotransport mentioned above, high-frequency electron dynamics has attracted much attention. In particular, cyclotron resonance (CR) measurements with electromagnetic waves in the millimeter-wave frequency range $f(=\omega/2\pi)$ of 30 – 100 GHz are powerful tools for determining the cyclotron mass m_c and the relaxation time τ directly [492]. L.P. Gor'kov and A.G. Lebed [558] proposed a mechanism of CR for a corrugated open FS in a quasi-1D metal. The idea, which is physically similar to that of the so-called Azbel-Kaner CR [559] for a closed orbit[8], is explained as follows.

In a quasi-1D system in an applied magnetic field $\boldsymbol{B}$, e.g., parallel to the c-axis, electrons moving along the most conducting a-axis with v_F follow an oscillatory path along the b-axis with the amplitude[9]

$$X_b = \frac{2t_b}{ev_F B} = \frac{2t_b}{\hbar v_F} \cdot \lambda_B^2 . \tag{5.15}$$

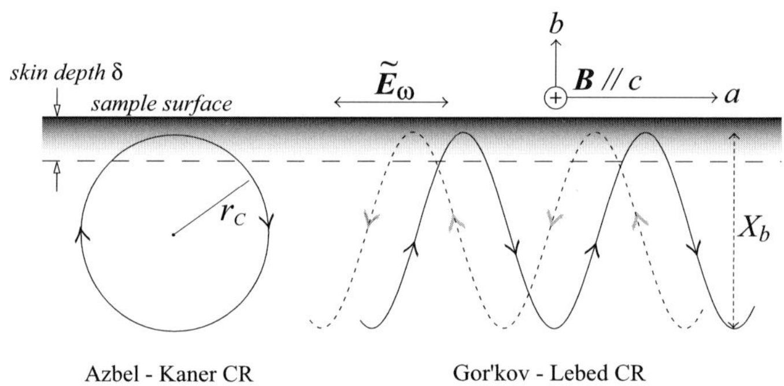

Fig. 5.3. Schematic configurations for Azbel-Kaner and Gor'kov-Lebed cyclotron resonances

Figure 5.3 illustrates the trajectory and the configuration of the oscillating electric-field component $\boldsymbol{E}_\omega$ parallel to the a-axis which decays within the skin-depth δ assumed to be less than X_b. For comparison, the circular cyclotron orbit in the Azbel-Kaner CR is also shown. Since the electron enters the skin-depth regime periodically with the frequency $\omega_c(= v_F b/\lambda_B^2)$[10]

---

[8] This prediction was verified by surface impedance measurements on Cu [560].

[9] This relation derived from (5.4) and (5.5) means that the electron trajectory becomes less corrugated (*field-induced one-dimensionalization*), which is what is essentially responsible for the field-induced SDW (Sect. 5.3.1) [561].

[10] Derived by substituting $\theta = 0$ in (5.8).

to absorb the energy from $\boldsymbol{E}_\omega$, the resonance occurs when both periodicities become commensurate,

$$\omega = n\omega_c = n\frac{v_F b}{\lambda_B^2}. \tag{5.16}$$

To observe the resonance that would be detected by the surface impedance as in the Azbel-Kaner CR, the condition $\delta < X_b < \ell$ must be satisfied. It is difficult, however, to meet the condition experimentally and the observation has not been made yet.[11]

Alternatively, millimeter-wave measurements on LDMM have played an important role in investigating the electrodynamics regarding AMRO. Measurements without any electrical contacts usually involve placing a tiny single crystal at various locations within either rectangular or cylindrical cavities [563, 271]. The dynamical magnetoconductivity $\sigma_{ij}(\omega)$ was calculated [517] by the kinetic form of the Boltzmann equation assuming a constant relaxation time τ:

$$\sigma_{ij}(\omega) = \frac{2e^2}{V}\sum_{\boldsymbol{k}}\left(-\frac{\mathrm{d}f(\boldsymbol{k})}{\mathrm{d}E(\boldsymbol{k})}\right) v_i(\boldsymbol{k},0)\int_{-\infty}^{0} v_j(\boldsymbol{k},t)\exp\left(\frac{1}{\tau}-\mathrm{i}\omega\right)t\mathrm{d}t. \tag{5.17}$$

Here the real parts are expressed[12] as

$$\begin{pmatrix}\sigma_{yy} & \sigma_{yz}\\ \sigma_{zy} & \sigma_{zz}\end{pmatrix} = N(E_F)\sum_{p,q}\left(\frac{et_{pq}}{\hbar}\right)^2\begin{pmatrix}p^2b^2 & pqbc\\ pqbc & q^2c^2\end{pmatrix}\cdot\frac{\tau}{1+[(\omega - v_F G_{pq})\tau]^2}. \tag{5.18}$$

Here t_{pq} represents the effective interchain transfer-integral associated with the lattice vector $\boldsymbol{R}_{pq} = (0, pb, qc)$ and

$$G_{pq} = (pb\cos\theta - qc\sin\theta)\frac{1}{\lambda_B^2}. \tag{5.19}$$

The conductivity in the dc limit ($\omega = 0$) gives a series of peaks at $G_{pq} = 0$ exactly equal to the commensurability condition (5.10), while the high-frequency conductivity exhibits a series of resonance-like peaks at

$$\omega = v_F G_{pq}. \tag{5.20}$$

---

[11] At liquid-helium temperatures, $\delta = (2/\sigma\mu\omega)^{1/2}$ of in-plane currents induced by millimeter-waves is in the μm range, while X_b is estimated to be 0.1 μm or less using $v_F \sim 10^7$ cm/s and $B = 1$ T. The mean free path ℓ in conventional single crystals of LDMM is in the order of 0.1 μm at most. For general considerations on the experimental configurations for millimeter-wave electrodynamics, refer to [562].

[12] Other elements like σ_{xy} and σ_{xz} are zero — no Hall effect —, while $\sigma_{xx}(\equiv 1/\rho_{xx}) = N(E_F)(ev_F)^2\tau/(1+(\omega\tau)^2)$ has no field dependence as already described in Sect. 5.1.1.

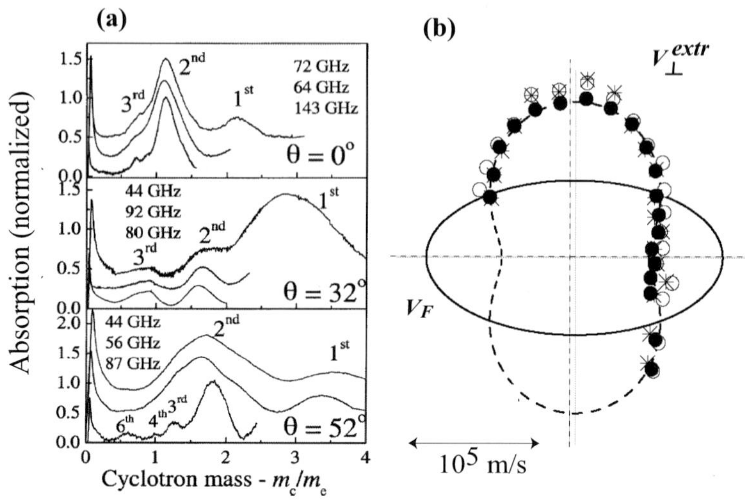

Fig. 5.4. (a) Millimeter-wave absorption in κ-$(ET)_2I_3$ as a function of magnetic field. The horizontal axis corresponds to the cyclotron mass ($m_c = eB/\omega$). The tilt angle θ of the magnetic field is measured from the interlayer direction (a-axis). The sharp peaks at low fields are due to a Josephson plasma resonance [569], after [570]. (b) Polar plot of $v_\perp^{\rm extr}$ and v_F in κ-$(ET)_2I_3$; after [571].

This idea for a novel type of cyclotron resonance was extended more generally to a corrugated FS, either open or closed. As electrons traverse such a curved surface, the oscillating velocity component can couple directly to the applied electromagnetic waves. Thus, S. Hill [564] called it *periodic-orbit resonance* or, according to S.J. Blundell, A. Ardavan and J. Singleton [565, 566, 567] it is also known as *FS-traverse resonance.* The effects of the higher harmonics and interlayer coherency were discussed by R.H. McKenzie and P. Moses [568].

Figure 5.4(a) shows the millimeter-wave absorption in κ-$(ET)_2I_3$ [570] obtained by S. Hill and his collaborators. The FS is similar to that of κ-$(ET)_2Cu(NCS)_2$ as shown in Fig. 4.5(b). The broad resonances, measured at different frequencies, line up at the same m_c and become narrower at higher frequencies. These resonances appear harmonically, e.g., at $m_c/m_0 = 2.2$, 1.1 and 0.71, and also shift to higher m_c/m_0 values following a $1/\cos\theta$ dependence expected for a cylinder. These features are consistent with the periodic-orbit resonance due to a corrugated cylindrical FS. Particularly interesting is the observation that the peak intensity of the fundamental resonance at $\theta = 0$ is smaller than that of the second harmonics. It may be consistent with the numerical simulation [564] based on the two-fold symmetry of the elliptical corrugated cylinder where the interlayer component of the Fermi velocity oscillates twice during one cycle of the cyclotron period. For the same materials, furthermore, A.E. Kovalev et al. [571] mapped $\boldsymbol{v}_F$ onto the basal plane from millimeter-wave absorption measurements at 54 GHz (corresponding to $\omega\tau \sim 2$) with rotating $\boldsymbol{B}$ in the basal bc-plane as shown in Fig. 5.4(b). By applying formula (5.17) for the dynamical conductivity to a corrugated cylinder, the interlayer conductivity $\sigma_{zz}(\omega)$ is found to be predominantly determined by the extremal perpendicular velocity $v_\perp^{\rm extr}$, leading to a resonance at $\omega = v_\perp^{\rm extr} a/\lambda_B^2$. The Fermi velocities along $c(x)$ and $b(y)$ were obtained as 1.3 and 0.62×10^5 m/s, respectively, and the effective mass of $2.5\, m_0$ along the c-axis on the β-orbit is consistent with those estimated from other measurements [570, 572, 573].

The described dynamical phenomena opened the way for angle-resolved FS spectroscopy of low-dimensional metals and have been extensively studied for LDMM like $(TMTSF)_2ClO_4$ [574, 575], κ-$(ET)_2Cu(NCS)_2$ [576], α-$(ET)_2KHg(SCN)_4$ [577, 578, 579, 580], α-$(ET)_2NH_4Hg(SCN)_4$ [581, 582, 583], β"-$(ET)_2SF_5CH_2CF_2SO_3$ [576] and θ-$(ET)_2I_3$ [584], as well as for Sr_2RuO_4 [585].

5.2 Magnetic Quantum Oscillations

So far, we have given an overview of magnetic-field effects on electron trajectories and their dynamics on a corrugated open or closed FS. In this section, we will focus on the magnetic quantum oscillations and their intriguing properties caused by magnetic breakdown, dimensionality, superconducting vortices and localized spins.

5.2.1 Landau Levels and Lifshitz-Kosevich Theory

Landau levels

The Bohr-Sommerfeld quantization rule can be applied to an electron moving on a topologically closed FS under the influence of a magnetic field

$$\oint \mathbf{p} \mathrm{d}\boldsymbol{r} = \oint \mathrm{d}\boldsymbol{r}\left(\hbar\boldsymbol{k} - e\boldsymbol{A}\right) = (n+\gamma)h, \tag{5.21}$$

where n is an integer and $\gamma = \frac{1}{2}$ for a free electron.[13] The total magnetic flux $\Phi \equiv BS_r$ penetrating the real space area S_r encircled by electrons is quantized as

$$\Phi \equiv BS_r = (n+\frac{1}{2})\Phi_0, \tag{5.22}$$

[13] In general the phase constant γ departs slightly from 1/2 depending on E and B [586].

where $\Phi_0 = h/e = 4.14 \times 10^{-11}\,\mathrm{Tcm}^2$.[14] The corresponding k-space area $S_k = S_r/\lambda_B^4$ derived from (5.4), that is exactly the cross-sectional area of the underlying FS perpendicular to $\boldsymbol{B}$, is also quantized as

$$S_k = (n + \frac{1}{2})(\frac{2\pi e}{\hbar})B = \frac{2\pi}{\lambda_B^2}(n + \frac{1}{2}). \tag{5.23}$$

This is known as Onsager's quantization relation [588].

The orbital quantization leads to the energy quantization,

$$E_n(k_z) = (n + \frac{1}{2})\hbar\omega_c \pm \frac{1}{2}g\mu_B B + \frac{\hbar^2 k_z^2}{2m_z^*}. \tag{5.24}$$

The third term is the kinetic energy of electrons moving along $\boldsymbol{B} \parallel z$ with the effective mass m_z^*. The first term represents the energy quantization for electrons in the plane perpendicular to $\boldsymbol{B}$. The discrete levels with the quantum number n, as illustrated in Fig. 5.5(a), are called Landau levels[15] [589], being equally separated by $\hbar\omega_c$, where

$$\omega_c = \frac{2\pi e B}{(\partial S_k/\partial E)_{k\perp}} = \frac{eB}{m_c} \tag{5.25}$$

is the cyclotron frequency and $m_c = (2\pi)^{-1}(\partial S_k/\partial E)_{k\perp}$ is the cyclotron mass. Moreover, as expressed in the second term of (5.24), each Landau level becomes split into spin-up and spin-down Zeeman levels separated by $g\mu_B B$, see Fig. 5.5(b). In the presence of magnetic exchange interactions, as, e.g., due to localized d-spins, these levels are modified as shown in Fig. 5.5(c), and will be discussed in Sect. 5.2.5.

For simplicity, consider a 2D system with $E(k_x, k_y) = \hbar^2(k_x^2 + k_y^2)/2m_\parallel$. The density of states[16] $N_{\mathrm{2D}}(E)$ in the absence of B becomes *a constant*, independent of E,

$$N_{\mathrm{2D}}(E) = \frac{2}{(2\pi)^2}\int \mathrm{d}^2\boldsymbol{k} \cdot \delta\left(E - \frac{\hbar^2(k_x^2 + k_y^2)}{2m_\parallel}\right) = \frac{m_\parallel}{\pi\hbar^2}. \tag{5.26}$$

The factor of 2 in the numerator comes from the spin degeneracy. In a finite magnetic field at $T = 0$, the continuous spectrum (occupied up to E_F in zero field) turns into an equally separated series of delta functions of Landau levels with the density-of-state spectrum

---

[14] This concept for the magnetic flux quantization is also applied to superconducting vortex states with a quantum flux $\phi_0 = h/2e = 2.068 \times 10^{-11}\,\mathrm{T\,cm}^2$ due to a Cooper pair with $2e$, see, e.g. [587, 365].

[15] In a 3D k-space each Landau level forms a hollow cylinder called a Landau tube.

[16] For 1D, $N_{1D}(E) \propto 1/\sqrt{E}$ being divergent at the band edges, while, for 3D, $N_{3D}(E) \propto \sqrt{E}$.

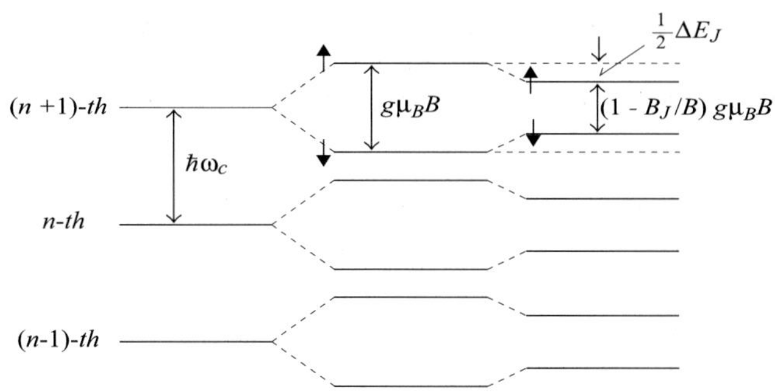

Fig. 5.5. Quantized energy levels in a magnetic field. (a) Landau levels with indices $n-1$, n and $n+1$, (b) spin-splitting Zeeman levels with the energy difference $g\mu_B B$ and (c) the exchange field (B_J)-induced modification of the Zeeman levels with $\Delta E_J = g\mu_B B_J$.

$$N_{2D}(E) = \frac{1}{\pi\lambda_B^2}\sum_n \delta(E - E_n), \tag{5.27}$$

where $E_n = (n+1/2)\hbar\omega_c$. Each Landau level accommodates electronic states within $\pm(1/2)\hbar\omega_c$, the number of which is given by

$$\hbar\omega_c N_{2D}(E) = eB/\pi\hbar = 1/\pi\lambda_B^2. \tag{5.28}$$

Here $m_c = m_\parallel$ is assumed. Since $\pi\lambda_B^2$ is the *real-space* area, the magnetic length λ_B (5.5) is exactly the same as the cyclotron radius $r_c(= v_F/\omega_c) = (2n+1)^{1/2}\lambda_B$ for $n = 0$.[17]

Quantum Oscillations and Lifshitz-Kosevich theory

According to (5.24), for a 3D electronic system, the higher B is, the wider the separation, and thereby the smaller the number of Landau levels involved below E_F. Since E_F (or the chemical potential) is assumed to be constant, the highest occupied Landau level crosses the Fermi level E_F periodically in $1/B$. Whenever this crossing occurs, electrons at the outmost Landau level are forced to redistribute on lower levels. Therefore, the electronic free energy changes periodically with the inverse magnetic field, resulting in quantum oscillations in all the quantities concerned with electronic states near E_F. For a FS of general shape, the phase of oscillations is cancelled out except those originating from extremal orbits, the periods of which are stationary with respect to a small change of k_z. In particular, the oscillatory behavior of the magnetization and magnetoresistance are called dHvA and SdH effects, respectively.

[17] For the derivation, use (5.23) and $S_r = \lambda_B^4 S_k = \pi r_c^2$.

With the cross-sectional area S_{extr} of extremal orbits, the periodicity of the quantum oscillations becomes

$$\frac{1}{B_{n-1}} - \frac{1}{B_n} = \frac{2\pi e}{\hbar S_{extr}}, \tag{5.29}$$

while the frequency F is given by the inverse,

$$F = \frac{1}{1/B_{n-1} - 1/B_n} = \frac{\hbar}{2\pi e} S_{\mathrm{extr}} = \frac{\phi_0}{(2\pi)^2} S_{extr}. \tag{5.30}$$

Taking into account effects of finite temperature, electron scattering and the Zeeman splitting, I.M. Lifshitz and A.M. Kosevich [590] formulated the oscillatory magnetization M_{osc} for one extremal FS as follows:

$$M_{\mathrm{osc}} \propto \left(\frac{B}{\left|\frac{\partial^2 S_{\mathrm{extr}}}{\partial k^2}\right|} \right)^{1/2} \sum_{r=1}^{\infty} \frac{(-1)^r}{r^{3/2}} R_T R_D R_S \cdot \sin\left[2\pi r \left(\frac{F}{B} - \frac{1}{2}\right) \pm \frac{\pi}{4}\right]. \tag{5.31}$$

The summation over r accounts for higher harmonic contributions and the sign $+(-)$ in the argument of the sine function corresponds to the minimum (maximum) cross-sectional area. From this expression, we can directly determine the cross-sectional area S_{extr} of an extremal orbit by measuring F; S_{extr} (in cm^{-2}) $= 9.545 \times 10^{11} F$ (in Tesla). The denominator of the prefactor in (5.31) measures the curvature of the extremal FS, giving an extraordinary enhancement in the oscillatory amplitude for a weakly corrugated cylinder, as in the present quasi-2D metals where the curvature is expected to be quite small.

The reduction factors R_T and R_D describe the phase smearing resulting in a broadening of the Landau levels. First, a Fermi-Dirac distribution function near E_F at $T \neq 0$ gives the temperature reduction factor R_T,

$$R_T = \frac{\alpha r \mu_c T/B}{\sinh(\alpha r \mu_c T/B)} \approx \frac{2\alpha r \mu_c T}{B} \exp\left(\frac{-\alpha r \mu_c T}{B}\right), \tag{5.32}$$

where $\alpha = 2\pi^2 k_B m_0/e\hbar = 14.69\,\mathrm{T/K}$ and $\mu_c = m_c/m_0$ is the cyclotron mass on the extremal orbit in units of the free electron mass m_0. To note, the approximation at the right side of (5.32) holds when the argument is sufficiently larger than 2. Next, the scattering of electrons by impurities, which predominates at low temperatures, induces a Landau-level broadening that is measured by the so-called Dingle temperature T_D [591],

$$k_B T_D = \frac{\hbar}{2\pi < \tau >}. \tag{5.33}$$

Here $< \tau >$ is the average of the electron-scattering time on an extremal orbit. The Dingle reduction factor R_D in (5.31) is given by

$$R_D = \exp\left(\frac{-\alpha r \mu_c T_D}{B}\right). \qquad (5.34)$$

As an experimental routine, μ_c is first determined with (5.32) by measuring the temperature dependence of the oscillatory amplitudes (or the spectral power of the Fourier transform) and then T_D is determined from the magnetic field dependence (5.34). It is noted that both parameters m_c (sometimes called de-Haas mass) and T_D, derived from these reduction factors, are renormalized by many-body effects in interacting Fermi liquids [497], see Sect. 4.5.

Finally, for the spin reduction factor R_S due to the Zeeman splitting (Fig. 5.5), $\pm(1/2)g\mu_B B$, one must take into account that the two sets of spin-dependent Landau levels (tubes) cross E_F at different fields. The effect is exactly equivalent to a spin-splitting phase,

$$\phi = 2\pi g \mu_B B/\hbar\omega_c = \pi g \mu_c, \qquad (5.35)$$

between the oscillations coming from the spin-up and spin-down electrons (or r times as much for the r-th harmonic). The superposition of these oscillations, therefore, provides R_S to the net amplitude of $M_{\rm osc}$ (5.31) [592, 591],

$$R_S = \cos(\frac{1}{2}r\phi) = \cos(\frac{1}{2}r\pi g\mu_c). \qquad (5.36)$$

Here the spin-splitting factor g and the effective cyclotron mass μ_c are renormalized in general by many-body effects such as electron-phonon and electron-electron interactions [497, 256]. In particular, in addition to many-body interactions, g is considerably modified by spin-orbit coupling. Therefore, once μ_c is determined from the temperature dependence of the amplitude (5.32), g can be evaluated by measuring the spin-splitting zeros ($R_S = 0$) as will be described below. Thus, g so obtained is, in general, different from that determined by electron-spin-resonance measurements which do not detect many-body effects.

The Lifshitz-Kosevich (LK) theory has been successfully used to determine the FS topology, cyclotron mass, relaxation time and many-body effects. Basically, the theory equally holds for oscillations in other thermodynamic quantities such as the electronic specific heat and the elastic constants, the so-called magnetothermal and magnetoacoustic oscillations, respectively.[18] On the other hand, the transport quantities such as magnetoresistance, ultrasonic attenuation, thermopower and so on are, in general, rather difficult to

[18] Quantum oscillations in the elastic constants were investigated for α-$(ET)_2NH_4Hg(SCN)_4$ by longitudinal ultrasound-velocity measurements [593]. For Sr_2RuO_4, refer to [594]. This technique, which requires a sample thickness of 1 – 2 mm at least – a single crystal of this size is exceptionally large in LDMM – can provide an evaluation of the electron-lattice coupling and its dependence on acoustic sound modes [595, 596]. Magnetothermal oscillations were observed in κ-$(ET)_2Cu(NCS)_2$ [597].

treat fully quantum-mechanically due to the complicated details of electron-scattering processes in magnetic fields [598]. A.B. Pippard pointed out [497], however, that the probability of scattering is proportional to the number of states into which the electrons can be scattered, and the resistivity determined by this probability ($1/\tau$) should oscillate in harmony with the oscillation of the density-of-states at the Fermi level.[19] Therefore the LK theory has been routinely used for analyzing transport data.

Unlike in semimetals and semiconductors, the SdH effect is often difficult to observe in ordinary metals. In the present LDMM, however, the SdH effect in magnetoresistance has been extensively studied, even more widely than the dHvA effect. This may be partly due to the experimental advantage that the helium-temperature resistance of tiny single crystals, available in the form of thin plates or needles, is rather high in the range of $0.01 - 1\,\Omega$, even in the most conducting plane or along the highly conducting chain. To note, a traditional field-modulation technique [501], with a phase-sensitive detection to obtain the field-derivatives of the resistance or magnetization, has been rarely employed. Alternatively, a conventional four-terminal resistance or magnetic-torque measurements ($\boldsymbol{M} \times \boldsymbol{B}$) with a minute mechanical cantilever [599] have been successfully performed.[20]

A case study in κ*-(ET)$_2$Cu(NCS)$_2$*

Detailed FS parameters of the superconductor κ-$(ET)_2Cu(NCS)_2$ with $T_c =$ 9.4 K have been obtained from quantum oscillations [96, 601, 602, 447, 603, 604]. Figure 5.6(a) shows the in-plane resistance as a function of B applied perpendicular to the basal bc-plane at $T = 0.65$ K. The superconductivity starts to be suppressed around 4 T, followed by a sharp recovery of the resistance. The SdH oscillations with $F_\alpha = 625$ T are clearly superimposed on the normal-state resistance for $B > 6$ T. Upon further increase of B, additional oscillations with a higher frequency $F_\beta = 3800$ T are discernible. The inset shows these periodic oscillations as a function of $1/B$. The two oscillations are identified as coming from α- and β-orbits, as shown in Fig. 5.6(b). These FSs were calculated by a 2D tight-binding model using an extended Hückel method [96]. The α-FS is identified as being an almost perfect cylinder giving

---

[19] An Einstein relation for the conductivity is given by $\sigma = 2e^2N(E_F)D$, where $D = (1/3)v_F\ell$ is the diffusion constant. In a metal, the current $\boldsymbol{J}$ is given by $\boldsymbol{J} = -eD\nabla n + \sigma\boldsymbol{E}$, where n is the number of carriers. At equilibrium, $n = n_o - 2eVN(E_F)$ where V is the potential; $\boldsymbol{E} = \nabla V$. Setting $\boldsymbol{J} = 0$, the above relation is obtained [587].

[20] For magnetic fields larger than $10 - 15$ T available in many laboratories, so-called hybrid-magnets installed in various facilities in Tallahassee, Grenoble, Nijmegen, and Sendai have been actively used. Even higher fields are supplied by pulse magnets generating up to $50 - 60$ T and which are installed in combination with ^{3}He refrigerators to cool samples down to $0.3 - 0.4$ K. A technique for contactless resistance measurements in pulsed fields is noticeable [600].

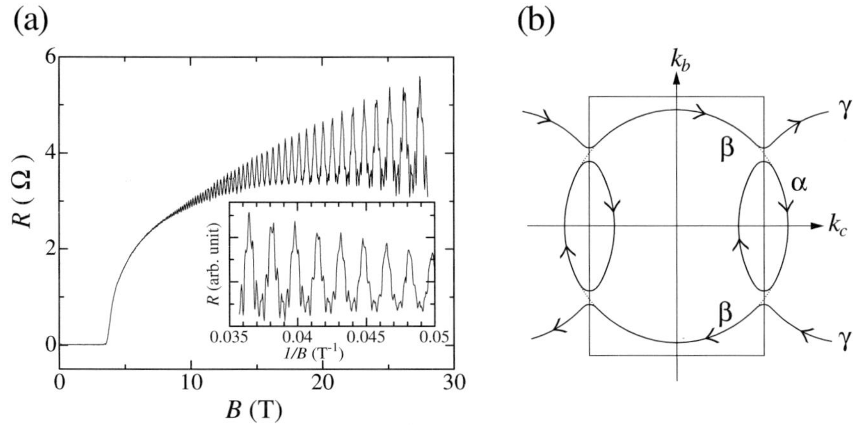

Fig. 5.6. (a) In-plane resistance as a function of B applied perpendicular to the basal plane in κ-$(ET)_2Cu(NCS)_2$ with $T_c = 9.4\,K$. The inset shows the SdH oscillations periodic in $1/B$, after [447]. (b) The FS projected to the $k_b k_c$-plane. The closed and open orbits are designated α and γ, respectively. The closed tunneling orbit β interconnecting between them is called magnetic-breakdown orbit.

a $1/\cos\theta$ dependence [604]. The β-orbit is caused by a magnetic-breakdown (MB) effect between the α- and γ-orbits (a pair of the corrugated open FSs) [602, 447, 603], see Sect. 5.2.2. The k-space areas encircled by these orbits are $S_\alpha = 5.97 \times 10^{14}$ and $S_\beta = 3.63 \times 10^{15}\,cm^{-2}$, occupying 16.3 % and 100.9 % of the first Brillouin zone.[21]

These observations proved that (1) exactly two carriers exist in the unit cell accommodating four ET donors — as expected from the stoichiometric charge-transfer (Sect. 2.2) —, leading to a quarter-filled band, and (2) the band structure based on the tight-binding approximation is highly reliable. The former conclusion is exactly consistent with the Luttinger theorem [607, 608] stating that the FS volume is necessarily conserved as far as a Fermi-liquid picture works, no matter what many-body interactions exist [609].

The cyclotron mass was determined from SdH and dHvA oscillations: $m_c^\alpha = (3.5-3.3)\,m_0$ and $m_c^\beta = (6.9\pm0.8)\,m_0$ for the α- and β-orbits [447, 603, 250, 610, 611], see also Table 4.3. The Dingle temperature, which is a sample-dependent, extrinsic quantity determined by defects in the crystals, was in the range of 0.4 – 1.0 K. J. Caulfield et al. carried out SdH measurements under pressure which suppresses superconductivity [250, 610]. With pressures up to 16 kbar[22], m_c^α decreases sharply from 3.5 down to 1.5 m_0 as shown in Fig. 4.30,

[21] Using the lattice parameters at 12 K [605], the 100.9 % occupation, obtained from the parameters at 104 K in [447], approaches exactly 100 % within an error of ± 0.2 %, via a sizable lattice contraction [606].

[22] A pressure of 5 – 6 kbar is sufficient to suppress superconductivity completely in this material [217, 612].

while T_D increases from 0.57 K up to 1.5 K. Therefore, the product $m_c^\alpha T_D$ is only weakly dependent on pressure which was ascribed to renormalization due to many-body effects[23], as discussed for several quasi-2D molecular metals [609]. For many-body effects, refer to Sect. 4.5.

A different approach to the renormalization effect was taken by J. Wosnitza et al. who observed the spin-splitting zeros $R_S = 0$ in (5.36) from the angular dependence of the dHvA oscillations [604]. Assuming $g = 2$, they obtained for the band mass $m_b = 2.6\, m_0$ giving the $1/\cos\theta$ dependence and an angular-independent electron-phonon contribution of about $0.7\, m_0$ and discussed T_c using the BCS formula. It must be noted, however, that, in the analysis for the product $g\mu_c$ (5.36), some ambiguity may remain in the separation of the different contributions.

From far-infrared transmission measurements, which suggested the existence of a much smaller mass of $m_c^\alpha = 1.25\, m_0$, J. Singleton et al. [613] pointed out that the large discrepancy could be understood by a renormalization effect due to short-range Coulomb interactions, which, according to Kohn's theorem [614], does not affect the CR frequency.[24] However, those measurements were later clarified to be more closely related with the AMRO effect as described above [564, 566].

At this point one may ask about the size of the band mass. In general, absolute values of band-electron energies in metals are more or less equivocal. For LDMM, this is partly because the estimation for the ionization energies of relevant molecular orbitals is empirical at the present state in band-structure calculations, see Sect. 2.2.

So far we have described magnetic quantum oscillations and then derived FS parameters for κ-$ET_2Cu(NCS)_2$. There have been many studies for other molecular metals[25], see [208, 209, 205]. Hereafter, we will focus on peculiar phenomena observed in quantum oscillation studies on LDMM.

[23] With the renormalization constant $Z \geq 1$, the bare (band) mass, the relaxation time and the Fermi velocity are renormalized like Zm_b, $Z\tau$ and v_F/Z, while quantities such as resistivity ($\propto m/\tau$), the mean free path ($\ell = v_F\tau = \hbar k_F \tau/m$) and the product $m_c T_D$ remain unrenormalized due to the cancellation, see, e.g. [497, 609].

[24] Kohn's theorem states that a cyclotron resonance between Landau levels in a metal irradiated by electromagnetic waves with the electric field perpendicular to $\boldsymbol{B}$ gives a resonance frequency independent of short-range electron-electron interactions. To note, Umklapp processes being important for a large FS may violate this theorem [615]. For CR in metals and semiconductors, refer to [497, 616].

[25] The single-component molecular metal $Ni(tmdt)_2$, for example, exhibits a conventional dHvA effect, indicative of a conventional 3D metal [135].

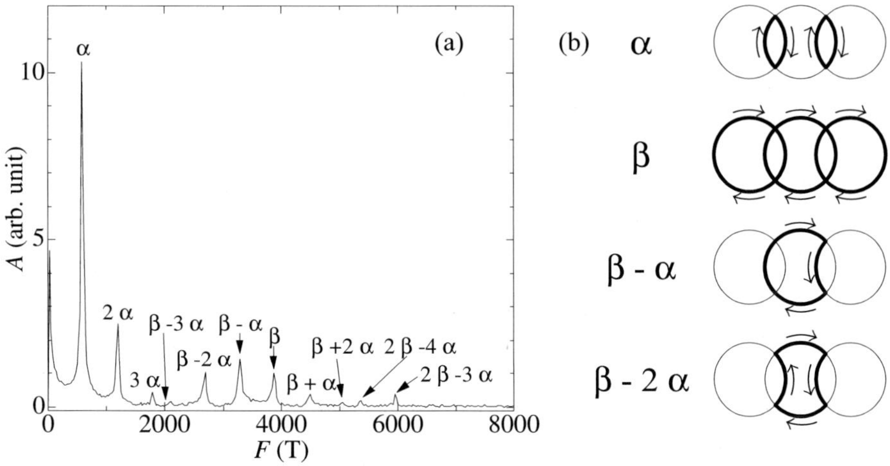

Fig. 5.7. (a) Fourier-transform spectrum of the SdH oscillations in κ-$(ET)_2Cu(NCS)_2$ shown in Fig. 5.6(a). (b) Schematic illustrations for the typical orbits corresponding to the spectum's peaks at F_α, F_β, $F_{\beta-\alpha}$ and $F_{\beta-2\alpha}$. After [447].

5.2.2 Magnetic Breakdown and Quantum Interference

As described in the preceding section, the β-orbit in Fig. 5.6(b) results from a magnetic breakdown between α- and γ-orbits. Here the magnetic breakdown is an electron-tunneling phenomenon, i.e. electrons accelerated by the Lorentz-force (5.1) can tunnel from an orbit on one part of FS to an orbit on another part separated from the first one by a small energy gap E_G, see, e.g. [617, 497].

The tunneling probability P is given by

$$P = \exp\left(-\frac{\pi E_G^2}{4\hbar\omega_c E_F \sin 2\theta}\right) = \exp\left(-\frac{B_{\mathrm{MB}}}{B}\right), \tag{5.37}$$

where 2θ is the Bragg reflection angle and $B_{\mathrm{MB}} = \pi m_c E_G^2/4e\hbar E_F \sin 2\theta$ is a characteristic field for the phenomenon to be visible. From numerical simulations with the parameters $E_F = \hbar^2 k_F^2/2m_c^\beta = (740 \pm 100)\,\mathrm{K} \times k_B$ and $2\theta = \pi/2$, E_G/k_B and B_{MB} were estimated to be about $(54 \pm 10)\,\mathrm{K}$ and $16\,\mathrm{T}$, respectively [618].[26]

Figure 5.7(a) shows the power spectrum of the Fourier transform for the data in Fig. 5.6(a) [447]. In addition to the fundamental frequencies F_α and F_β and the higher harmonics $2F_\alpha$ ($r = 2$) and $3F_\alpha$ ($r = 3$), many combinations of these are clearly discernible. T. Sasaki et al. [618] pointed out that

[26] The gap is expected to be smeared out for $T > E_G/k_B$ via thermally assisted tunneling, in terms of which an anomalous increase in the Hall coefficient below $60 - 70\,\mathrm{K}$ [233, 619] was discussed [620].

β, $\beta + \alpha$ and $\beta + 2\alpha$ orbits could be allowed by Pippard's network model [497, 621, 622, 623], while other orbits labeled $\beta - \alpha$, $\beta - 2\alpha$, $2\beta - 3\alpha$ would be forbidden as illustrated in Fig. 5.7(b). This semiclassical model assumed *a priori* that electrons should traverse along several paths across neighboring FSs in the definite direction determined by the Lorentz force (5.1). On the contrary, taking into account the property of coherence of electron wave functions, W.G. Chambers [624], and H. Shiba and H. Fukuyama [625] predicted quantum-mechanical interference phenomena between wave functions on neighboring FSs.[27] This quantum interference effect associated with the magnetic breakdown phenomena, was proved by SdH measurements on Mg [627, 628, 629]. J. Caulfield et al. [630] pointed out that the above semiclassically forbidden orbits could be allowed by this effect, which was later investigated extensively [631, 632, 633].

Figure 5.8 shows the k-space geometry for this effect. As described in the caption, the transmission probability from P_1 to P_2 is proportional to $\cos(\lambda_B^2 S_{k,\alpha}) = \cos(2\pi F_\alpha / B)$. Here $S_{k,\alpha}$ is the k-space area enclosed by the interference path, consequently oscillating as a function of B^{-1} with a frequency equal to F_α. The double Bragg reflections make such forbidden oscillations like $F_{\beta-\alpha}$ observable, as illustrated using a semiclassical (wave-packet) picture in Fig. 5.7(b). In contrast to semiclassical breakdown phenomena, this interference effect has two characteristic features as described below. Firstly, this interference oscillation, which is not caused by the Landau quantization, is expected to be much less sensitive to temperature than SdH (and dHvA) oscillations. For a semiclassical breakdown orbit, the effective cyclotron mass can be expected to be a sum of the mass on relevant orbits, e.g., $m_c^{\beta+\alpha} = m_c^\beta + m_c^\alpha$ for the $(\beta + \alpha)$-orbit. Indeed, it has been confirmed that the observed mass $11\,m_0$ [630] is in agreement with the sum yielding $10 - 11\,m_0$. For the forbidden orbits due to the interference effect, M. Kartsovnik et al. [632] derived the effective mass from the energy dependence of the oscillation phase: $m_c^{\beta-n\alpha} = m_c^\beta - n m_c^\alpha$ for $(\beta - n\alpha)$-orbits, and made a crucial test for $n = 2$ by observing SdH oscillations in the inter-plane resistance at $B < 15\,\mathrm{T}$. While the F_α oscillations are highly damped at $T > 2.4\,\mathrm{K}$, the rapid oscillations with 2630 T corresponding to $F_\beta - 2F_\alpha$ remain observable at higher temperatures. The mass was evaluated to be $(0.9 \pm 0.1)\,m_0$ being consistent with the above derivation for $n = 2$. Furthermore, a pressure of 8.5 kbar has been applied in order to tune the mass to a value as small as possible. As shown in Fig. 5.9(a), the rapid oscillations continue to be observable even at 9 K, which was ascribed to the quite small mass of $0.3\,m_0$ from the LK analysis as shown in Fig. 5.9(b). Though there remain some unclear points concerning, e.g., the assumption of a temperature-independent relaxation time at these high temperatures, the quantum interference effect can be considered responsible for the forbidden magnetic breakdown orbits.

[27] This prediction for the k-space interference is essentially the same as the Aharonov-Bohm effect [626], see Fig. 5.8(b) and the caption.

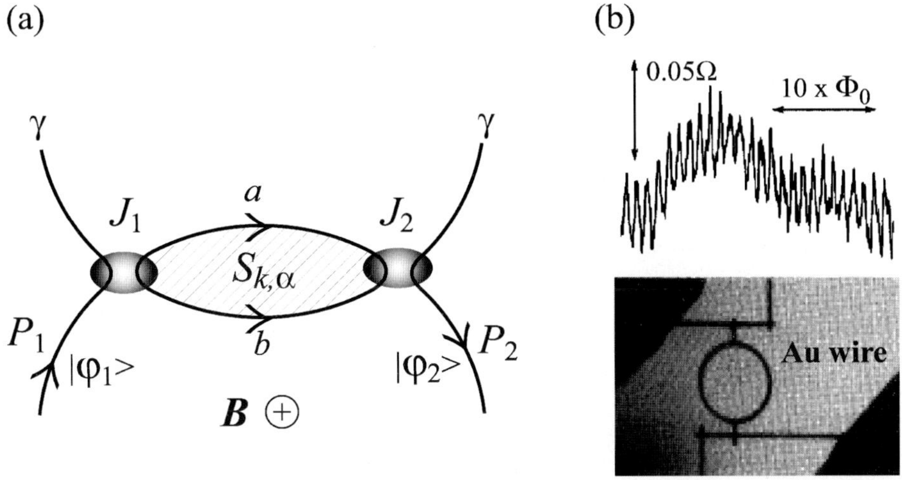

Fig. 5.8. (a) The k-space geometry for the quantum interference effect. Consider an electron with the wave function φ_1 of unit amplitude and zero phase starting from P_1 on the γ-orbit and passing through two equivalent junctions J_1 and J_2 with breakdown gaps eventually to P_2 on the counterpart orbit with the final wave function φ_2. The tunneling probability is given by $P_{1,2} = | < \varphi_2 | \varphi_1 > |^2$ and there are two possible paths a and b on the α-orbit between J_1 and J_2. Transmitted (to a) and reflected (to b) electrons at J_1 can be assumed to have wave functions written as p and $(1-p^2)^{1/2}\exp(\mathrm{i}\pi/2)$. (The relative phase difference should be $\pi/2$ due to an orthogonality relation.) Here the transmission probability p^2 is given by (5.37). Those electrons traveling on different paths receive phases from the vector potential $\boldsymbol{A}$ just before arriving at J_2; the wave functions are evolved as $p\exp(\mathrm{i}\theta_1)$ and $(1-p^2)^{1/2}\exp[\mathrm{i}(\pi/2+\theta_2)]$. The second tunneling then occurs at J_2, resulting in $\varphi_2 = p^2\exp(\mathrm{i}\theta_1) - (1-p^2)\exp(\mathrm{i}\theta_2)$. Thus $P_{1,2} = \{p^4 + (1-p^2)^2\} - 2p^2(1-p^2)\cos(\theta_2-\theta_1)$, where $\theta_2 - \theta_1 = (e/\hbar)(\int_{J_1}^{J_2} \boldsymbol{A}\mathrm{d}\boldsymbol{r}_a - \int_{J_1}^{J_2} \boldsymbol{A}\mathrm{d}\boldsymbol{r}_b) = (e/\hbar)\oint \boldsymbol{A}\mathrm{d}\boldsymbol{r}_{ab} = \lambda_B^{-2} S_{r,\alpha} = \lambda_B^2 S_{k,\alpha}$. (b) This effect is essentially the same as the Aharonov-Bohm effect [626] which was experimentally proved by magnetoresistance measurements on a multiply-connected Au wire demonstrating the conductance quantization with Φ_0. After [634].

Secondly, since the interference is caused by oscillations between the electric current and the density-of-states, it should not be observed in thermodynamic quantities which are solely due to density-of-states effects [625]. The dHvA measurements [635, 611], however, revealed the existence of the forbidden oscillations as $F_{\beta-\alpha}$.[28] Notably, the power intensity is smaller by a factor of ~ 3

[28] The same situation has been found for Mg [636] and recently in Sr_2RuO_4 [637, 551, 552, 553], GaAs/AlAs [638], and $In_xGa_{1-x}As$ [639], which showed forbidden-orbit dHvA oscillations. To note, in the field-modulation method conventionally used for dHvA measurements, the induced eddy-current may cause the interference effect.

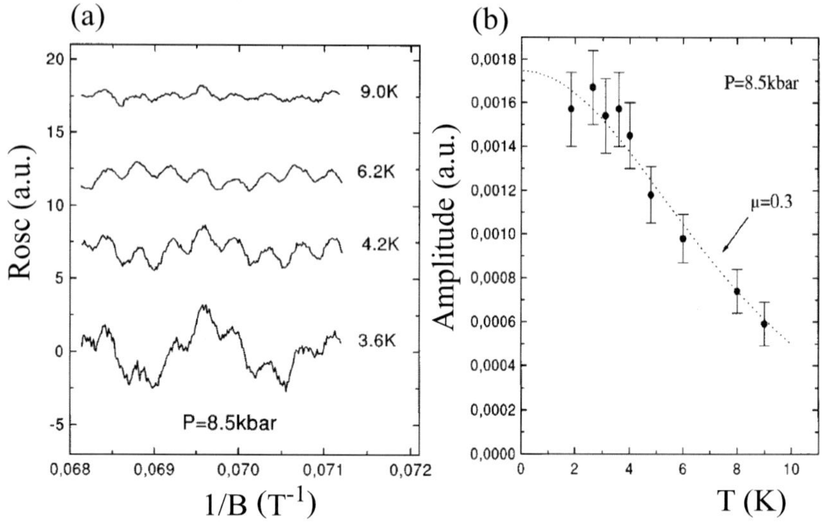

Fig. 5.9. (a) Oscillatory part in the interplane resistance in κ-(ET)$_2$Cu(NCS)$_2$ at temperatures up to 9 K, at pressure of 8.5 kbar. (b) The temperature dependence of the $\beta - 2\alpha$ oscillation amplitude. The dashed line represents the theoretical fit of the reduction factor (5.32) to the amplitude in (a). After [632].

than that of F_β, which is in strong contrast to the power spectrum as shown in Fig. 5.7; the power intensity of $F_{\beta-\alpha}$ is even larger than that of F_β. This has been the subject of controversy in a number of numerical calculations [633, 640, 641, 642, 643, 644, 645, 646, 647, 648, 649, 650, 651]. Particularly important among these is a possible chemical-potential oscillation, which is related to the 2D effect on the waveshape in the oscillatory magnetization described below.

5.2.3 Two-Dimensionality Effect

In an ideal 2D metal, all the electrons on the perfectly cylindrical FS can coherently contribute to quantum oscillations. D. Shoenberg derived a 2D version of the LK theory by assuming a field-independent constant chemical potential $\mu = E_F$ [652] for a grand canonical ensemble. The oscillatory magnetization $M_{\rm osc}$ is given by

$$M_{\rm osc} \propto \sum_{r=1}^{\infty} \frac{1}{r} R_T(r) R_D(r) R_S(r) \cdot \sin\left[2\pi r \left(\frac{F}{B} - \frac{1}{2}\right)\right] . \qquad (5.38)$$

The difference to the 3D-LK formula (5.31) is as follows. While the three reduction factors expressed by (5.32), (5.34), and (5.36) remain unchanged, the $B^{1/2}$-dependence in the prefactor and $\pi/4$ in the phase factor in (5.31) are omitted, and the harmonic content becomes modified by r^{-1} (instead of $r^{-3/2}$). When the reduction factors are close to unity (guaranteed by very high B and low T for small T_D), the waveform in 2D can resemble an *inverse saw-tooth* as a function of B due to an increased harmonic content. To note here, the condition with a fixed chemical potential can be realized, provided that there exists an electron reservoir as in a multiband system to allow a change in the number of carriers on the Landau levels near E_F.[29]

J. Wosnitza and his collaborators [254] measured dHvA oscillations in β"-$(ET)_2SF_5CH_2CF_2SO_3$. Figure 5.10 shows the data that can be completely reproduced by the 2D-LK formula (5.38) using $m_c = 1.9\,m_0$, $gm_c/m_0 = 3.9$ and $T_D = 0.4\,\mathrm{K}$. Since a less-quality sample with $T_D = 1.3\,\mathrm{K}$ and, hence, higher scattering rates did exhibit less pronounced signals, the authors ascribed the observed inverse saw-tooth waveforms as predicted by theory to the exceptionally high sample quality. This was discussed in terms of a fixed chemical potential or, equivalently, the existence of an electron reservoir. According to band-structure calculations [653], see Fig. 4.5(d), a pair of quasi-1D electron sheets exist in addition to the 2D hole pocket that gives rise to the dHvA signals. This quasi-1D band may act as the charge-carrier reservoir causing the chemical potential to be pinned. For a single-band 2D system without any reservoirs, on the other hand, a canonical ensemble where the number of charge carriers is fixed and hence an oscillation of the chemical potential should be allowed, may be reasonably adequate. For this case it

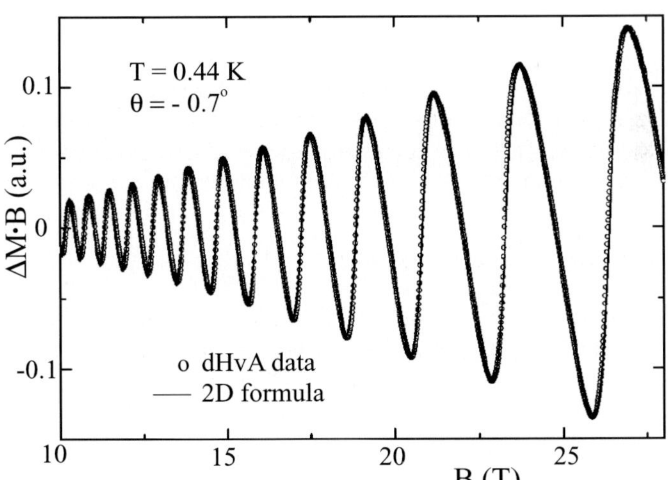

Fig. 5.10. Comparison of the measured dHvA waveforms (open circles) in a high-quality sample of β"-$(ET)_2SF_5CH_2CF_2$ with the 2D formula (5.38) (solid line) lying virtually on top of the experimental data. After [254].

[29] This is the reason why the LK formula holds quite well in a 3D system.

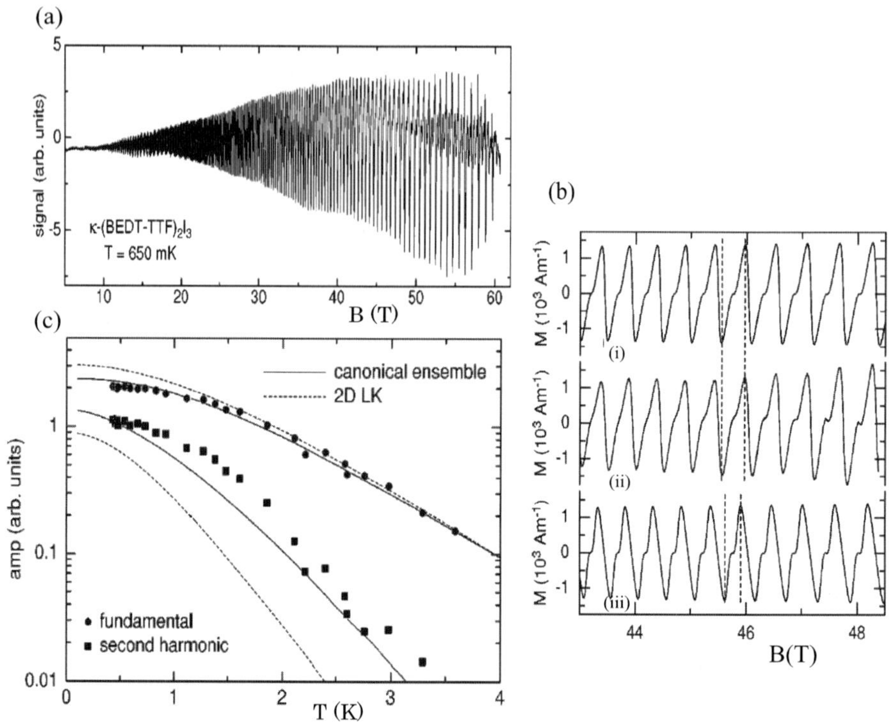

Fig. 5.11. (a) An example of a raw voltage signal proportional to the time-derivative of magnetization in κ-(ET)$_2$I$_3$. (b): (i) The calculated magnetization using the canonical ensemble with the parameters $m_c^\beta = 3.9\,m_o$, $g\mu_c = 8.63$ and $\tau^{-1} = 0.3 \times 10^{12}\,\mathrm{s}^{-1}$. The vertical dashed lines mark the interval between successive minima and maxima. (b): (ii) The magnetization obtained by integrating the induced voltage. (b): (iii) The magnetization calculated by means of the 2D LK theory using the parameters above. (c) A comparison of the temperature dependencies of the measured dHvA amplitudes of the fundamental frequency and of the second harmonic with results using the numerical model (solid lines) and the 2D LK theory (dashed lines). After [573].

was theoretically predicted that a *normal saw-tooth* waveform might appear associating sharp drops of both the magnetization and chemical potential with B [654, 655].

N. Harrison et al. [573] studied the dHvA oscillations in another highly-2D material, κ-(ET)$_2$I$_3$, in high fields up to 60 T. (For the SdH oscillations, refer to [656, 657].) As described in Sect. 5.1.3 above, the overall FS shape [43] is similar to that in the κ-(ET)$_2$Cu(NCS)$_2$ salt, Fig. 5.6(b), except the magnitude of the magnetic breakdown gap E_G [658]. In κ-(ET)$_2$I$_3$ with a centrosymmetric space group, E_G is much smaller and thereby the magnetic breakdown β-orbit with $m_c^\beta = (3.9 \pm 0.1)\,m_0$ [659, 446] is easily observed above 2.5 T being much smaller than 16 T in κ-$(ET)_2Cu(NCS)_2$ [618]. Figure 5.11(a) shows the dHvA oscillations in the time-derivative of the magnetization in pulse-field experiments. In addition to F_α and $F_{\beta-\alpha}$, oscillations with a quite small spectral power occur[30]; the spectrum has a clear series of F_β oscillations of 3875 T with higher harmonics up to $r = 9$, indicating the high sample quality with a small value of $T_D = (0.26 \pm 0.06)$ K. The temperature dependence of the amplitudes of both dHvA and SdH oscillations with F_β clearly deviates from the 3D LK formula (5.31), suggesting the importance of strong two-dimensionality at low temperatures and high fields. Taking into account the double peaks above 45 T in Fig. 5.11(a) that are induced by the Zeeman spin-splitting in high fields, the oscillatory waveform of the magnetization was successfully reproduced, as shown in Fig. 5.11(b), by the 2D model [641] within the framework of a canonical ensemble with a constant number of carriers. There, the waveform like *normal saw-tooth* is considered to be associated with the signature of the spin-splitting into Zeeman levels. These modifications of the waveform correspond to a change in the relative spectral weight of the fundamental and its harmonics. Figure 5.11(c) compares the fundamental and the second harmonic data with the numerical calculations. For the fundamental, the data below 1 K, which show a discernible deviation from the 2D LK theory within the grand canonical ensemble, agree quite well with calculations based on the canonical ensemble. Furthermore, the second harmonics, which, with the use of the 2D LK theory, gives a strikingly small value for $m_c^\beta = 2.4m_0$, is also well explained by the canonical ensemble. Thus both the waveform and the amplitude of β-oscillations are highly influenced by the two-dimensionality effect and the oscillation of the chemical potential.

5.2.4 Superconducting Vortex States

The dHvA effect in the superconducting vortex state has been a subject of interest since its discovery in $NbSe_2$ [660] and other type-II superconductors with high upper critical fields B_{c_2} such as A15 compounds [661, 662, 663], borocarbides [664], and some heavy fermion compounds [665]. See, e.g. [668] for a recent review. The phenomena, commonly associated with a sizable damping effect of the vortex state on the oscillatory amplitudes, have been also studied in κ-$(ET)_2Cu(NCS)_2$ [669, 670, 671, 667, 666] and β"-$(ET)_2SF_5CH_2CF_2SO_3$ [253].

Figure 5.12 demonstrates the SdH [666] and dHvA [667] oscillations of F_α in κ-$(ET)_2Cu(NCS)_2$, both of which are visible above 4 T at low temperatures. The field and temperature range was ascribed to quantum vortex liquid (QVL) states (Sect. 6.5) that were distinguished from thermal vortex liquid (TVL) states [672]. The quasiparticle scattering by the vortex lattice with

[30] In the SdH spectrum, the (forbidden) $F_{\beta-\alpha}$ power becomes much more enhanced [573], which, as described in the preceding section, is similar to the case in κ-$(ET)_2Cu(NCS)_2$.

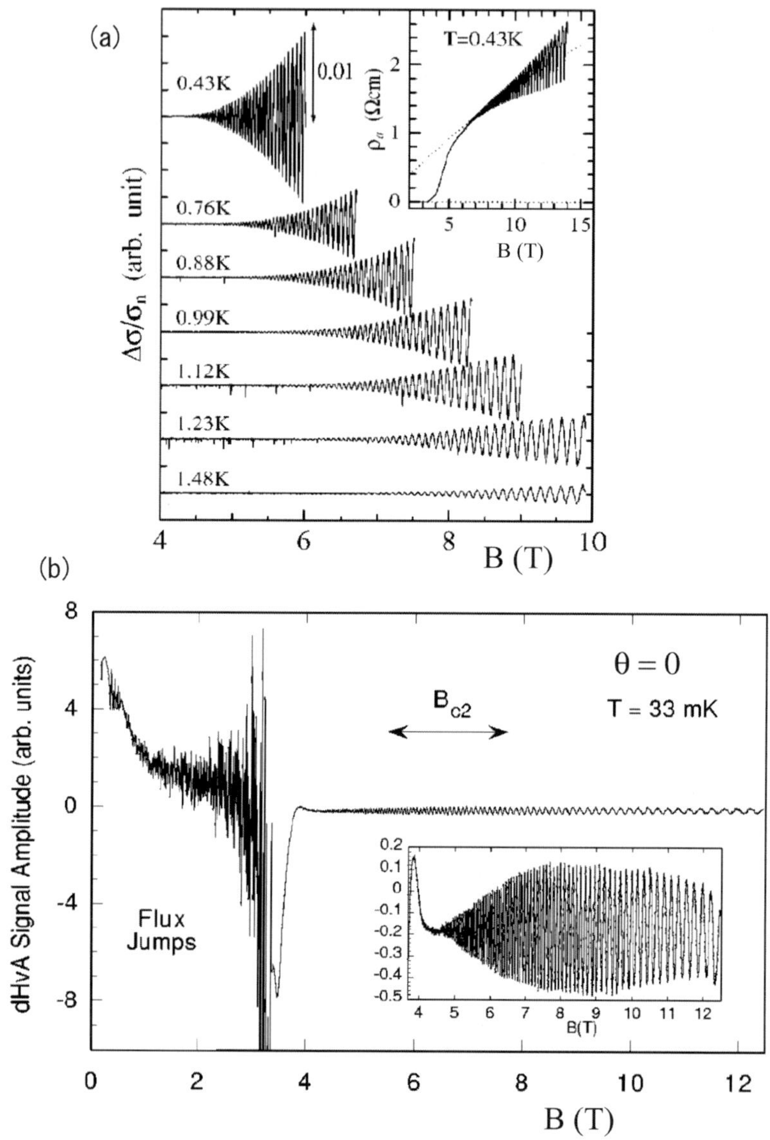

Fig. 5.12. (a) SdH oscillations in the interlayer resistivity ρ_a. Here $\Delta\sigma$ is the oscillatory part of the conductivity. The inset shows the overall magnetic-field dependence at $T = 0.43\,\mathrm{K}$. The dotted line indicates the normal-state resistivity. After [666]. (b) Oscillatory magnetization in κ-$(ET)_2Cu(NCS)_2$ as measured by the second harmonic pick-up, with the magnetic field applied perpendicular to the *bc*-plane at $T = 33\,\mathrm{mK}$. The dHvA oscillations with $F_\alpha = 601\,\mathrm{T}$ are observed above 4 T. The increasing noise below 3.5 T is due to flux jumps occurring within the sample. After [667].

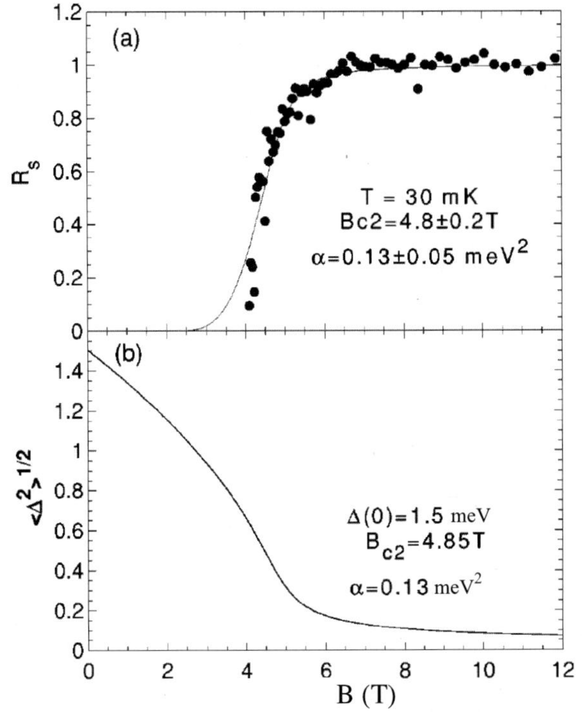

Fig. 5.13. (a) Damping R_{SC} of the dHvA signal at 30 mK in a magnetic field applied perpendicular to the *bc*-plane. The solid line is a fit to (5.39). (b) Root-mean-square order parameter as given by (5.40). After [667].

large superconducting fluctuations near the mean-field B_{c_2} must be taken into account. The reduction factor, which should be included in (5.31) as a fourth damping mechanism, is given [673] by

$$R_{SC} = \exp\left[-\pi^{3/2}\frac{<\Delta^2>}{(\hbar\omega_c)^2}\left(\frac{B}{F}\right)^{1/2}\right], \tag{5.39}$$

where $<\Delta^2>$ is the mean square of the fluctuating superconducting order parameter. The latter is approximated as

$$<\Delta^2>=\frac{\Delta(0)^2}{2}\left(1-\frac{B}{B_{c_2}}\right)+\sqrt{\left[\frac{\Delta(0)^2}{2}\left(1-\frac{B}{B_{c_2}}\right)\right]^2+\alpha(T)^2}, \tag{5.40}$$

where both $\Delta(0)$ and B_{c_2} are mean-field values, and $\alpha(T)$ is a temperature-dependent scaling parameter for the fluctuations [667]. At B_{c_2} where $<\Delta>=0$, it is a measure for the strength of the fluctuations. As shown in Fig. 5.13 where the magnetic field dependencies of R_{SC} and $<\Delta^2>^{1/2}$ calculated using the BCS gap of 1.5 meV ($2\Delta = 3.5k_BT_c$) are plotted, the above model can reproduce the dHvA data quite well using the parameters given. The damping observed in SdH oscillations (not shown here), however, is a little stronger than in dHvA oscillations. It is noted here that the current dependence due to the nonlinear transport is important [666] and further studies may shed light on the phase coherence of vortices in the quantum vortex state.

5.2.5 Magnetic Exchange-Interaction

Quantum oscillations have been extensively studied in magnetic metals, in which the conduction electrons interact with each other via an exchange coupling (itinerant magnetism), in systems with diluted magnetic impurities (Kondo systems), or in materials with periodically localized *d*- or *f*-spins. For these systems, magnetic quantum oscillations have proved to be a powerful tool for spin-resolved spectroscopy [497, 674, 675].

For conduction electrons exposed to an exchange field B_J from localized moments, the Zeeman levels are modified as illustrated in Fig. 5.5(c). The up (down)-spin level lowers (rises) by $\Delta E_J/2$ where $\Delta E_J = g\mu_B B_J$ and consequently the Zeeman splitting becomes $(1-B_J/B)g\mu_B B$ [674]. Therefore the spin-splitting phase ϕ (5.35) is modified as

$$\phi = 2\pi \frac{g\mu_B B(1 - B_J/B)}{\hbar\omega_c} = \pi g\mu_c \left(1 - \frac{B_J}{B}\right) . \tag{5.41}$$

Placing this equation into (5.36) and (5.31), the oscillatory magnetization M_{osc} with frequency F_0 is given by the sum of spin-dependent terms with different amplitudes ($A_\downarrow$ and $A_\uparrow$) and phases[31],

$$M_{\text{osc}} \propto A_\downarrow \cos\left[2\pi\left(\frac{F_0}{B} - \frac{1}{2}\right) + \frac{\phi}{2}\right] + A_\uparrow \cos\left[2\pi\left(\frac{F_0}{B} - \frac{1}{2}\right) - \frac{\phi}{2}\right] \tag{5.42}$$

$$= A_\downarrow \cos\left[2\pi\left(\frac{F_\downarrow}{B} - \frac{1}{2}\right) + \frac{\pi g\mu_c}{2}\right] + A_\uparrow \cos\left[2\pi\left(\frac{F_\uparrow}{B} - \frac{1}{2}\right) - \frac{\pi g\mu_c}{2}\right] , \tag{5.43}$$

where

$$F_\downarrow = F_0 - \frac{1}{4} g\mu_c B_J \tag{5.44}$$

$$F_\uparrow = F_0 + \frac{1}{4} g\mu_c B_J . \tag{5.45}$$

This well-established formula was applied [677] to the splitting of the dH frequency of the α-orbit in π-d interacting mixed salts of λ-(BETS)$_2$Fe$_x$Ga$_{1-x}$Cl$_4$ [125]. Figure 5.14 shows the antiferromagnetic insulator-to-paramagnetic metal transitions and SdH-oscillations observed in the interlayer resistance under

[31] For simplicity, only the fundamental ($r = 1$) is considered here.

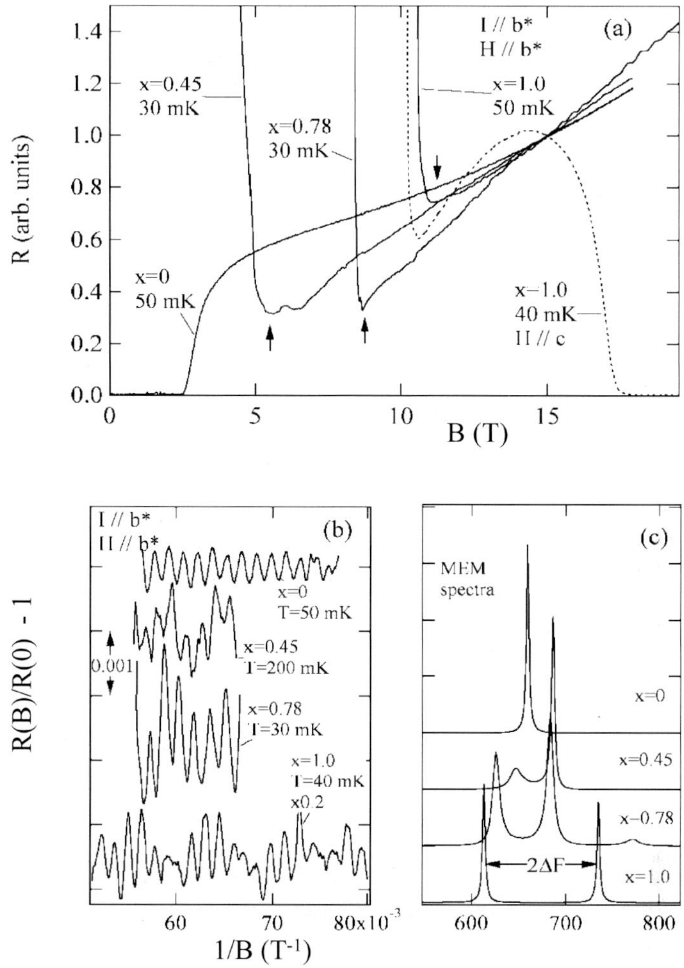

Fig. 5.14. (a) Interlayer resistance ($I, B \parallel b^*$) in a high magnetic field region for λ-(BETS)$_2$Fe$_x$Ga$_{1-x}$Cl$_4$. The antiferromagnetic insulator-to-paramagnetic metal transitions are indicated by arrows. (b) SdH oscillations as a function of inverse magnetic field. $R_0(H)$ is the non-oscillatory background. (c) Fast-Fourier-Transformation spectra calculated by a maximum entropy method. After [676].

magnetic fields perpendicular to the conducting plane. From the above equations using the observed split frequencies and the cyclotron mass in the SdH effects, the exchange field B_J was directly obtained: $B_J = 32$, 16 and 10 T, respectively, for $x = 1.0$, 0.78 and 0.45. As described in Sect. 7.2.6, these systems show field-induced superconductivity due to the Jaccarino-Peter compensation effect. The superconductivity is most enhanced for $B = B_J$ (see Fig. 7.18), which consistently shifts to lower external fields with decreasing Fe concentration.

5.3 High-Magnetic Field States

5.3.1 Field-Induced Cascade Transitions to SDWs and Anomalous High-Field States

As described in Sect. 4.4.1, the SDW insulator $(TMTSF)_2PF_6$ turns into a superconductor below 1 K upon applying pressure in excess of 6 kbar. In the metallic state at $B > B_{c_2}$, J.F. Kwak et al. [678] observed SdH-like oscillations in the resistance along the most conducting a-axis, followed by quite similar observations in the ClO_4 salt [679, 680]. Since these quasi-1D salts lack any closed FSs, the results had been quite puzzling before it was proved by NMR [681, 682], magnetization [683], and specific heat [684] measurements to be a cascade of thermodynamic phase transitions. In particular, the disappearance of the ^{77}Se-NMR signals strongly suggested that the transitions are due to SDW states.

Figure 5.15(a) shows the magnetization at 60 mK as a function of magnetic field applied nearly parallel to the least-conducting c-axis of relaxed $(TMTSF)_2ClO_4$ [683]. Successive jumps in the magnetization associated with hysteresis clearly indicate a cascade of first-order phase transitions. These anomalies are absent when $B \parallel b$. The B-T phase diagram is shown in Fig. 5.15(b). The lines correspond to the positions of maxima in the magnetization curves. The lower inset indicates that the peak index n — with $n = 1$ arbitrarily chosen for the magnetization jump around 8 T — is periodic in $1/B$, while the upper inset shows an enhancement of the amplitude of the $n = 1$ phase with decreasing temperature. This phase diagram is consistent with that obtained from magnetoresistance measurements [679, 680].

This so-called field-induced SDW (FISDW) phenomenon can be conceptually understood within a semiclassical picture (Sect. 5.1.3) in which a magnetic field enhances the one-dimensionality of the quasi-1D electron orbits as expressed in (5.15), thereby enhancing the DW instability simply by improving the nesting condition (Sect. 4.4.1). Extensive theoretical studies have been performed on FISDW [561, 685, 686, 687, 688]. (For details, refer to [86].) Following the model calculations by G. Montambaux et al. [685] who extended the idea of L.P. Gor'kov and A.G. Lebed [561], we outline the mechanism explaining how the metallic state becomes periodically unstable against forming FISDW condensates as a function of $1/B$.

Figure 5.16(a) illustrates the Zeeman splitting $g\mu_B B$ of the linearized 1D band expressed by the first term in (4.23). The SDW (CDW) correlations couple the bands with anti-parallel (parallel) spins and hence the nesting vector $Q_{\uparrow\downarrow} = Q_{\downarrow\uparrow}$ ($Q_{\uparrow\uparrow} < Q_{\downarrow\downarrow}$). Therefore a magnetic field, which does not change the $B = 0$ nesting condition for a SDW, can make the SDW state more stable via the field-enhanced 1D character of the electron orbits. (For CDW, the nesting condition in the ground state deteriorates continuously with magnetic fields, as described in Sect. 5.3.2.) The quasi-1D electron band (4.23) is extended as

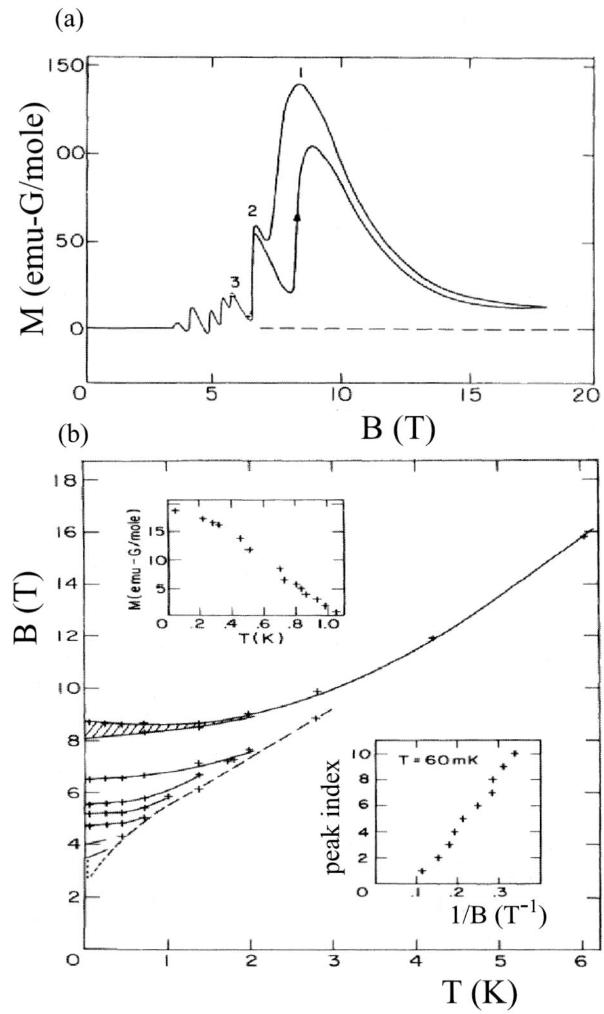

Fig. 5.15. (a) Magnetization vs magnetic field at 60 mK for $(TMTSF)_2ClO_4$. (b) Phase diagram showing peak-maxima field positions vs temperature. The shaded region shows hysteresis of the $n = 1$ phase (the upper trace obtained by increasing B). The dashed line represents the threshold fields. The upper inset shows M vs T for $n = 1$ and the lower inset shows the indexed peaks vs $1/B$. After [683].

$$E(k_x, k_y) = \hbar v_F(|k_x| - k_F) - 2t_b \cos(k_y b) - 2t_b' \cos(2k_y b). \tag{5.46}$$

The last term represents an additional corrugation along k_y. Figure 5.16(b) illustrates the FS of this quasi-1D band with the corrugation $4(t_b + t_b')/\hbar v_F$ along k_y, where the best nesting vector is given by $\boldsymbol{Q}$ connecting the inflection points on the two sheets. The nesting is perfect when $t_b' = 0$, as shown in Fig. 4.22. In the case of $t_b' \neq 0$, however, it becomes imperfect leaving a hatched area S_k between the solid line on the right-hand sheet and the dotted line obtained by the translation of the left-hand band. The size is characterized by $t_b'/\hbar v_{\mathrm{F}}$ along k_x.

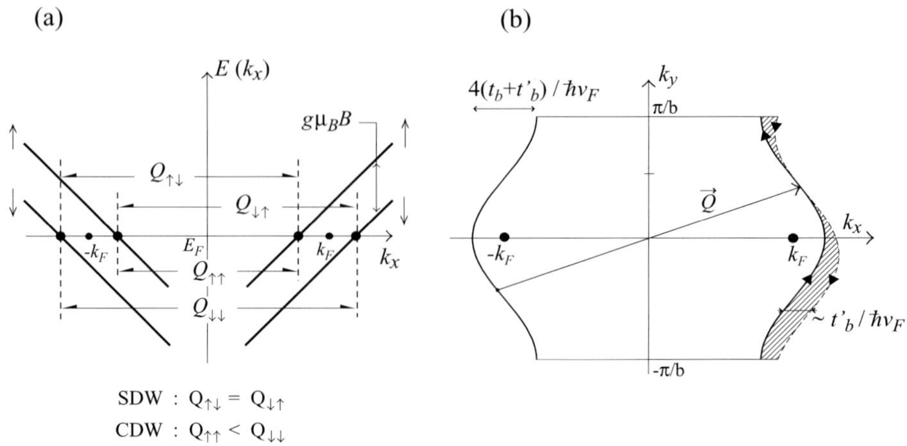

Fig. 5.16. (a) The Zeeman splitting $\pm(1/2)g\mu_B B$ of the linearized 1D band $E(k_x)$ (the first term in (5.46)). The $Q_{\uparrow\downarrow}$ and $Q_{\downarrow\uparrow}$ ($Q_{\uparrow\uparrow}$ and $Q_{\downarrow\downarrow}$) are nesting vectors for SDW (CDW). (b) The corrugated FS derived by (5.46). At $B = 0$, the best nesting vector $\boldsymbol{Q}$ connects the inflection points on the two sheets. After [685].

The electronic response function $\chi(\boldsymbol{Q}(Q_x, Q_y), B, T)$, depending on magnetic field and temperature, exhibits a series of maxima when $S_k = 2\pi q_{\parallel}/b$ is quantized; $S_k = nS_0$ where $S_0 = 2\pi eB/\hbar = 2\pi/\lambda_B^2$ [685]. This Landau quantization occurs necessarily at

$$q_{\parallel,n} = Q_x - 2k_F = n(b/\lambda_B^2). \tag{5.47}$$

This means that the metallic state in magnetic fields becomes unstable against the formation of a cascade of SDW states periodically in $1/B$.

A different theoretical approach was taken with an anisotropic 2D Hubbard model by K. Yamaji [686]. The total energy was found to be lowered when the above quantization occurs; the highest Landau level below E_F is fully occupied and the next level just above E_F is empty. Upon sweeping B, $q_{\parallel,n}$ adjusts itself in order to optimize the energy gain of each SDW subphase. This self adjustment occurs discontinuously and successively from n to $n-1$, and the last SDW state with $n = 0$ in this model, i.e. $\boldsymbol{Q}_o = (2k_F, \pi/b)$ is reached eventually. The top curve in Fig. 5.15(b) should be ascribed to the most stable SDW phase with $n = 0$. Thus, the cascade of first-order phase transitions of FISDW was understood at least qualitatively. With more realistic refined models [689, 690, 691, 692, 693, 694], comparisons with experiments were made in detail. (For reviews, refer to [86, 695] and references therein.) The experiments on the PF_6 salt are fairly well explained, while those on the ClO_4 salt are rather complicated due to the anion ordering responsible for the band folding [693, 694].

To note, a step-wise field dependence of the Hall voltage was observed in the ClO_4 [696, 697] and PF_6 salts [698, 699, 695], which was reasonably attributed to an integer quantum Hall effect, see, e.g. [503, 504, 505, 507]; $R_{xy,\nu=1} = h/2e^2 = 12.9\,\mathrm{k\Omega}$, where the factor 1/2 is due to the spin degeneracy. This is because the FISDW subphases contain completely filled Landau levels.

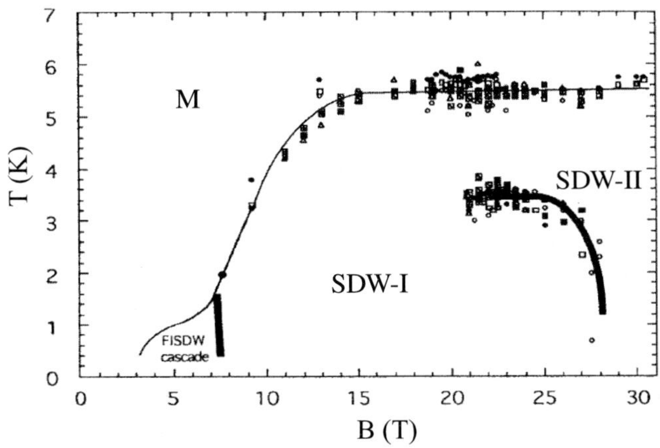

Fig. 5.17. Temperature-magnetic field phase diagram for $(TMTSF)_2ClO_4$. Thin and heavy lines indicate second- and first-order phase transitions, respectively. After [700].

For the high-field SDW phase assigned to $n = 0$ in the ClO_4 salt, McKernan et al. [700] obtained the phase diagram as shown in Fig. 5.17 from magnetization and resistance measurements at $T > 0.5\,\mathrm{K}$ and $B(\parallel c) < 30\,\mathrm{T}$. The thin line is the second-order phase boundary between the metal and FISDW phases. The thick line around 7 T indicates the first-order transition to the last SDW-I state with $n = 0$, where the above-mentioned quantized Hall resistance appears. In this SDW-I state, another first-order phase boundary exists, which seems to terminate at a critical point at 20 T and 3.5 K.

From measurements in higher fields up to 45 T, S. Uji et al. [701] observed in the SDW-II state that the Hall resistance oscillates changing its sign with increasing amplitude as a function of B, as shown in Fig. 5.18. This oscillation has the same frequency of 260 T that already was known as a rapid oscillation [405].[32] Both R_{xx} and R_{zz} sharply increase around 26 T accompanying the rapid oscillations, well consistent with the phase diagram (Fig. 5.17). The sign-reversal oscillations in R_{xy} in the SDW-II state, which was discussed in terms of intra-subband nesting vectors, may shed more light on the physical mechanism both for the rapid oscillations and the nature of the *almost insulating* state.

[32] The phase diagram and the rapid oscillation have been the subject of theoretical controversy [694, 702, 703, 704, 705, 706, 707].

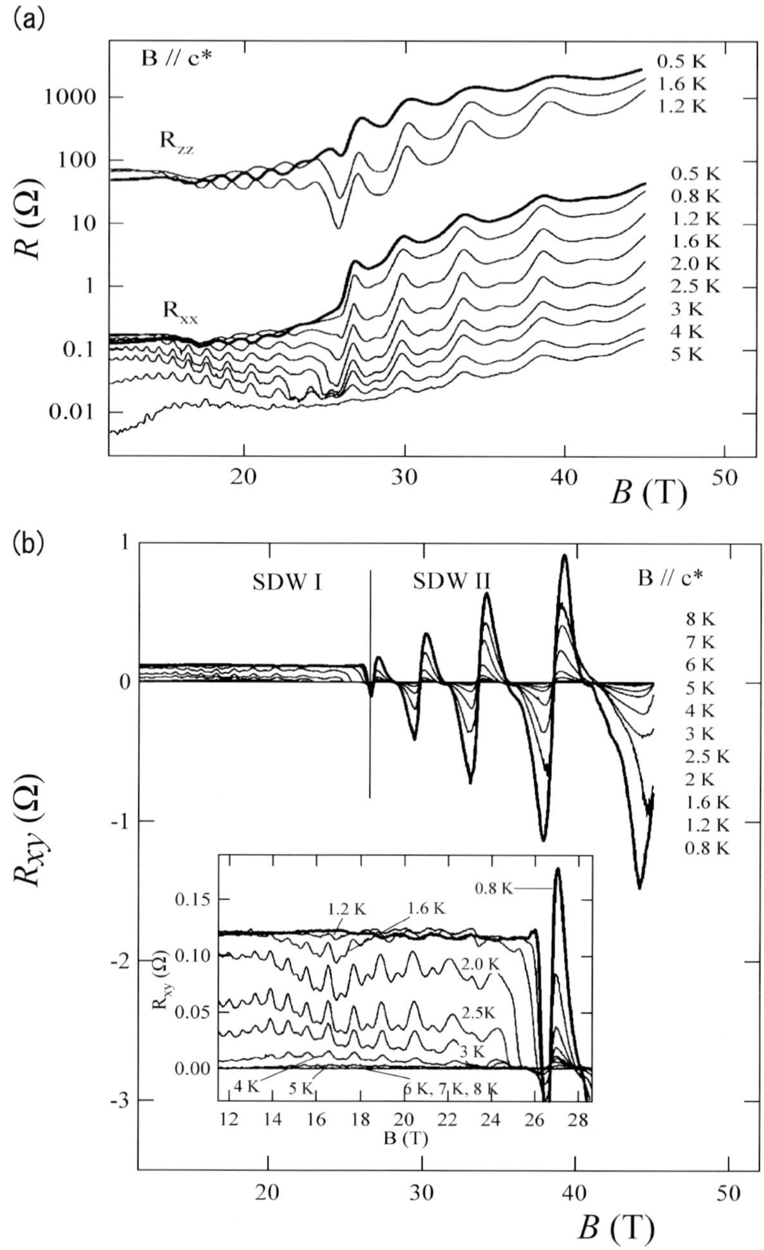

Fig. 5.18. (a) Magnetoresistance R_{xx} and R_{zz} along the a- and c-axis of $(\mathrm{TMTSF})_2\mathrm{ClO}_4$, respectively under magnetic fields along c^*. (b) Hall resistance R_{xy} at various temperatures. The inset shows the data below 28 T. After [701].

5.3.2 Field-Induced Destruction of CDW and Anomalous High-Field States

As well as for the FISDW states in the quasi-1D TMTSF salts, high magnetic field effects have been extensively studied for quasi-2D α-$(ET)_2MHg(SCN)_4$ (M = K, Tl, Rb). (For the crystal structure and electronic states, see Sects. 3.2.2 and 4.2.)

The debate regarding the ground state in the $KHg(SCN)_4$ salt is summarized as follows. A scenario of a SDW metal was first proposed by T. Sasaki et al. from observations of anomalies near $T_p = 8\,\mathrm{K}$, i.e. a hump in the temperature-dependent resistance [708] and an anisotropic behavior in the magnetic susceptibility suggesting an antiferromagnetic-like ordering [200]. The SDW scenario was supported by muon-spin-rotation experiments [201] which detected a magnetic anomaly below T_p associating a small magnetic moment of about $10^{-3}\,\mu_B$. However, ESR [709] and NMR [710] measurements did not detect any anomalies characteristic for a SDW transition, but indicated a reduction of the density-of-states at the Fermi level. On the other hand, T. Osada et al. [711] revealed a so-called kink anomaly around $B_k = 23\,\mathrm{T}$ in the magnetoresistance accompanying SdH oscillations for B perpendicular to the basal ac-plane, followed by similar observations for M = Tl (Rb) with $T_p = 9\,\mathrm{K}$ ($10\,\mathrm{K}$) and $B_k = 25\,\mathrm{T}$ ($32\,\mathrm{T}$) [712, 645].

On the other hand, from AMRO studies suggesting that a pair of open sheets tilts by about 20° from the k_c direction, M.V. Kartsovnik [712] proposed a FS reconstruction by the nesting vector $\boldsymbol{Q} = (1/8, 1/8, 1/6)$ due to a Peierls instability (CDW)-induced band folding. The CDW ground state, in contrast to the above SDW, has been considered to be highly plausible, since it fits better to the experimental fact because a CDW can be easily suppressed by a magnetic field as described in Sect. 5.3.1, Fig. 5.16. To note, slightly different tilt angles were pointed out from subsequent AMRO studies [713, 714]. The associated superlattice formation, which is a crucial test for this CDW scenario, indeed was observed by X-ray diffraction measurements [715].[33] However, the derived $\boldsymbol{Q} = (0.13(2), 0.1(1), 0.42(2))$ is inconsistent with the above nesting vector but highly consistent with $\boldsymbol{Q} = (1/5, \zeta, 2/5)$ (ζ is an unknown interplane component) as proposed by more detailed AMRO effects observed by N. Harrison et al. [645], see [209] and references cited therein. Recently K. Maki and his coworkers [716, 717, 718] have discussed the ground state and the colossal Nernst effect [719] from the theoretical viewpoint of an unconventional density-wave with nodal excitations, in analogy to unconventional superconductivity.

From magnetic torque measurements up to 22 T applied nearly parallel to the planes, P. Christ et al. [721] found that the magnetization simply

---

[33] The superlattice reflections with a factor-of-two lower intensity than that below T_p were observed even above T_p up to room temperature [715]. This is reminiscent of the strong fluctuations in the $2k_F$ and $4k_F$ Peierls instabilities in the metallic state of TTF-TCNQ (Sect. 4.4.1).

decreases with decreasing temperature below T_p even at 22 T. This is not what is expected for an antiferromagnetic (SDW) transition which is usually associated with a spin-flop, metamagnetic transition at some lower field. Moreover, T_p decreases linearly with B^2, which is also supported by specific heat measurements [722]. The field dependence is described as

$$\frac{T_p(B) - T_p(0)}{T_p(0)} = -A\left[\frac{\mu_B B}{k_B T_p(0)}\right]^2, \tag{5.48}$$

where A is a band parameter-dependent constant in the order of unity. To note, the Pauli (Zeeman) splitting-induced quadratic decrease of T_p was first predicted by W. Dietrich and P. Fulde [723] and then proved in TTF-TCNQ [724]. Thus, a consensus has been obtained that the metallic ground state is of CDW nature, not SDW. Hereafter we will focus on the issue of high-field effects on the metallic state in coexistence with CDW condensates, which have been extensively studied on the α-$(ET)_2KHg(SCN)_4$ salt.

Figure 5.19 shows the isothermal magnetoresistance for B almost perpendicular to the basal plane [720] showing a maximum around 10 T, followed by the so-called kink transition field $B_k = 23$ T, which reproduces earlier results [711, 200]. At $B > B_k$ and $T > 3$ K, the amplitude of the SdH oscillations grows rapidly, which can be understood as follows. At integral Landau-level filling, where the chemical potential is pinned in the gap between adjacent Landau levels, the oscillatory resistance maxima increase with decreasing temperature in an insulating-like manner. Conversely, at half-integral filling

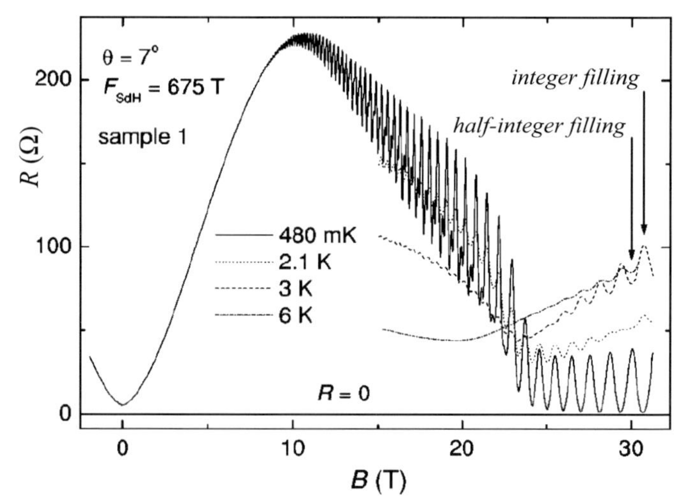

Fig. 5.19. Magnetoresistance of α-$(ET)_2KHg(SCN)_4$, with SdH oscillations of F_α = 675 T at $\theta = 7°$. The *phase-inversion* effect is pronounced at fields above the kink transition field (23 T), with the inverted SdH minima having resistivity values almost 4 times lower than the minimum resistivity at zero field. After [720].

factors, the resistivity minima behave in a metallic manner [725]. This is totally consistent with the theoretically predicted behavior of a quasi-2D metal in a magnetic field [641, 726]. The drastic change in the field dependence of the magnetoresistance at $B > B_k$ between above and below 3 K was ascribed to a new metallic phase [727, 728].

Magnetic-torque measurements have been performed using the high-field facilities at Grenoble [202] and Tallahassee [720]. The results are quite consistent with each other, leading to a very similar B-T phase diagram. The solid lines in Fig. 5.20(a) show the steady part of the isothermal torque, and the dotted lines the total one at 5 K exhibiting dHvA oscillations in a magnetic field slightly tilted from the perpendicular direction. At $T = 10\,\mathrm{K} > T_p$, an almost temperature-insensitive torque proportional to B^2 is seen. With the temperature decreasing below T_p, the quadratic term increases at low B, but above 4 T the B dependence becomes subquadratic. At higher fields, the curves bend to merge with the high-T quadratic dependence. The field at which the torque returns to its normal behavior coincides with B_k, as determined in other experiments [113, 729, 730, 727]. The curve at 3.2 K does not return back to the high-T part at B_k but stays below. At T below 3 K, however, the dHvA amplitude becomes so strong that the steady torque can no longer be reliably extracted. Figure 5.20(b) shows a window of the field sweep from 18 T to 28 T and back at 0.4 K indicating a clear transition from

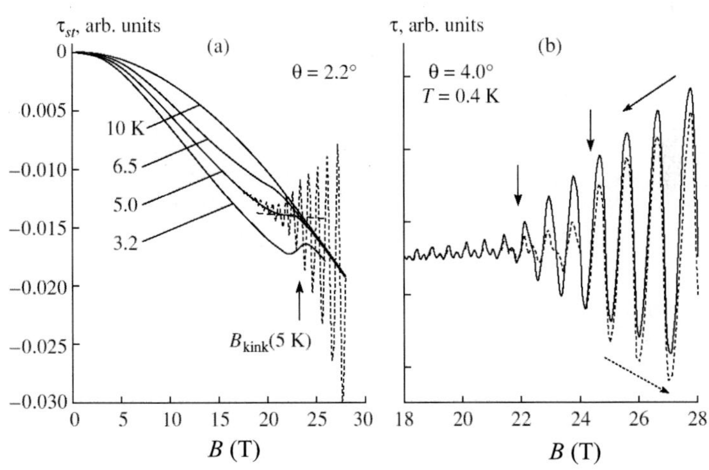

Fig. 5.20. Magnetic torque of α-$(ET)_2KHg(SCN)_4$ as a function of the magnetic field applied nearly perpendicular to the basal (ac) plane: (a) steady part of the torque at different temperatures; the dotted curve represents the total signal from the sample, with the dHvA oscillations at $T = 5.0\,\mathrm{K}$; (b) up (dotted line) and down (solid line) field sweeps of the torque at 0.4 K. After [202].

a low-B state with a double-peak structure[34] to a high-B state without any splitting but with a hysteresis in the field interval indicated by arrows.

Figure 5.21 shows the B-T phase diagram [732] with B, almost perpendicular to the basal plane from both groups [720, 202], which agrees with specific heat data [722] indicating the second-order phase boundary. With increasing B, T_p decreases continuously down to about 3 K, below which it becomes so insensitive to B that the critical magnetic field is almost divergently enhanced. Moreover, there is a first-order phase boundary below 4 K. The kink transition at B_k divides the low-temperature phase into low- and high-B states, which are named CDW_0 and CDW_x, respectively, after the theoretical prediction by D. Zanchi, A. Bjeliš and G. Montambaux [733]. According to the magnetization data, with fields up to 28 T applied almost parallel to the basal plane, the second-order phase boundary is globally in coincidence with the perpendicular case, as shown in the figure. This may indicate that the second-order critical field is rather isotropic, suggesting the importance of the Zeeman interaction being orientation-independent. In contrast, the B_k transition inside the low-T phase is quite anisotropic, suggesting the importance of orbital effects as well, which was inferred from simultaneous measurements of the interplane resistance and the magnetic torque by D. Andres et al. [734].

In the preceding section, we have seen that the orbital effect of a magnetic field is responsible for the cascade of FISDW. Similarly, it may induce some CDW subphases, where, in contrast to the orbital effect in FISDW, the Pauli effect may play an important role [733, 532]. The qualitative discussion [734]

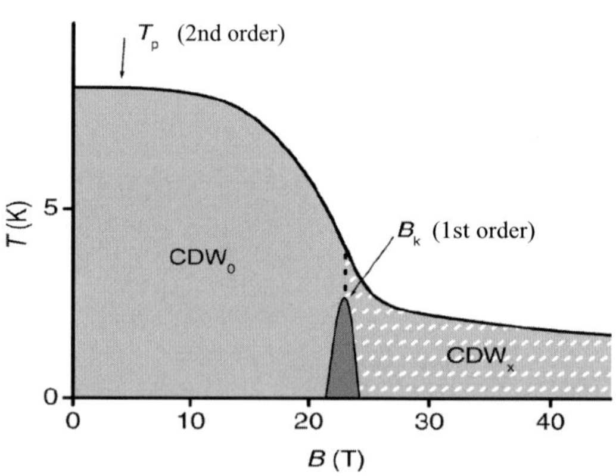

Fig. 5.21. Magnetic field vs temperature phase diagram of α-$(ET)_2KHg(SCN)_4$. The thick solid line represents the second-order transition at T_p into a CDW ground state depicted in light grey. The data from specific heat measurements up to 14 T coincide with the transition line. The vertical dotted line represents the first-order kink transition field B_k between the CDW_0 and CDW_x regimes, with the region of hysteresis depicted in dark grey. After [732].

---

[34] In the SDW scenario, this was erroneously interpreted as an exchange-induced modification of the spin-splitting factor [200, 731, 729]. An alternative idea was proposed that a pinned CDW (and also SDW) may induce a frequency doubling when the chemical potential oscillates [645, 633, 646].

based on the theory [733] including both CDW and SDW correlations in quasi-1D systems is summarized as follows.

We take a quasi-1D CDW-system with the corrugated tight-binding band expressed by (5.46), in which t'_b is assumed to be smaller than some critical value t'^*_b. From Fig. 5.16(a), we can express the optimal nesting conditions as

$$Q^{opt}_{\downarrow\downarrow(\uparrow\uparrow),x}(B) = Q_{0x} \pm \frac{2\mu_B B}{\hbar v_F}, \tag{5.49}$$

where the sign + (−) stands for the spins antiparallel (parallel) to B. However, the system tends to maintain the original nesting vector $\boldsymbol{Q}_0$ in order to keep the condensation energy $-(1/2)N(E_F)\Delta^2_{CDW}$ as large as possible and therefore both subbands may remain fully gapped up to the critical field $B_k = \Delta_{CDW}/2\mu_B$. Above B_k, however, $\boldsymbol{Q}_0$ is no longer a good nesting vector since both subbands can be expected to be ungapped. Then, at low enough temperatures, $T < T^*$, a CDW$_x$ state becomes stable with the field-dependent nesting vector,

$$Q_x(B) = Q_{0x} + q^{Pauli}_x(B), \tag{5.50}$$

where $q^{Pauli}_x(B)$ asymptotically approaches the value $2\mu_B B/\hbar v_F$ (5.49). It is noted that the nesting condition may be improved for one of the spin-down subbands at the cost of an additional unnesting of the other spin-up subband. This marginal state may be possibly stabilized just as in FISDW by the orbital quantization of the small pocket FS due to the unnested subband, see (5.47). The nesting vector is then expressed as

$$Q_{x,n} = Q_{0x} - \frac{2\mu_B B}{\hbar v_F} + \frac{nb}{\lambda_B^2}. \tag{5.51}$$

This predicts a new quantum phase combining both spin and orbital interactions, which is called a FICDW state. So far, no cascades of FICDW transitions have been observed for an almost perpendicular magnetic field except the kink transition asserted to the CDW$_0$-CDW$_x$ transition. D. Andres et al. discussed the rather complicated B vs θ phase diagram in terms of this model [734].

As mentioned above, the magnetic-field effect on CDW states is closely reminiscent of that of a type-II superconductor with the orbital effect-induced vortex state and the Pauli limiting field. In particular, a Fulde-Ferrell-Larkin-Ovchinnikov state [735, 736] is a theoretical analogue to CDW$_x$. The highly conducting state above B_k, as shown in Fig. 5.19, has been discussed in relation to superconductivity [720, 737, 738], the quantum Hall effect [725, 739, 740, 741, 742], and unconventional quantum liquid [743]. It is interesting to note that the CDW metal turns into a superconductor with $T_c = 0.1\,\mathrm{K}$ under pressures of about 2.5 kbar [744].

5.3.3 Approaching the Quantum Limit

With a further increase of the magnetic field B, the system can eventually enter the so-called quantum limit; then the Landau filling factor $\nu = F/B = 1$, corresponding to $n = 0$, where $B = (\hbar k_F^2/2e)$ if $S_{k,extr} = \pi k_F^2$. As is well known, this condition is easily satisfied at conventional fields in the order of several Tesla for a 2D electron gas in artificial semiconductor heterostructures like GaAs/GaAlAs where the low carrier density is responsible for an extremely small FS.[35] In contrast, the present LDMM with F in the order of 10^2–10^3 T or k_F in the order of $10^7\,\mathrm{cm}^{-1}$ as described in this chapter, cannot satisfy the condition, even with pulse magnets providing fields up to about 60 T [746].[36] However, some peculiar situations can place the system close to the quantum limit, as demonstrated, for example, in Fig. 5.18, for the FISDW caused by an imperfect nesting and its anomalous high-field states in TMTSF salts. So far, several experimental groups have tried to explore novel quantum-limit phenomena in quasi-2D salts, which also have rather small FSs.

At around $B = 20$ T and $T = 0.48$ K, T. Konoike et al. [747] found SdH oscillations of $F = 50$ T with quite large amplitudes due to the $n = 2$ Landau level in the interplane resistance of a quasi-2D metal τ-(EDO-S,S-DMEDT-TTF)$_2$(AuBr$_2$)$_{1+y}$($y \sim 0.75$) [748]. (Here EDO represents the ethylenedioxy group.) To note, this phenomenon was observed in the semiconducting-like state with a resistance upturn of a factor of ~ 2 below 50 K. In other salts [749, 750, 751] with the pyradio group instead of EDO, J. Brooks and his collaborators [752, 753] found field-induced metal-insulator transitions above 35 T and below 14 K from a combination of resistance, magnetic-torque and thermopower measurements. These features have been ascribed to a bulk thermodynamic phase transition of first-order with a slight field-orientation dependence. This novel phase transition was discussed in terms of a field-induced change in the electronic structure with a small band width of 10 meV or less that would open a gap at the Fermi level.

On the other hand, J. Wosnitza et al. [755] pointed out a field-induced-MI transition in the quasi-2D superconductor β''-(ET)$_2$SF$_5$CH$_2$CF$_2$SO$_3$. The SdH oscillations with $\nu \geq 7$ were superimposed on the background inter-plane resistance ρ_{zz} ($B \parallel z < 30$ T). In comparison with the dHvA oscillations in the magnetic torque, they extracted the background resistance not following Kohler's rule but some dynamical scaling, suggesting a MI transition.

---

35 For α-(ET)$_2$I$_3$ under pressure, N. Tajima [745] asserted, on the assumption of a two-band semimetal, that the temperature-dependent carrier density and mobility change from $10^{21}\,\mathrm{cm}^{-3}$ and $1\,\mathrm{cm}^2/\mathrm{Vs}$, respectively, at room temperature to $10^{15}\,\mathrm{cm}^{-3}$ and $10^5\,\mathrm{cm}^2/\mathrm{Vs}$ at Helium temperature. To note, however, no quantum phenomena such as SdH or quantum Hall effect have been observed.

36 *Destructive* magnets installed, for example, at the Institute of Solid State Physics (University of Tokyo) can produce fields far exceeding 100 T within the extremely short duration.

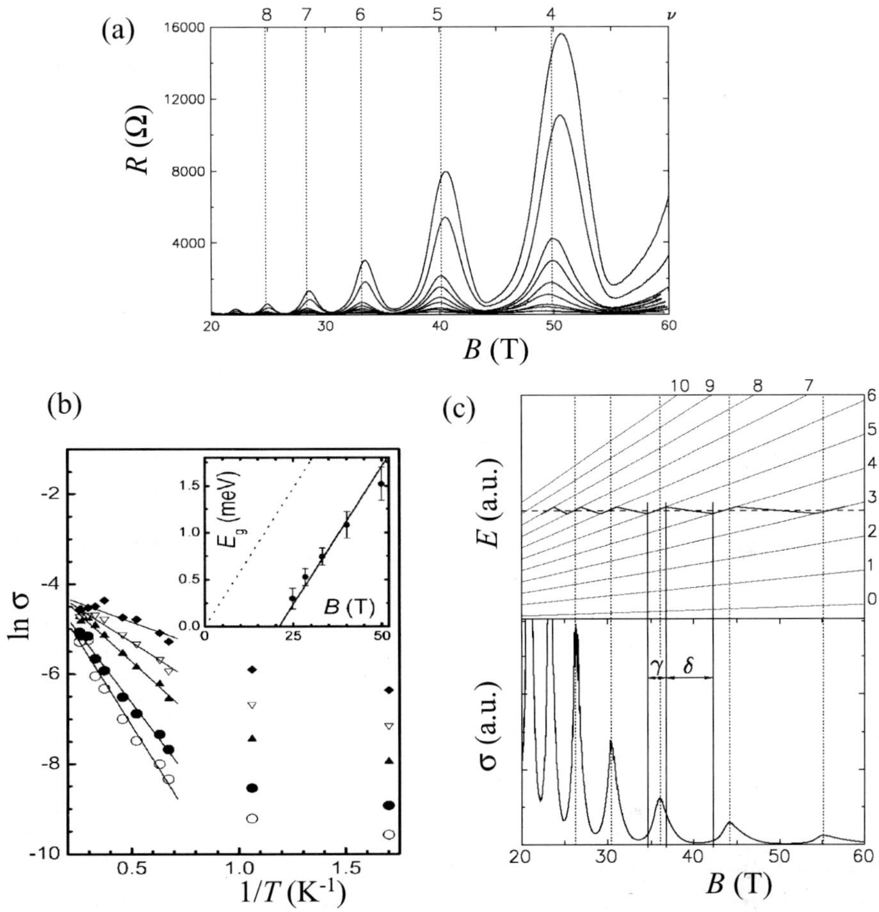

Fig. 5.22. (a) Temperature-dependent magnetoresistance in β''-$(ET)_2SF_5CH_2CF_2SO_3$ (from the top, 0.59, 0.94, 1.48, 1.58, 1.91, 2.18, 2.68, 3.03, 3.38, 3.80, and 4.00 K). The dotted lines and numbers indicate integer Landau-level filling factors. (b) σ_{zz} against $1/T$ at the magnetoresistance peaks close to integer filling factors; filled diamond $\nu = 8$, open triangle $\nu = 7$, filled triangle $\nu = 6$, filled circle $\nu = 5$, and open circle $\nu = 4$. The lines are fits used to extract Δ. The inset shows the magnetic-field dependence of $E_G = 2\Delta$. The dotted line shows $E = \hbar\omega_c = e\hbar B/m_c$ for comparison, and the solid line shows $E = \hbar\omega_c - E_0$, where E_0 is a constant offset energy (see text). (c) The upper figure shows the Landau fan diagram with a schematic view of the motion of the chemical potential μ where $T = T_D = 0$. The solid line indicates the motion of μ when it is alternately pinned in quasi-1D and quasi-2D states. The dashed line shows μ held constant by, e.g., an incommensurate density wave. The lower part shows the magnetoconductivity oscillation at 500 mK; the conductivity peaks whenever μ is at the center of a Landau level. After [754].

Alternatively, from the same experiments up to 60 T, M. Nam et al. [754] arrived at a different conclusion suggesting that the conductivity minima can be explained simply by an activation-type mechanism across the Landau gap. Figure 5.22(a) shows the magnetoresistance curves for temperatures down to 0.59 K, which associate the SdH oscillations with $F = 196\,\mathrm{T}$ for $\nu = 4 - 9$ (upper horizontal axis). The parameters, such as $m_c = 1.96\,m_0$ and $T_D = 0.82\,\mathrm{K}$, obtained from the low-field data below 6 T, result in the half-width Landau level broadening of 0.44 meV due to the finite lifetime and the Landau-level spacing $\hbar\omega_c = 2.94\,\mathrm{meV}$ at 49.85 T ($\nu = 4$). At these high fields, the Landau levels are sharp enough for the energy gap to be well defined. Figure 5.22(b) shows a logarithmic plot of the peak values in (a) as a function of $1/T$ which gives an activation-type dependence, $\sigma_{zz} \propto \exp(-E_G/k_\mathrm{B}T)$. As shown in the inset, E_G is proportional to B; $E_G = \hbar\omega_c - E_0$, where $E_0 = 1.23\,\mathrm{meV}$. The Landau-level broadening, caused by both the smearing effect and the tail of the Fermi-Dirac distribution function ($\sim 6k_\mathrm{B}T = 0.3\,\mathrm{meV}$), induces a reduction of the effective energy gap. This result strongly indicates that $E_G(B)$ is directly related to the gap between the Landau-level centers. This behavior is explained by the characteristic movement of the Landau levels with respect to the chemical potential μ as a function of magnetic field, as illustrated in Fig. 5.22(c). For simplicity, we assume $T = 0$ and that there is no band smearing ($\tau = \infty$). With increasing B, μ moves up in a peculiar Landau level until the degeneracy of the levels below has increased sufficiently for them to accommodate all the quasiparticles, just where μ drops discontinuously into the Landau level which lies below [497]. However, the presence of quasi-1D FS sheets, which give a continuous dispersion in a magnetic field, modifies this behavior [641, 209]. There, μ is alternately pinned, as B increases, to a Landau level or to the quasi-1D density-of-state. While the system acts as a metal if μ is pinned to a Landau level (the region of γ in the figure), it behaves as an insulator in the region δ when μ is located in between two adjacent Landau levels. The data in (b) suggest that quasi-1D FS sheets may not contribute to σ_{zz} and quasiparticles are expected. The reason was ascribed to a possible DW formation around 140 K where the resistance minimum appears. Quantum transport calculations by T. Champel and V.P. Mineev [756] reproduced the SdH oscillations in the quantum limit such that the pseudo-gap is centered on the integer ν and consequently a thermally activated behavior of the conductivity minima appears.

6 Superconductivity

6.1 Occurrence of Superconductivity in Molecular Metals

About 25 years after the discovery of superconductivity in the Bechgaard salt $(TMTSF)_2PF_6$ by D. Jérome and coworkers in 1980 [33], a large and varied family of organic superconductors has been developed due to the continued efforts from chemists and physicists working in this exciting field. Among the more than 100 charge-transfer superconductors known until now, the majority of systems consists of donor molecules which are derivatives of the prototype TMTSF. This includes the BEDT-TTF (ET) and their selenium and oxygen substituted variants BEDT-TSF (BETS) and BEDO-TTF, respectively, as well as various combinations of these molecules which form asymmetric hybrids such as DMET or MDT-TTF, see Fig. 3.1. Table 6.1 gives a selection of organic superconductors with their transition temperatures. For some materials the application of pressure is necessary to stabilize the superconducting ground state.[1]

As Table 6.1 indicates, superconductivity in these charge-transfer salts is still a low-temperature phenomenon with T_c values of several degrees Kelvin at maximum. Despite considerable efforts in the last decade the search for materials with higher transition temperatures has remained unsuccessful leaving the κ-$(ET)_2Cu[N(CN)_2]Br$ salt with a T_c value of about 11.5

[1] In the early stage of ET superconductivity the β-$(ET)_2I_3$ salt ($T_c = 1.4 - 1.5\,\mathrm{K}$) was studied extensively because of the occurrence of a pressure-induced "high-T_c" phase. Upon increasing hydrostatic pressure, which has been applied using the clamp-cell method, T_c increases discontinuously to $7 - 8\,\mathrm{K}$ near 1 kbar before it starts to decrease again at higher pressures [47, 757]. This observation is related to a structural phase transition associated with an incommensurate lattice modulation below 200 K which is due to the interaction of the ET molecules' ethylene endgroups and the triiodide anion. Upon applying pressure of about 1 kbar the incommensurate lattice modulation disappears and the high-T_c state emerges, which has been discussed in terms of a suppression of superconductivity by a random scattering potential induced by the lattice distortion due to the incommensurate structure. A detailed discussion is given in [86].

Table 6.1. Selection of organic superconductors and their superconducting transition temperatures T_c which may vary depending on the experimental technique employed. p_c denotes the critical pressure necessary to stabilize superconductivity. See also [758] for a recent compilation of organic superconductors so far discovered.

	T_c (K)	p_c (kbar)	Reference
$(TMTSF)_2PF_6$	0.9	12	[33]
$(TMTSF)_2AsF_6$	1.1	12	[759]
$(TMTSF)_2ClO_4$	1.4		[759]
$(ET)_2ReO_4$	2	4.5	[313]
β_L-$(ET)_2I_3$	1.4		[760]
β_H-$(ET)_2I_3$	8	0.5	[47, 757]
κ-$(ET)_2I_3$	3.6		[42]
β-$(ET)_2IBr_2$	2.8		[129]
κ-$(ET)_4Hg_{2.89}Cl_8$	1.8	12	[761]
α-$(ET)_2(NH_4)Hg(SCN)_4$	1.15		[111]
κ-$(ET)_2Cu(NCS)_2$	9.4		[45]
κ-$(ET)_2Cu[N(CN)_2]Br$	11.5		[46]
κ-$(ET)_2Cu[N(CN)_2]Cl$	12.8	0.3	[105]
κ-$(ET)_2CuCN[N(CN)_2]$	11.2		[762]
κ-$(ET)_2Cu_2(CN)_3$	2.8	1.5	[158]
β''-$(ET)_2SF_5CH_2CF_2SO_3$	5.3		[118]
β'-$(ET)_2ICl_2$	14.2	82	[763]
λ-$(BEDT\text{-}TSF)_2GaCl_4$	8		[764]
β_m-$(BEDO\text{-}TSF)_3Cu_2(NCS)_3$	1.1		[52]
$(BEDO\text{-}TSF)_2ReO_4 \cdot H_2O$	2.5		[53]
$(DMET)_2Au(CN)_2$	0.8	5	[49]
$(DMET)_2AuCl_2$	0.83		[765]
$TTF[Ni(dmit)_2]_2$	1.6	7	[54]
$(CH_3)N[Pd(dmit)_2]_2$	6.2	6.5	[766]

K as the (ambient-pressure) record holder since its discovery in 1990 [46].[2]

[2] The highest transition temperature of 14.2 K (onset) among the present organic superconductors has been oberved in β'-$(ET)_2ICl_2$ under a pressure (generated

This material, together with its isomorphic counterparts κ-(ET)$_2$X with X = Cu[N(CN)$_2$]Cl [105] and Cu(NCS)$_2$ [45], has since then attracted strong interest not only because of the high transition temperatures but also due to the quasi-2D character and unusual properties of the normal state above T_c, see Chapt. 4. Since these κ-(BEDT-TTF)$_2$X salts represent the best characterized and most intensively studied materials, we will discuss their superconducting properties in detail. Other materials of recent interest which will be discussed here are the archetype (TMTSF)$_2$X salts as well as the two-component π-d materials of the (BEDT-TSF)$_2$X family.

6.2 Ginzburg-Landau Theory

Before discussing the superconducting properties of the present molecular materials, we recall some basic results of the Ginzburg-Landau theory which examines the problem of superconductivity from the view point of the thermodynamics of phase transitions.

Based on L.D. Landau's theory of second-order phase transitions, see, e.g., [767], V.L. Ginzburg and L.D. Landau [768] introduced the idea of a superconducting order parameter $\Psi(\boldsymbol{r}) = |\Psi(\boldsymbol{r})|e^{i\varphi}$, which is a complex and position-dependent function characterized by a phase $\varphi(\boldsymbol{r})$ and a modulus $|\Psi(\boldsymbol{r})|$. In this formalism the free energy F of the system is expressed in powers of the order parameter which measures the extent of macroscopic phase coherence, or in other words, the difference in the symmetry between the ordered and non-ordered phases. This order parameter is assumed to gradually grow upon cooling below the superconducting transition and to vanish for temperatures above T_c, which implies that the order parameter is small near T_c. In their phenomenological description, the local density of superconducting carriers $n_s(\boldsymbol{r})$ is given by the square of the order parameter $|\Psi(\boldsymbol{r})|^2$ indicating that $\Psi(\boldsymbol{r})$ has the character of a wave function. In the absence of a magnetic field, one then finds for the free energy of the superconductor

$$F = F_n + \int \left(\frac{\hbar^2}{4m} |\nabla \Psi|^2 + a|\Psi|^2 + \frac{b}{2}|\Psi|^4 \right) \mathrm{d}V, \tag{6.1}$$

where F_n is the free energy of the normal state where $\Psi = 0$ and m is the electron mass. For a uniform superconductor (where the gradient term vanishes) of volume V it follows

$$F = F_n + aV|\Psi|^2 + \frac{bV}{2}|\Psi|^4. \tag{6.2}$$

Minimization of F yields the equilibrium value for the order parameter. Choosing Ψ to be a real function, the solutions of the equation $\partial F/\partial \Psi = 0$

in a diamond anvil cell) of 8.2 GPa [763]. The system is a Mott insulator under ambient conditions and becomes superconducting under extreme pressures above 7 GPa.

are $\Psi = 0$ and $|\Psi|^2 = -a/b$, corresponding to the normal and superconducting state, respectively. At a given temperature, the phase with minimal free energy is stabilized. For F to be a minimum, the second derivative $\partial^2 F/\partial\Psi^2 = 2aV$ for the normal state and $\partial^2 F/\partial\Psi^2 = -4aV$ for the superconducting state has to be positive. This implies that the coefficient a must be positive for $T > T_c$ and negative for $T < T_c$. According to the Landau theory of second-order phase transitions, the coefficient a can be expanded near T_c in integral powers of $(T - T_c)$. Considering only the first term in the Taylor expansion, one may write $a = \alpha(T - T_c)$, where α is a positive function and $|T - T_c| \ll T_c$. For $T < T_c$ one gets $|\Psi|^2 = \alpha(T_c - T)/b$. Substituting this into equation 6.2 yields

$$F_s = F_n - V\frac{a^2}{2b} = F_n - V\frac{\alpha^2}{2b}(T_c - T)^2. \tag{6.3}$$

The coefficients α and b can be linked to the thermodynamic critical field, $B_{c_{th}}$ by considering the Gibbs free energy G, which, for superconductors, is the most convenient thermodynamic potential to work with. From a thermodynamic integration of G along the field axis in the B-T plane one obtains for the condensation energy

$$G_s - G_n = -\frac{V}{2\mu_0}B_{c_{th}}^2, \tag{6.4}$$

with μ_0 the magnetic permeability of free space. By expanding the Gibbs free energy in the same way as done in (6.3), one finds a linear temperature dependence of $B_{c_{th}}$ close to T_c: $B_{c_{th}}(T) = B_{c_{th}}(0)\cdot(1 - T/T_c)$. Differentiating (6.4) twice with respect to temperature and using $B_{c_{th}}(T = T_c) = 0$ yields the jump in the specific heat at T_c of

$$(C_s - C_n)_{T=T_c} = \frac{VT_c}{\mu_0}\left(\frac{\partial B_{c_{th}}}{\partial T}\right)^2_{T=T_c}. \tag{6.5}$$

If we consider a non-uniform configuration with a positional-dependent wave function $\Psi(\boldsymbol{r})$, (6.1) indicates that a significant change in energy will result as soon as the gradient term $(\hbar^2/4m)|\nabla\Psi|^2$ is of the same order as $a|\Psi|^2$. This defines a natural length scale

$$\xi_{GL} = \sqrt{\frac{-\hbar^2}{2ma}}, \tag{6.6}$$

the Ginzburg-Landau coherence length which gives the typical distance over which the wave function may change. By using the temperature dependence of a one gets

$$\xi_{GL}(T) = \frac{\xi_{GL}(0)}{\sqrt{1 - T/T_c}}, \tag{6.7}$$

which diverges for $T \to T_c$.

In the presence of a magnetic field, the magnetic energy $\boldsymbol{B}^2/2\mu_0$ must be added to the integrand in (6.1). Moreover, since in a superconductor the particles are charged, the order parameter $\Psi(\boldsymbol{r})$ couples to the vector potential $\boldsymbol{A}(\boldsymbol{r})$ in a gauge-invariant way, where $\boldsymbol{A}(\boldsymbol{r})$ is related to the magnetic field $\boldsymbol{B}$ via $\boldsymbol{B} = \nabla \times \boldsymbol{A}$. To restore the invariance, the gradient term $|\nabla\Psi|^2$ must be replaced by $|[\nabla - i(2e/\hbar)\boldsymbol{A}]\Psi|^2$. Minimization of F with respect to variations in $\Psi(\boldsymbol{r})$ and $\boldsymbol{A}(\boldsymbol{r})$ then leads to the two well-known Ginzburg-Landau (GL) equations which can be solved assuming appropriate boundary conditions. For the simple case of a uniform superconducting carrier density, the second Ginzburg-Landau equation reduces to the London equation

$$\nabla \times \boldsymbol{j_s} = -\frac{e^2 n_s}{m}\boldsymbol{B}, \tag{6.8}$$

where $\boldsymbol{j_s}$ is the supercurrent density. Together with the Maxwell equations

$$\nabla \times \boldsymbol{B} = \mu_0 \boldsymbol{j_s}, \tag{6.9}$$

and

$$\nabla \boldsymbol{B} = 0, \tag{6.10}$$

this results in

$$\Delta \boldsymbol{B} = \frac{1}{\lambda_L^2}\boldsymbol{B}. \tag{6.11}$$

Equation 6.11 defines a second length scale, the London penetration depth λ_L, which shows the same temperature dependence as the Ginzburg-Landau coherence length (6.7):

$$\lambda_L(T) = \frac{\lambda_L(0)}{\sqrt{1 - T/T_c}}, \tag{6.12}$$

with its $T = 0$ limit

$$\lambda_L(0) = \sqrt{\frac{m^*}{\mu_0 n_s e^2}}. \tag{6.13}$$

Within the framework of the above GL equations, the so-called GL parameter defined by

$$\kappa = \frac{\lambda_L}{\xi_{GL}}, \tag{6.14}$$

is temperature independent.

J. Bardeen, L.N. Cooper and J.R. Schrieffer (BCS) developed a microscopic description in 1957 which is now considered the standard theory of superconductivity — the BCS theory [364]. It was soon recognized that this theory was able to explain a number of important experimental observations.

Among them are the isotope effect, i.e. the changes of the transition temperature with the mass M of the crystal lattice ions, and the existence of an energy gap 2Δ at the Fermi level. In a normal metal, the electronic states are filled up to the Fermi level, E_F, with a finite density of states $N(E_F)$. The situation is dramatically changed for a superconductor, where below T_c the electron density of states acquires a gap 2Δ at the Fermi energy.

Shortly after the publication of the BCS theory, L.P. Gor'kov [769] was able to demonstrate that near T_c, the phenomenological Ginzburg-Landau equations can be directly derived from the microscopic BCS theory. By this he was able to give a microsocopic explanation of the order parameter Ψ introduced by V.L. Ginzburg and L.D. Landau. He not only found that $\Psi(\mathbf{r})$ can be directly related with the wave function for the Cooper pairs, but that it is also proportional to the energy gap Δ. Both quantities thus measure the degree of long-range phase coherence in the superconducting state.

The BCS theory is based on the following three main insights:
(i) Under certain circumstances there can be an effective attractive interaction between electrons. Such a situation can arise due to the coupling between the electrons and the phonons of the underlying crystal lattice. (ii) In the presence of a fully occupied Fermi surface, a pair of electrons form a stable bound state irrespective of how weak the attractive force is. The stability of such a Cooper pair is maximum for the electrons having opposite values of their momenta and spins. (iii) A many-particle wave function can be constructed in which all electrons near the Fermi surface are paired up. In this situation an energy of 2Δ is required for breaking up a Cooper pair into two "free" electrons.

For the derivation of the BCS theory, which requires advanced techniques of many-body theory, we refer to, e.g. [770, 771, 365, 772, 773]. Here we will recall only a few central results which are useful for the general discussion and for analyzing experimental data.

The BCS theory is a mean-field theory and becomes nearly exact in the limit of weak coupling. This requires that the dimensionless coupling parameter λ, which is defined by

$$\lambda = |g_{eff}|^2 \cdot N(E_F), \tag{6.15}$$

is much less than 1. Here $-|g_{eff}|^2 = -V^*$ is the effective (attractive) interaction for electrons with energies in the range of $\pm\hbar\omega_D$ (ω_D the Debye frequency) of the Fermi surface, and $N(E_F)$ the density of states at the Fermi level. The central result of the BCS theory is the gap equation

$$1 = \lambda \int_0^{\hbar\omega_D} d\epsilon \frac{1}{\sqrt{\epsilon^2 + |\Delta|^2}} \tanh\left(\frac{\sqrt{\epsilon^2 + |\Delta|^2}}{2k_B T}\right), \tag{6.16}$$

which implicitly determines the gap $\Delta(T)$ at any temperature T. The temperature dependence of $\Delta(T)$ is shown in Fig. 6.1.

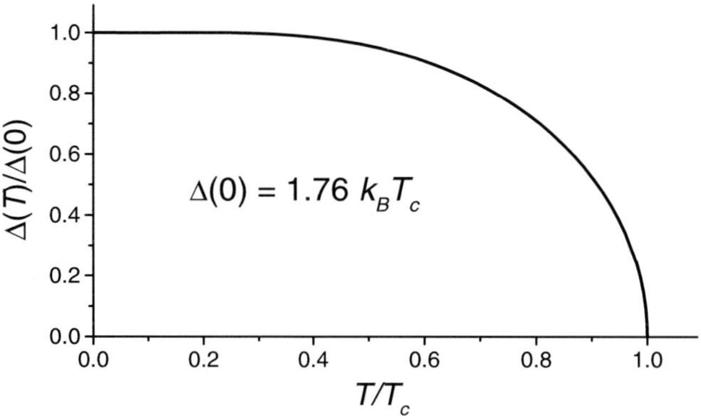

Fig. 6.1. Temperature dependence of the gap function in the BCS theory. After [774].

From (6.16) the transition temperature T_c can be obtained in the limit $\Delta \rightarrow 0$, resulting in the well-known BCS formula for T_c,

$$k_B T_c = 1.13\hbar\omega_D \cdot \exp\left(-\frac{1}{\lambda}\right). \tag{6.17}$$

Integration of (6.16) at $T = 0$ gives the famous BCS result

$$2\Delta(0) = 3.52 k_B T_c. \tag{6.18}$$

The BCS theory has been extended to allow for strong coupling, i.e. for λ ranging from about $0.2 - 0.5$. In this parameter range, the effects of the phonons on the electrons and also the effects of the electrons on the phonons have to be taken into account self-consistently. An appropriate theoretical framework has been developed by G.M. Eliashberg (see [771]), where the coupling to the phonons is characterized by a single function $\alpha^2(\omega)F(\omega)$, with $F(\omega)$ the phonon density of states, and $\alpha(\omega)$ the effective electron-phonon matrix element resulting in a mean electron-phonon-coupling constant

$$\lambda = 2\int_0^\infty \frac{\alpha^2(\omega)F(\omega)}{\omega} d\omega. \tag{6.19}$$

W.L. McMillan derived the following approximate expression for the critical temperature [775]:

$$k_B T_c = \frac{\hbar\omega_D}{1.45} \cdot \exp\left[\frac{1.04(1+\lambda)}{\lambda - \mu^*(1+0.62\lambda)}\right]. \tag{6.20}$$

The parameter μ^* is the Coulomb pseudopotential, which takes into account the screened Coulomb repulsion between the electrons.

6.3 The Superconducting Phase Transition

The superconductors discussed here are characterized by a highly anisotropic electronic structure, a low charge-carrier concentration and unusual lattice properties. The latter are a result of the molecular nature with rather rigid molecular units bound together by weak van der Waals bonds. The combination of these unique material parameters gives rise to a variety of unusual properties of the superconducting state such as pronounced thermal and quantum fluctuations of the superconducting order parameter, a variety of anomalies in the superconducting mixed state as well as an extraordinarily high sensitivity to external pressure.

6.3.1 Anisotropy and Superconducting Fluctuations

In standard three-dimensional superconductors, the transition into the superconducting state manifests itself in a sharp, step-like disappearance of the electrical resistivity with the 50 % point usually lying close to the transition temperature determined by thermodynamic measurements. This holds true also when a magnetic field B, lower than the upper critical field B_{c_2}, is applied to the superconductor which shifts the transition to lower temperatures while leaving its widths almost unaffected. The situation is significantly altered for strongly anisotropic materials, where pronounced effects of fluctuations of the superconducting order parameter may cause a substantial broadening of the transition, in particular when exposed to a magnetic field. The field-induced broadening is illustrated in Fig. 6.2, where the resistive transition on single crystalline κ-$(ET)_2Cu[N(CN)_2]Br$ is shown for various magnetic fields aligned parallel and perpendicular to the conducting planes [776].

Similar effects have been observed for the various members of the $(TMTSF)_2X$ and $(ET)_2X$ salts, see, e.g. [777, 35], and also for the high-T_c cuprates [778, 779]. Depending on the actual anisotropy of the material, the resistive transitions may broaden over a substantial fraction of T_c. In the strongly anisotropic cases, the comparison with the transition width observed in thermodynamic experiments, such as the specific heat, clearly shows that the $R = 0$ point in finite fields may lie deep within the superconducting mixed state and thus is no longer a good measure of the transition temperature, see also Sect. 6.5. In other words, there is a substantial temperature range below T_c where a dissipative mechanism is active although the material is already in the superconducting state. A less problematic determination of the transition temperature, even in magnetic fields, is provided by thermodynamic measurements such as magnetization, specific heat or thermal expansion.

To account for the effect of moderate anisotropies in the superconducting state properties, the phenomenological Ginzburg-Landau and London models have been extended by introducing an effective-mass tensor [780]. For the extreme case of quasi-2D superconductors characterized by a superconducting coherence length perpendicular to the planes, $\xi_\perp$, being much shorter than the spacing between the conducting layers, s, these models are no longer valid. Instead, the material has to be described by explicitly taking into account the layered structure. Such a model has been proposed by W.E. Lawrence and S. Doniach [781] who considered a set of superconducting layers separated by

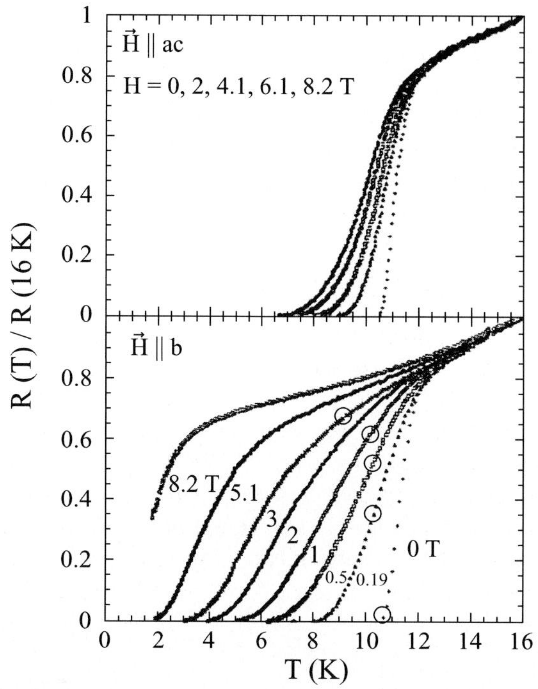

Fig. 6.2. Resistive transition of κ-$(ET)_2Cu[N(CN)_2]Br$ in various magnetic fields parallel (upper panel) and perpendicular (lower panel) to the conducting planes. The circles denote the nucleation temperatures as determined from magnetization measurements. After [776].

thin insulating sheets. According to this work, the three-dimensional phase coherence is maintained by Josephson currents running across the insulating layers. In fact, the presence of an intrinsic Josephson effect has been demonstrated for the present κ-$(ET)_2Cu(NCS)_2$ superconductor and for the most anisotropic high-T_c superconductors, see, e.g. [782, 783].

The above models for anisotropic superconductors can be used to determine the degree of anisotropy for a given material. For layered systems such as the present $(ET)_2X$ compounds, it is customary to use the effective-mass ratio

$$\Gamma = m^*_\perp / m^*_\parallel, \tag{6.21}$$

where $m^*_\perp$ and $m^*_\parallel$ denote the effective masses for the superconducting carriers moving perpendicular and parallel to the conducting planes, respectively. In the London and Ginzburg-Landau model, Γ is directly related to the anisotropies in the magnetic penetration depth λ and coherence length ξ by

$$\sqrt{\Gamma} \equiv \gamma = \frac{\lambda_\perp}{\lambda_\parallel} = \frac{\xi_\parallel}{\xi_\perp}.^3 \tag{6.22}$$

[3] γ is called anisotropy parameter of the system.

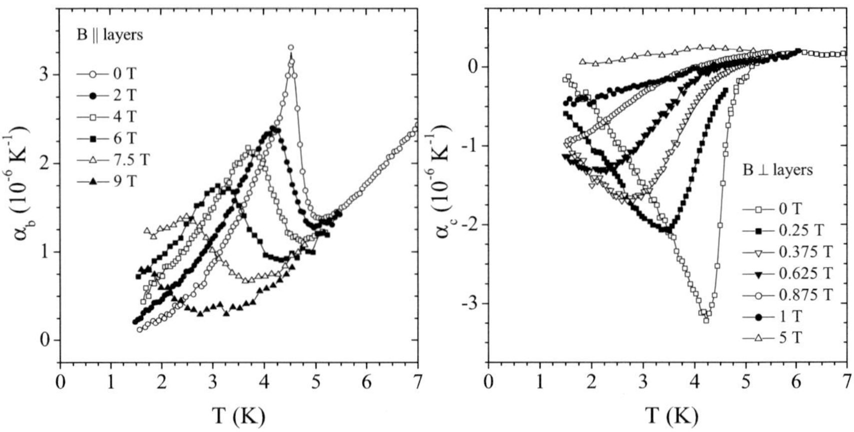

Fig. 6.3. Coefficient of thermal expansion of β''-$(ET)_2SF_5CH_2CF_2SO_3$ measured parallel (left panel) and perpendicular (right panel) to the conducting planes in varying fields applied along the measuring directions. After [348].

Figure 6.3 shows the results of the uniaxial thermal expansion coefficient at the superconducting transition in various magnetic fields for the quasi-2D salt β''-$(ET)_2SF_5CH_2CF_2SO_3$, which is one of the most anisotropic materials in this class of superconductors. While fields aligned parallel to the planes (left panel) result in an only moderate reduction of T_c accompanied by a slight broadening, a much stronger effect is observed for fields perpendicular to the planes (right panel). Here the phase-transition anomaly becomes substantially reduced and rounded already in a small field of only 0.25 T. This broadening can be attributed to a field-induced dimensional crossover: with increasing fields, the quasiparticles become progressively confined to their lower Landau levels, this process being accompanied by a reduction of the effective dimensionality from quasi-2D in small fields to quasi-0D at high fields [784, 785]. As a result, the order-parameter fluctuations grow with increasing fields [786] leading to the observed broadening of the transition.

A measure of the strength of thermal order-parameter fluctuations is provided by the Ginzburg number [787]

$$G_i = \frac{1}{2}\left[\frac{k_B T_c}{B_{c_{th}}^2(0)\xi_{\|}^2\xi_{\perp}}\right]^2 , \tag{6.23}$$

where $B_{c_{th}}(0)$ is the thermodynamic critical field. G_i measures the ratio of thermal energy to the condensation energy per coherence volume. For classical 3D superconductors, the Ginzburg number amounts to about $G_i \sim 10^{-8} - 10^{-11}$ and therefore, deviations from a mean-field behavior are unimportant. In contrast, for the present $(ET)_2X$ compounds and some

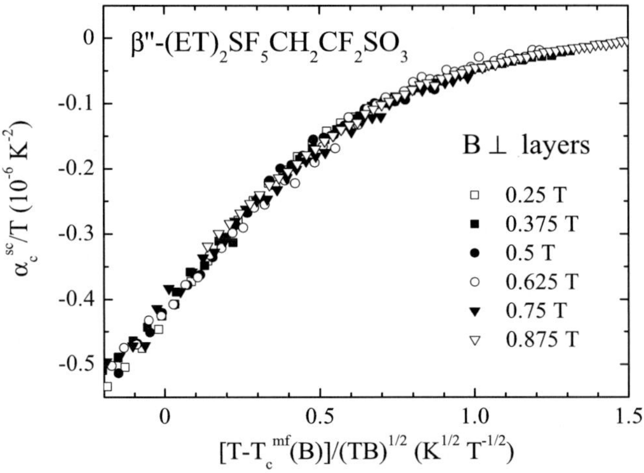

Fig. 6.4. Scaling behavior of the superconducting contribution to the linear coefficient of thermal expansion of β''-$(ET)_2SF_5CH_2CF_2SO_3$ [348] in magnetic fields applied perpendicular to the planes. The raw data are shown in Figure 6.3.

generic high-T_c cuprates such as $YBa_2Cu_3O_{7-\delta}$ (YBCO) or the Bi- and Tl-based materials (e.g., $Bi_2Sr_2CaCu_2O_{8+\delta}$, BSCCO), one finds several orders of magnitude larger numbers $G_i \sim 10^{-2} - 10^{-3}$ [788, 789, 787, 790, 791, 779]. The main reason for this is the short coherence lengths $\xi \propto \hbar v_F/(k_B T_c)$ for these superconductors as a consequence of the relatively high transition temperatures and the low charge-carrier concentration. The latter results in a low Fermi velocity v_F.

The effect of thermal fluctuations on transport and thermodynamic properties has been studied by several authors [792, 793], see also [794] for a recent review. Assuming the lowest-Landau-level approximation and taking into account only non-interacting Gaussian fluctuations, S. Ullah and A.T. Dorsey obtained an expression for a scaling function of various thermodynamic quantities such as the magnetization M or the specific heat C

$$\Xi_i = F_i \left(A \frac{T - T_c^{\mathrm{mf}}(B)}{(TB)^n} \right), \tag{6.24}$$

with $\Xi_i = M/(TB)^n$ or C/T [788]. F_i is an unknown scaling function, A a temperature- and field-independent coefficient characterizing the transition width and $n = 2/3$ for anisotropic 3D materials and $n = 1/2$ for a 2D system. Thus from a scaling analysis, both the actual dimensionality as well as the mean-field-transition temperature $T_c^{\mathrm{mf}}(B)$ can be determined.

Figure 6.4 shows the thermal expansion data of Fig. 6.3 for fields perpendicular to the planes in the scaling form α_c^{sc}/T vs $(T - T_c^{\mathrm{mf}}(B))/(TB)^{1/2}$, where α_c^{sc} denotes the superconducting contribution to the coefficient of thermal expansion.[4] As Fig. 6.4 demonstrates, the various field curves $\alpha_c^{sc}(T, B)$ follow the 2D scaling over a rather wide temperature and field range. A similar scaling analysis of high-field magnetization [789] and electrical conductivity [795] data on the κ-(ET)$_2$Cu(NCS)$_2$ revealed that this compound is at the threshold from being a strongly anisotropic 3D to a 2D superconductor. On the other hand, a distinct 2D behavior has been claimed from a scaling analysis of low-field magnetization data by H. Ito et al. [777]. For κ-(ET)$_2$Cu[N(CN)$_2$]Br, the magnetization curves have been found to scale according to the two-dimensional fluctuation theory in a high magnetic field [796]. The above scaling analysis can be used to determine the mean-field phase transition temperatures $T_c^{\rm mf}(B)$ and by that, the upper critical fields. From the initial slopes parallel, $B_{c2}^{\parallel'}$, and perpendicular, $B_{c2}^{\perp'}$, to the conducting planes the anisotropy parameter γ can be determined by the following relations [797, 365]:

$$B_{c2}^{\perp'} = \left|\frac{\mathrm{d}B_{c2}^{\perp}}{\mathrm{d}T}\right| = \frac{\phi_0}{2\pi\xi_{\parallel}^2 T_c} \quad \text{and} \quad \frac{B_{c2}^{\perp'}}{B_{c2}^{\parallel'}} = \frac{\xi_{\perp}}{\xi_{\parallel}} = \frac{1}{\gamma}, \tag{6.25}$$

where ϕ_0 denotes the flux quantum.[5] The analysis of the data of Fig. 6.3 for β''-(ET)$_2$SF$_5$CH$_2$CF$_2$SO$_3$ reveals $\xi_{\parallel} = (144 \pm 9)$ Å, $\xi_{\perp} = (7.9 \pm 1.5)$ Å and $\gamma \approx 18$ [348] highlighting the quasi-2D character of the superconducting state in this material.

For κ-(ET)$_2$Cu(NCS)$_2$values of $\gamma \sim 10 - 50$ have been extracted from the initial slopes of the upper critical fields in resistivity, magnetization, and specific heat measurements, see, e.g., [799, 217, 800, 801, 802, 789].

Here we note that the anisotropy parameter γ of the present quasi-2D compounds differs significantly depending on the experimental method employed to its determination. The values, extracted from measurements of the upper critical fields, are usually lower than those from magnetic torque measurements which can be considered as a more sensitive probe to estimate the real anisotropy parameter. For κ-(ET)$_2$Cu(NCS)$_2$ for example, in torque measurements, γ values of about 200 have been reported [803, 804] which place this material in the same class of quasi-2D superconductors as the

[4] Since the volume coefficient of thermal expansion, $\beta(T)$, is related to the specific heat $C_V(T)$ via the Grüneisen relation $\beta(T) = \Gamma_G \cdot \frac{\kappa_T}{V_{\rm mol}} \cdot C_V(T)$, where κ_T denotes the isothermal compressibility, $V_{\rm mol}$ the molar volume and Γ_G a field- and temperature-independent Grüneisen parameter, the scaling form (6.24) holds also for α/T.

[5] Alternatively, the value of the upper critical field at zero temperature can be used to estimate the coherence lengths via $B_{c2}^{\perp}(0) = \phi_0/(2\pi\xi_{\parallel}^2(0))$. Using these relations — both are based on a mean-field treatment — can lead to different coherence lengths. A discussion of the limitations of deriving the coherence lengths based on conventional mean-field theory is given in [798].

most anisotropic high-T_c cuprates with $\gamma \sim 140 - 420$ for $Bi_2Sr_2CaCu_2O_{8+x}$ [805, 806].

6.3.2 Pressure Dependence of T_c

The application of external pressure to a superconductor enables the study of the volume dependence of the pairing interaction.[6] At the same time, the observed pressure shifts of T_c may serve as a guideline for the search for new materials with potentially higher transition temperatures. Indeed, the discovery of $YBa_2Cu_3O_7$ [808], containing the small Y^{3+}-ions compared to La^{3+}, was initially stimulated by the large enhancement of T_c found for the first La-based high-T_c superconductor under hydrostatic pressure [809]. For the majority of classical superconductors, external pressure causes a reduction of T_c indicating a pressure-induced weakening of their pairing interaction, the electron-phonon coupling constant $\lambda_{el-ph} \propto \Theta_D^{-2}$, where Θ_D is the Debye temperature [775]. This has been understood as a pressure-induced stiffening of the crystal lattice and a corresponding enhancement of Θ_D.

For the molecular superconductors, such as the $(TMTSF)_2X$ and $(ET)_2X$ families, one generally finds also a reduction of T_c with pressure but with an extraordinarily large pressure coefficient. In Fig. 6.5 the pressure dependence of T_c is shown for a selection of β- and κ-type $(ET)_2X$ salts under

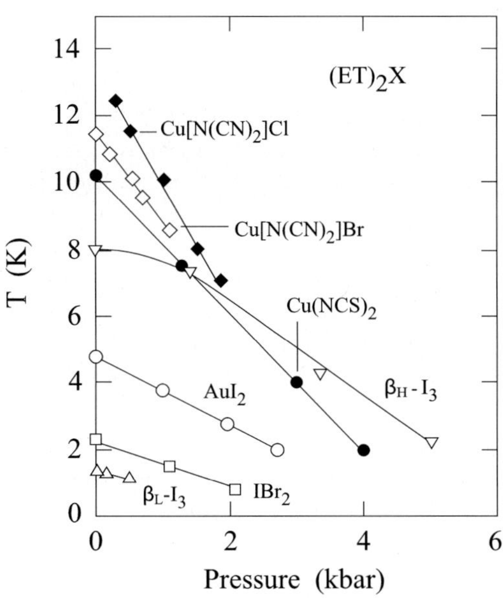

Fig. 6.5. Hydrostatic pressure dependence of T_c for various β- and κ-type $(ET)_2X$ superconductors. After [86].

[6] See [807] for a recent review of high-pressure studies on organic conductors and superconductors.

hydrostatic-pressure conditions.[7] The initial slope of the pressure dependence of T_c, $(\partial T_c/\partial p)_{p\to 0}$, determined from resistivity measurements ranges from -0.25 K/kbar for α-$(ET)_2NH_4Hg(SCN)_4$ ($T_c = 1$ K) [113] to -3.2 K/kbar for κ-$(ET)_2Cu[N(CN)_2]Cl$ ($T_c = 12.8$ K at 0.3 kbar) [219]. For $(TMTSF)_2PF_6$, one finds $-(0.08 \pm 0.01)$ K/kbar ($T_c = 1.1$ K at 6.5 kbar) [811]. At first glance a strong pressure dependence of T_c appears not surprising in view of the weak van der Waals bonds between the organic molecules, giving rise to a highly compressible crystal lattice. In fact, the value of the isothermal compressibility $\kappa_T = -\partial \ln V/\partial p$ for κ-$(ET)_2Cu(NCS)_2$, for example of $\kappa_T = (122\,\text{kbar})^{-1}$ [172, 150], exceeds the values found for ordinary metals by about a factor of five. To account for this "lattice effect" one should, therefore, consider the physically more meaningful volume dependence of T_c:

$$\frac{\partial \ln T_c}{\partial \ln V} = \frac{V}{T_c} \cdot \frac{\partial T_c}{\partial V} = -\frac{1}{\kappa_T \cdot T_c} \cdot \frac{\partial T_c}{\partial p}. \tag{6.26}$$

Using the above isothermal compressibility, one finds $\partial \ln T_c/\partial \ln V \approx 40$ for κ-$(ET)_2Cu(NCS)_2$[348] which exceeds the values found for a rather soft ordinary metallic superconductor such as Pb with $\partial \ln T_c/\partial \ln V = 2.4$ [812] by more than one order of magnitude.

For strongly anisotropic superconductors, like the $(TMTSF)_2X$ or $(ET)_2X$ salts, uniaxial-pressure studies provide supplementary information which may help to identify the relevant microscopic couplings responsible for superconductivity in these systems. Different techniques have been employed to determine the uniaxial-pressure coefficients of T_c including measurements under uniaxial strain or stress [8] [814, 114, 815, 816, 817] or by using a thermodynamic analysis of ambient-pressure thermal expansion and specific heat data [818, 819, 348, 155]. The latter approach is based on the Ehrenfest relation (4.22) which connects the pressure coefficients of T_c for uniaxial pressure along the i-axis (in the limit of vanishing pressure) to the phase-transition anomalies at T_c in the coefficient of thermal expansion, $\Delta\alpha_i$, and specific heat, ΔC via

$$\left(\frac{\partial T_c}{\partial p_i}\right)_{p_i \to 0} = V_{\text{mol}} \cdot T_c \cdot \frac{\Delta\alpha_i}{\Delta C}, \tag{6.27}$$

with V_{mol} being the molar volume. Figure 6.6 shows results of the linear thermal expansion coefficients along the three principal axes of β''-$(ET)_2SF_5CH_2CF_2SO_3$, κ-$(ET)_2Cu(NCS)_2$, and κ-$(ET)_2Cu[N(CN)_2]Br$.

[7] In [810] it has been demonstrated for the κ-$(ET)_2Cu(NCS)_2$ compound that cooling in a frozen pressure medium may cause deviations from hydrostatic conditions due to the strongly anisotropic contraction of the sample.

[8] In [813] the uniaxial strain method, where a crystal is compressed along any direction with no deformation in the transverse directions, and its application to many kinds of molecular conductors is reviewed. Besides results on κ-$(ET)_2Cu(NCS)_2$ and κ-$(ET)_2Cu[N(CN)_2]Br$, e.g., the band structure control of α-$(ET)_2MHg(SCN)_4$ (M = K, NH_4) by uniaxial strain to induce or enhance superconducting states is discussed.

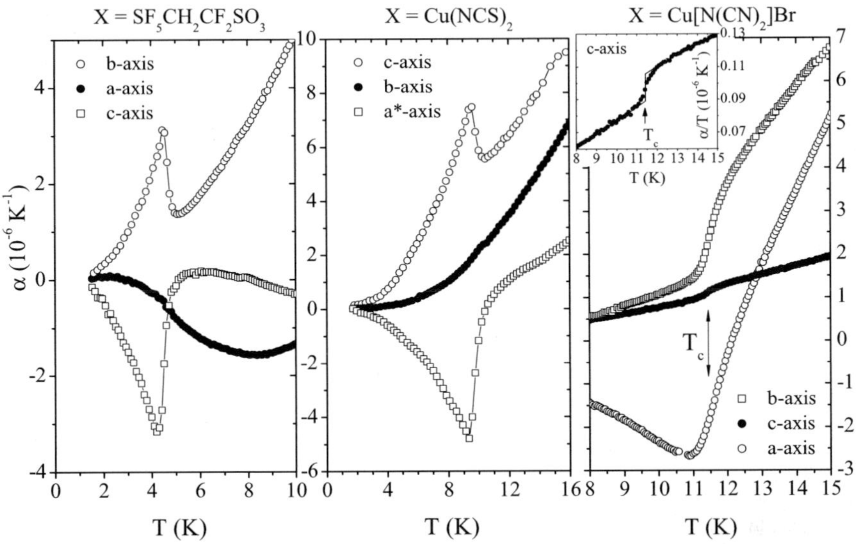

Fig. 6.6. Phase transition anomalies at the superconducting transition in the uniaxial coefficients of thermal expansion, α_i, for β''-$(ET)_2SF_5CH_2CF_2SO_3$ (left), κ-$(D_8$-$ET)_2Cu(NCS)_2$ (middle) and κ-$(ET)_2Cu[N(CN)_2]Br$ (right), after [348, 155, 115]. Open squares indicate α data perpendicular to the planes; open and closed circles correspond to the in-plane expansion coefficients.

These data disclose strong anisotropies in the uniaxial expansion coefficients for all three superconductors with striking similarities in the α_i's for the β''-$(ET)_2SF_5CH_2CF_2SO_3$ and κ-$(ET)_2Cu(NCS)_2$ salts. Table 6.2 collects the uniaxial-pressure coefficients derived from these data using equation (6.27) [348, 155] and literature results for the specific heat discontinuity, i.e. [459] for the β''-system and [820, 801, 458] for the κ-phase compounds. To check for consistency, the hydrostatic-pressure dependencies can be calculated by summing up the uniaxial-pressure coefficients, cf. right column in Table 6.2. These values are found to be in good agreement with the results obtained from hydrostatic-pressure experiments yielding $-1.43\,$K/kbar for β''-$(ET)_2SF_5CH_2CF_2SO_3$ [821], $-3\,$K/kbar for κ-$(ET)_2Cu(NCS)_2$ [469] and $-(2.4 \sim 2.8)\,$K/kbar for κ-$(ET)_2Cu[N(CN)_2]Br$ [470, 234].

An obvious step towards a deeper microscopic understanding of superconductivity in these molecular materials would be to trace out those uniaxial-pressure effects which are common to a certain class of substances. As Fig. 6.6 demonstrates, all three above-discussed superconductors share an extraordinarily large negative out-of-plane pressure effect on T_c. Apparently, it is this huge uniaxial-pressure component which predominates the enormously large hydrostatic-pressure effect. At the same time, the above results clearly demonstrate that even among the two κ-phase $(ET)_2X$ salts the systems behave quite non-uniformly concerning the in-plane pressure ef-

Table 6.2. Uniaxial-pressure dependencies of T_c for various $(ET)_2X$ salts, after [348, 155]. From left to right, the in-plane and out-of-plane axes are the b-, a-, and c-axis, respectively, for X = $SF_5CH_2CF_2SO_3$, the c-, b-, and a^*-axis for X = $Cu(NCS)_2$, and the a-, c-, and b-axis for X = $Cu[N(CN)_2]Br$. Somewhat different uniaxial-pressure coefficients of T_c for κ-$(ET)_2Cu(NCS)_2$ have been reported from tensile- and uniaxial-stress studies [814, 817] as well as earlier thermal expansion experiments [819]. However, these studies all agree upon a rapid decrease of T_c under interplane stress, see [817]. For the calculation of the hydrostatic-pressure coefficients in the right column $\left(\frac{\partial T_c}{\partial p}\right)_{hydr} = \sum_i \frac{\partial T_c}{\partial p_i}$ has been used.

anion X	$\partial T_c/\partial p_i$ (K/kbar)			$(\partial T_c/\partial p)_{hydr}$
	in-plane	in-plane	out-of-plane	(K/kbar)
$SF_5CH_2CF_2SO_3$	$+(3.9 \pm 0.15)$	$+(0.39 \pm 0.1)$	$-(5.9 \pm 0.25)$	$-(1.6 \pm 0.5)$
$Cu(NCS)_2$	$+(3.44 \pm 0.15)$	$-(0.14 \pm 0.1)$	$-(6.2 \pm 0.25)$	$-(2.9 \pm 0.5)$
$Cu[N(CN)_2]Br$	$-(1.16 \pm 0.2)$	$-(0.12 \pm 0.05)$	$-(1.26 \pm 0.25)$	$-(2.54 \pm 0.5)$

fects. While for κ-$(ET)_2Cu(NCS)_2$ the in-plane pressure coefficients of T_c are either vanishingly small or positive, they are both negative for the related κ-$(ET)_2Cu[N(CN)_2]Br$ system [155]. This finding of a large negative uniaxial-pressure coefficient of T_c for pressure perpendicular to the planes, as the only feature common to the κ-$(ET)_2X$ family, is supported by results on the related κ-$(ET)_2I_3$ salt, see [822].[9]

A $T_c(p)$ dependence, determined solely by in-plane electronic processes, has been claimed by A.-K. Klehe et al. [810] based on their measurements on κ-$(ET)_2Cu(NCS)_2$ using different pressure media. These authors argued that the pressure-medium-dependent shifts of T_c can be correlated with the size of the quasi-2D Fermi surface pockets[10] and thus the quasi-2D carrier density, which have been determined simultaneously. This has been interpreted to indicate a pressure-induced transfer of charge carriers from the quasi-1D to the 2D sections of the Fermi surface [810, 823].

[9] A different situation is encountered for the α-$(ET)_2MHg(SCN)_4$ salt, where uniaxial pressure perpendicular to the planes is found to either induce superconductivity by suppressing an ambient-pressure density-wave ground state for M = K, or enhance T_c for M = NH_4 [114, 115, 813]. This behavior is most likely related to the exceptionally thick anion layers specific to this compound resulting in a strong decoupling of the conducting layers.

[10] For a 2D metal, the area of the 2D Fermi surface, S_{FS}^{2D}, is directly proportional to the 2D carrier density via $n_{2D} = 1/(2\pi^2) \cdot S_{FS}^{2D}$.

Since pressure-induced changes of the interatomic distances may affect the spectra of both the lattice and the electronic excitations, several factors have to be considered when interpreting the above uniaxial-pressure results for the present quasi-2D superconductors. Among them are pressure-induced changes in the interlayer interactions, i.e. the interlayer electronic coupling — the degree of two-dimensionality — as well as changes in intermolecular phonons. Likewise, changes in the lattice degrees of freedom could be of relevance for the intraplane-pressure effects on T_c. In addition, in-plane stress effectively modifies the electronic properties by changing the various transfer integrals t_{ij} as well as the inter- and on-site Coulomb repulsion, V and U, respectively, cf. (4.2) in Chapt. 4.1.

The important information contained in the above uniaxial-pressure results is the large negative out-of-plane pressure effect on T_c and the nonuniform behavior for the in-plane coefficients. This implies that models considering pressure-induced T_c shifts originating from purely in-plane processes are inappropriate for these materials, see also [824].

6.3.3 Disorder

As discussed in Sect. 3.3, disorder is an important issue for molecular conductors and has to be taken into account when discussing their superconducting properties, see, e.g. [825]. Disorder may arise from various sources such as impurities, contaminations or crystalline defects which can be influenced to some extent by the preparation process, cf. Sect. 4.3.1. In addition, defects can be introduced intentionally by alloying or irradiation with fast particles such as high-energy electrons. Besides these *extrinsic* factors, many of the molecular compounds discussed here tend to be *intrinsically* disordered. This kind of disorder is a consequence of the materials' particular crystalline structure which enables certain atoms or molecules to adopt one of two possible orientations which are almost degenerate in energy [139, 140]. This refers to the anion ordering in the $(TM)_2X$ salts with non-centrosymmetric anions and the glass-like transition of the ethylene groups in the κ-$(ET)_2X$ compounds. It has been demonstrated by a variety of experiments that rapid cooling through the ordering temperature increases the degree of frozen disorder, see e.g. [826, 827, 828, 829]. However, little is known about the residual disorder in materials which had been cooled *"slowly"* on the time scale of a usual experiment. For the κ-$(ET)_2Cu[N(CN)_2]Br$ compound, a fraction of 5 – 20 % disordered ethylene groups has been proposed even for the slowly cooled crystals [155, 148]. According to a more recent X-ray study on differently prepared crystals of this compound, the degree of anion ordering at 100 K has been refined to $(92 \pm 2)\%$ [160]. Likewise, the nature of imperfections resulting from certain preparation parameters — such as the choice of the solvent which may have a severe influence on the electrical resistivity [160] — has not been explored in detail so far.

In Sect. 3.3 it has been discussed that for the salts κ-$(ET)_2X$ with X = $Cu[N(CN)_2]Br$, $Cu[N(CN)_2]Cl$, and $Cu(NCS)_2$ a kinetic, glass-like transition in the conformational degrees of freedom of the ET molecule's ethylene endgroups occurs which has been investigated in detail by means of thermal expansion measurements [155, 115, 154]. Below the characteristic glass-transition temperature T_g, a certain degree of disorder (in the 8 % range) remains frozen-in depending on the cooling rate and pressure conditions at T_g. An influence of the thermal history on the electronic properties of the above compounds had been realized before by various authors and interpreted in different ways. A strong depression of T_c, as large as 1.2 K, when the sample is rapidly cooled, as well as a pronounced kink in the resistivity at 75 K accompanied by hysteresis between heating and cooling, had been reported in early studies on κ-$(ET)_2Cu[N(CN)_2]Br$ [776, 606, 218]. Similar behavior in the vicinity of T_g was observed also for κ-$(ET)_2Cu(NCS)_2$ [225]. In later studies, based on resistance measurements of structural relaxation kinetics on κ-$(ET)_2Cu[N(CN)_2]Br$, it has been claimed that the ethylene-endgroup ordering is associated with a sequence of first-order phase transitions around 75 K [165]. Besides that, other more recent resistivity measurements on κ-$(ET)_2Cu[N(CN)_2]Br$ yielded interesting time dependencies affecting not only the electronic properties around 70 – 80 K but also the properties at lower temperatures: X. Su et al. reported relaxation effects in $R(T)$ and a separation of the curves below about 80 K as a function of the cooling rate q_c. As shown in the inset of Fig. 6.7 (upper panel), the residual resistivity increases with increasing $|q_c|$ [826, 166]. The inset also shows that the way of cooling through 70 – 80 K does influence the superconducting properties such that T_c decreases on increasing $|q_c|$. In addition, magnetization measurements revealed that with increasing $|q_c|$, a growing amount of disorder is induced causing an enlarged penetration depth [828]. In [829] an ac-susceptibility investigation of the magnetic penetration depths and their dependence on the cooling-rate-dependent intrinsic disorder has been performed. The authors found that the superconducting-state properties are critically determined by the time scale of the experiment around T_g.

For the deuterated salt κ-$(D_8\text{-}ET)_2Cu[N(CN)_2]Br$, it has been reported that rapid cooling through 80 K drives the superconducting ground state into an antiferromagnetic insulating state [830, 831].[11] As shown in the lower panel of Fig. 6.7, the ac-susceptibility data reveal a strong suppression of

[11] As discussed in Sect. 4.6 the ground state of deuterated κ-$(D_8\text{-}ET)_2Cu[N(CN)_2]Br$ is strongly sample dependent, i.e. there are both superconducting as well as non-superconducting samples. In [830] it is suggested that the crystals always contain superconducting and non-superconducting components, the latter having a magnetic character possibly similar to that of κ-$(ET)_2Cu[N(CN)_2]Cl$. It is thus believed that κ-$(D_8\text{-}ET)_2Cu[N(CN)_2]Br$ is situated in the critical region of the phase diagram just between the antiferromagnetic insulating and superconducting phases, see Fig. 4.32, and that rapid cooling drives the system towards the magnetic side.

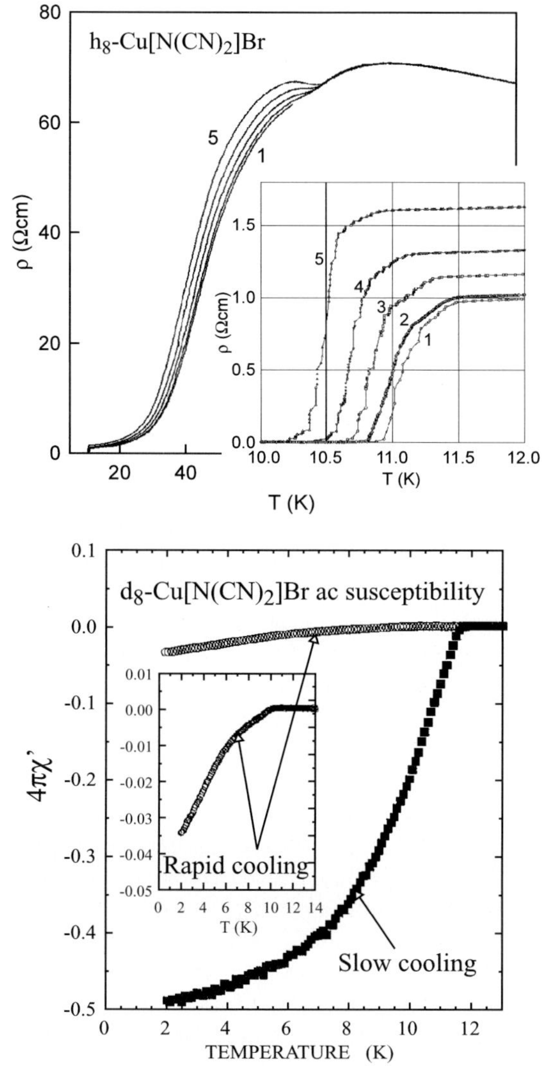

Fig. 6.7. Upper panel: resistivity as a function of temperature for a sample of κ-(H_8-ET)$_2$Cu[N(CN)$_2$]Br cooled at different rates ranging from about 0.5 K/min, lower curve (1), to 60 K/min upper curve (5). The inset shows an expansion of the data near the superconducting transition. Reproduced from [826]. Lower panel: ac-susceptibility of κ-(D_8-ET)$_2$Cu[N(CN)$_2$]Br after slow and rapid cooling. After [830].

the superconducting volume fraction with increasing cooling rate [830]. It is tempting to assign the apparent deterioration of superconductivity to the frozen disorder at T_g: besides the possibility of disorder-induced local moments or localization effects, it is conceivable that via the C-H$\cdots$donor and C-H$\cdots$anion contact interactions, disorder in the ethylene groups introduces a random potential that may alter the effective transfer integrals t_{eff} and, by this, may destroy superconductivity, see also [158, 154].

The effect of disorder, controlled by the cooling process, on the vortex pinning in the superconducting mixed state of κ-(H_8-ET)$_2$Cu[N(CN)$_2$]Br (see also Sect. 6.5) has been discussed by N. Yoneyama et al. [832]. In magnetization measurements, the zero-field-cooled (shielding) curves show almost the full Meissner volume and a negligible cooling-rate dependence indicating that the superconducting volume fraction is almost unchanged by the cooling process in contrast to the above findings for κ-(D_8-ET)$_2$Cu[N(CN)$_2$]Br. The magnetization hysteresis, however, is found to become smaller with increasing cooling rate [473, 832]. Also the critical current density decreases. These results indicate a *suppression* of the effective vortex pinning by increasing cooling rate. This finding is surprising because one would expect an *increase* of vortex pinning with increasing disorder. The authors suggest that in a slowly cooled sample, domains of ordered and disorder-including areas are formed where the domain boundaries act as extended pinning sites. In the quenched state on the other hand — although the amount of disordered ethylene endgroups is larger — many individual ethylene-disordered sites are randomly distributed leading to a quasi-homogeneous state on the length scale of the vortex cores $2\xi_{\parallel}(0) \sim 50$ Å [832] and thus a weaker pinning effect. The same group reported a partial suppression of superconductivity by fast cooling measured by magnetic susceptibility and scanning microregion infrared reflectance spectroscopy [833]. The latter technique revealed macroscopic insulating regions in the vicinity of the crystalline surface after fast cooling with diameters of $50 - 100\,\mu$m and depths of a few μm. The existence of such a shell-type insulating domain can explain the unusual large values of the in-plane magnetic penetration depth $\lambda_{\parallel}(0)$ obtained from measurements in magnetic fields $B < B_{c_1}$ after fast cooling, see discussion in Sect. 6.6.

Similar conclusions as in [832] have been drawn from a recent detailed study of magnetization measurements on fully deuterated κ-(D_8-ET)$_2$Cu[N(CN)$_2$]Br [167]. In this work, the superconducting properties have been studied as a function of annealing time (t_a) and temperature (T_a) around T_g with subsequent rapid cooling of the sample from T_a. Besides the consequences for the disordered superconducting ground state (diluted point-like defects vs. granular/cluster-like structure), the results also give additional insight into the dynamics of the glassy transition. It is found that the magnetic susceptibility χ reflects a decrease in both the superconducting volume and T_c with decreasing t_a, accompanied by a broadening of the transition. The quenched state exhibits hysteresis for $B \ll B_{c_1}$. Also, the (diamagnetic) peak magnetization M_p, which saturates with increasing t_a, increases with T_a approaching T_g. The relaxation is more rapid for χ than for M_p, see Fig. 6.8. These observations are explained by a percolation cluster model with two distinct time scales: at short times, clusters of superconducting regions in the sample grow with t_a following a power law (dominant at lower T_a) and at large time scales, the clusters grow towards a thermodynamic state following a stretched exponential law (dominant at temperatures T_a closer to T_g) $\propto \{1 - \exp[-(t/\tau)^{\beta}]\}$ with $\beta \approx 1$ for $T_a > 69$ K and $\beta \approx 0.5$ at 66 K which is close to the percolation threshold temperature. Since such a two-regime sce-

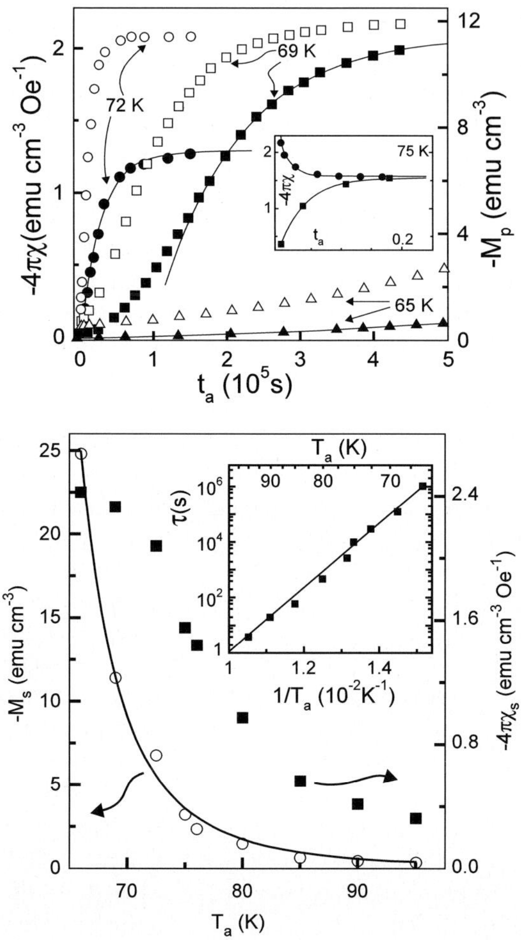

Fig. 6.8. Upper panel: peak magnetization M_p (filled symbols) and susceptibility χ (open symbols) at 2 K vs annealing time t_a at different temperatures $T_a = 65$, 69, and 72 K. The solid lines represent fitting curves to a stretched exponential law, see text. Lower panel: asymptotic values M_s and χ_s as functions of T_a. The solid line is a scaling law for M_s. The inset shows an Arrhenius plot of τ vs. T_a^{-1}, cf. Fig. 3.14. After [167].

nario is also found in spin glasses, the asymptotic magnetization indeed fits a scaling law consistent with an Ising spin-glass model [167]. T_g is estimated to be around 55 K which is considerably lower than T_g deduced from thermodynamic measurements (see Figs. 3.13 and 3.14). The equilibrium grain size has been determined to vary from $\sim 0.1\,\mu$m at $T_a \approx 100$ K to $10-30\,\mu$m at 66 K. These observations are explained by a cluster (or granular) structure which gives rise to Josephson-like effects characterized by the ratios λ/r and ξ/d where r and d are the average radius of the grains and the grain boundary thickness, respectively.

The above remarks indicate that disorder in the present molecular materials bears some additional aspects which are unknown from inorganic compounds. The high sensitivity of the electronic properties of these materials to disorder together with their material–specific problems have to be taken into account when interpreting experimental data in terms of theoretical models which consider idealized conditions. We have mentioned the reduction of the superconducting transition temperature of κ-$(ET)_2Cu[N(CN)_2]Br$ employing different cooling rates through the glass transition around 80 K. These measurements revealed an increase of the residual interlayer resistivity, accompanied by a reduction of T_c upon increasing the cooling rate, indicating a suppression of superconductivity by the cooling-rate induced disorder. By investigating the Shubnikov-de Haas effect under controlled cooling conditions, a rise in the Dingle temperature, T_D from 2.1 K to 2.4 K upon increasing the cooling rate has been found [164] indicating that the disorder acts as scattering centers for the electrons. The Dingle temperature (5.33) can be used to extract the electronic mean free path ℓ by

$$\ell = \frac{\hbar v_F}{2\pi k_B T_D}, \tag{6.28}$$

where v_F is the Fermi velocity. Employing (6.28) yields a decrease of the in-plane mean free path from 38 nm to about 30 nm on increasing the cooling rate in the above experiment.

In general, by studying the effect of disorder on T_c important information on the pairing state can be gained. Provided that the character of the disorder is known, the response of the superconductor to variations in the impurity concentration may help to distinguish between different pairing states. For a conventional singlet superconductor, Anderson's theorem [834] states that potential scattering by non-magnetic defects do not change the transition temperature (as far as the electronic structure near E_F does not change). The reason is that Cooper pairs in these s-wave superconductors are formed from time-reversed states — a symmetry which is conserved also in the presence of non-magnetic disorder. On the contrary, magnetic impurities strongly reduce T_c in singlet superconductors as they break the time-reversal symmetry [835]. A.A. Abrikosov and L.P. Gor'kov described the scattering of the conduction electrons by the impurity spin within the first Born approximation, i.e. to second order in the exchange interaction parameter [836]:

$$\ln\left(\frac{T_{c0}}{T_c}\right) = \psi\left(\frac{1}{2} + \frac{\hbar}{4\pi k_B T_c}\cdot\frac{1}{\tau_M}\right) - \psi\left(\frac{1}{2}\right), \tag{6.29}$$

where T_{c0} is the superconducting temperature in the pure system and $\psi(x)$ is the digamma function. τ_M is the quasiparticle lifetime due to scattering from magnetic impurities.

As a matter of fact (6.29) holds true also for the suppression of T_c due to non-magnetic impurities in a non-s-wave superconductor, when τ_M is replaced by τ_N, the corresponding lifetime for non-magnetic impurities [837]. For sufficiently low impurity concentrations, i.e. $\hbar/\tau \ll k_B T_c$, the Abrikosov-Gor'kov (AG) equation yields a linear reduction of T_c with the scattering rate $1/\tau$,

$$T_{c_0} - T_c \simeq \frac{\pi\hbar}{8k_B}\frac{1}{\tau}, \tag{6.30}$$

where $\tau = \tau_M$ for the magnetic impurities in singlet superconductors and $\tau = \tau_N$ for non-magnetic impurities in non-s-wave pairing states. The suppression of T_c upon increasing the cooling rate, reported by X. Su et al. [826], has been analyzed recently by B.J. Powell and R.H. McKenzie [838] and found to be consistent with the above Abrikosov-Gor'kov formula. It has been argued in [838] that the above cooling-rate dependent T_c suppression [826, 164] is a strong indication for a d-wave pairing, provided that the impurities can be clearly identified as non-magnetic scattering centers. Given the proximity of these salts to a magnetic insulator in the anion/pressure space, cf. the phase diagram in Sect. 4.6, however, it seems possible [838] that the disorder induces local magnetic moments. Indeed, for the nearby deuterated κ-(D_8-ET)$_2$Cu[N(CN)$_2$]Br compound, indications for an inhomogeneous phase coexistence between antiferromagnetic and superconducting regions have been claimed [839, 840]. Thus a careful investigation of the cooling-rate-induced disorder in terms of having a magnetic or non-magnetic character is necessary to settle this problem. As will be discussed in more detail in Sect. 6.6, the effect of disorder on the magnetic penetration depth λ has been investigated [828, 829, 841].

The cooling-rate induced disorder in the ethylene endgroups is rather similar to the anion disorder in the Bechgaard salts with non-centrosymmetric anions (TMTSF)$_2$ClO$_4$ and (TMTSF)$_2$ReO$_4$, where the anions can adopt two non-equivalent orientations. It has been shown that fast cooling leads to partially disordered domains [842] and to a reduction or even complete suppression of superconductivity in favor of a spin-density-wave. In [843], the effect of anion disorder on T_c has been studied in the solid solution (TMTSF)$_2$(ClO$_4$)$_{1-x}$(ReO$_4$)$_x$. Similar to the above results on the κ-(ET)$_2$Cu[N(CN)$_2$]Br salts, these studies revealed a suppression of T_c with increasing the residual resistivity in accordance with the Abrikosov-Gor'kov formula.

So far we have discussed two possible scenarios how disorder in the present materials can lead to a suppression of T_c: either non-magnetic impurities in a non-s-wave pairing state or magnetic impurities in a singlet state. In other words, these interactions can violate the time-reversal symmetry in superconductivity. An alternative mechanism has been proposed by Y. Hasegawa and H. Fukuyama [844]. They argued that a weak localization, due to non-magnetic disorder, may lead to the suppression of T_c in organic superconductors, in violation of Anderson's theorem. Their mechanism, however, reveals a quite different scattering-rate dependence as compared to the Abrikosov-Gor'kov formula.

6.4 Superconducting Parameters

In general, applying a magnetic field to a superconductor leads to a suppression of T_c. In a conventional superconductor, where the Cooper pairs are formed by electrons of opposite momentum $\boldsymbol{k}$ and spin $\boldsymbol{s}$, the magnetic field $\boldsymbol{B}$ can interact with the superconducting electrons in two ways: via the Zeeman interaction, i.e. $\boldsymbol{s} \cdot \boldsymbol{B}$, and via the electromagnetic interaction of the vector potential $\boldsymbol{A}$ (associated with $\boldsymbol{B}$) with the momentum $\boldsymbol{k}$ of the electron, $\boldsymbol{k} \cdot \boldsymbol{A}$. Since both of these interactions are favorable for one member of the electron pair $(\boldsymbol{k} \uparrow, -\boldsymbol{k} \downarrow)$ and unfavorable for the other, they break up the Cooper pairs and cause a suppression of T_c. The Zeeman interaction of the applied field with the spin of the electrons leads to the paramagnetic limiting field or Pauli field B_P

$$B_P = \left[\frac{N(E_F)}{\chi_n - \chi_s} \right]^{1/2} , \tag{6.31}$$

where $N(E_F)$ is the density of states at the Fermi level E_F, and χ_n and χ_s are the normal- and superconducting-state Pauli susceptibilities, respectively [845, 846]. The Pauli field can be expressed in terms of the zero-temperature energy gap

$$B_P = \Delta_0 \frac{1}{\sqrt{2}\mu_B} , \tag{6.32}$$

where μ_B denotes the Bohr magneton. Assuming a BCS relation for the energy gap

$$\Delta_0 = 1.76\, k_B T_c \tag{6.33}$$

and a Pauli susceptibility in the superconducting state $\chi_s = 0$ then yields

$$B_P \,(\text{in Tesla}) = 1.84 \times T_c \,(\text{in Kelvin}), \tag{6.34}$$

the Chandrasekhar-Clogston limiting field [845, 846]. Thus for a singlet superconductor B_P sets a natural upper limit for the upper critical field B_{c2}.

The electromagnetic interaction of the orbital momentum of the Cooper pairs with an external magnetic field gives rise to the orbital critical field

$$B^*_{c_2} = \frac{\phi_0}{2\xi^2} , \tag{6.35}$$

where $\phi_0 = h/2e$ is the flux quantum and ξ the superconducting coherence length [847, 848].

All known organic superconductors are strongly type-II superconductors, i.e. $\kappa = \lambda_L/\xi_{GL} \gg 1$, with lower critical fields, B_{c_1}, of the order of a few mT and upper critical fields, B_{c_2}, in the Tesla range. In accordance with the highly anisotropic electronic structure, these critical fields strongly depend on the direction of the applied field. Due to the pronounced fluctuation effects in these superconductors of reduced dimensionality, however, the determination of the upper critical fields is difficult and in many cases not free of ambiguities. This holds true in particular for resistivity measurements in finite fields, as phase fluctuations of the order parameter give rise to a resistive state which tends to descend far below the mean-field transition temperature, see Fig. 6.2.

6.4.1 $(\mathrm{TMTSF})_2\mathrm{X}$

The critical fields have been intensively studied first for the ambient-pressure $(\mathrm{TMTSF})_2\mathrm{ClO}_4$ salt and, more recently, also for pressurized $(\mathrm{TMTSF})_2\mathrm{PF}_6$, see also Sect. 6.6 below. For the former system, the Meissner and diamagnetic shielding effects have been examined for magnetic fields aligned along the three principal axes [849, 850]. From these experiments, the lower critical field values B_{c_1} (at 50 mK) have been determined to 0.2, 1 and 10 (in units of 10^{-4} T) along the *a*-, *b*- and *c*-axis, respectively [849]. From the condensation energy determined by specific heat measurements, the thermodynamic critical field has been estimated to $B_{c_{th}} = (44 \pm 2) \cdot 10^{-4}$ T [851]. Figure 6.9 shows the temperature dependence of the upper critical fields, $B_{c_2}(T)$, for $(\mathrm{TMTSF})_2\mathrm{ClO}_4$ as determined by early resistivity measurements [852]. Note that a midpoint criterium had been used to extract $T_c(B)$ from the rather broad resistive transitions, implying a considerable uncertainty, cf. the discussion in Sect. 6.3.1.

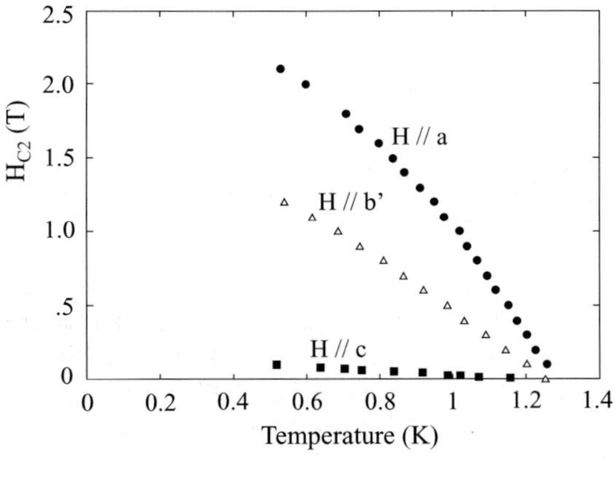

Fig. 6.9. Upper critical fields determined from resistivity measurements as a function of temperature for $(\mathrm{TMTSF})_2\mathrm{ClO}_4$ for the three principal crystal axes, after [852].

For the zero-temperature upper critical fields values of 2.8 T, 2.1 T and 0.16 T for fields along the *a*-, *b*- and *c*-axis, respectively, have been reported [849, 852]. The so-derived value for the field parallel to the *a*-axis, where B_{c_2} is the largest, is close to the Pauli-limiting field B_P, which for $(\mathrm{TMTSF})_2\mathrm{ClO}_4$ with $T_c = 1.25$ K is $B_P = 2.3$ T. The fairly good coincidence with the experimentally-derived maximum critical field has been taken as an indication for a spin-singlet pairing state [852, 853, 854]. As will be discussed in more detail in Sect. 6.6, more recent resistivity and torque- magnetization measurements on pressurized X = PF_6 and slowly cooled X = ClO_4 at ambient pressure revealed upper critical field curves which show an upward curvature for $T \to 0$ with no sign of saturation down to 0.1 K [35, 855]. It has been argued in [35, 855] that this unusual enhancement of B_{c_2}, exceeding the Pauli paramagnetic limit by a factor of $2-4$, indicates a spin-triplet ($S=1$) pairing state.

Yet from the initial slopes of the upper critical field curves in Fig. 6.9, $\mathrm{d}B^i_{c_2}/\mathrm{d}T$, the Ginzburg-Landau (GL) coherence lengths $\xi_i(0)$ can be derived using the following relation [797, 365]:

$$\frac{\mathrm{d}B^i_{c_2}}{\mathrm{d}T} = \frac{\phi_0}{2\pi\xi_j(0)\xi_k(0)T_c}, \tag{6.36}$$

where i, j and k can be a, b or c, see (6.25). The so-derived $\xi_c(0)$ value of about 20 Å, being much smaller than the numbers for the a- and b-axis coherence lengths of about 700 Å and 335 Å, respectively, but comparable to the lattice parameter $c = 13.5$ Å, indicates that superconductivity has, in fact, a more quasi-2D character. For the London penetration depth for B parallel to the a-axis, the axis of highest conductivity, a value of $\bar{\lambda} = 40\,\mu$m has been reported [850]. This number exceeds the GL coherence lengths by orders of magnitude indicating that the present system is an extreme type-II superconductor.

6.4.2 $(BEDT\text{-}TTF)_2X$ and $(BEDT\text{-}TSF)_2X$

Among the $(ET)_2X$ superconductors, most data on the critical fields are available for the κ-phase salts. The left panel of Fig. 6.10 shows the upper critical field curves for κ-$(ET)_2Cu[N(CN)_2]Br$, as determined from dc-magnetization measurements [789, 103] and employing a scaling analysis (6.24) to account for the fluctuation effects, see also [776]. The right panel shows B_{c_2} for fields applied parallel to the planes from magnetoresistance and specific heat data [856, 857]. The $B_{c_2}(T)$ curve for B aligned parallel to the highly conducting planes, as determined from resistivity measurements, is shown in the right panel of Fig. 6.10 over an extended field range. For the layered superconductors with negligible in-plane anisotropy, the expression (6.25) can be used to determine the GL coherence lengths perpendicular and parallel to the conducting planes.

Table 6.3 compiles the B_{c_2} values together with other superconducting parameters for the above ET compounds. For comparison, the Table also contains data for the BETS-based system λ-$(BETS)_2GaCl_4$. As indicated in the Table, the transition temperatures reported in the literature show a considerably large variation depending on the method and criterion used to determine T_c. This may partly be related to the fact that the superconducting transition is usually found to be relatively broad. Even for high-quality single

Table 6.3. Superconducting-state parameters of representative κ-phase $(ET)_2X$ salts with X = I_3, $Cu(NCS)_2$, $Cu[N(CN)_2]Br$, and β''-$(ET)_2SF_5CH_2CF_2SO_3$ and λ-$(BETS)_2GaCl_4$.
Note that $\lambda_\perp$ and $\lambda_\parallel$ describe the screening by supercurrents flowing perpendicular and parallel to the conducting planes, respectively, and *not* the screening of a magnetic field along those directions.
Note the wide range of values for the in-plane penetration depth $\lambda_\parallel(0)$. As discussed in Sect. 6.6, page 198, experiments performed in the shielding state, $B < B_{c_1}$, yield an in-plane penetration depth $\lambda_\parallel(0)$ ranging from $1.4 - 2\,\mu m$ for κ-$(ET)_2Cu(NCS)_2$, whereas experiments performed in the mixed state, $B_{c_1} < B < B_{c_2}$, consistently yield considerably lower values for $\lambda_\parallel(0)$; see also [858] for a recent compilation of values derived from μSR studies in the vortex state.

	$SF_5CH_2CF_2SO_3$	I_3	$Cu(NCS)_2$	$Cu[N(CN)_2]Br$	$GaCl_4$	References
T_c (K)	$4.5 - 5.3$	3.5	$8.7 - 10.4$	$11.0 - 11.8$	$5 - 6$	
$B_{c_1}^{\perp}(0)$ (mT)	2 ± 0.5	4 ± 1	6.5	3		[459, 859, 208, 860]
$B_{c_1}^{\parallel}(0)$ (mT)	$< 6\,\mu T$	0.05	0.2			[459, 859, 208]
$B_{c_2}^{\perp}(0)$ (T)	1.4	0.2	6	$8 - 10$	3	[348, 861, 859, 103, 862, 452, 863, 864]
$B_{c_2}^{\parallel}(0)$ (T)	$10.4 - 11.9$	7	$30 - 35$	> 30	12	[348, 861, 859, 865, 866, 856, 867, 863]
$B_{c_{th}}(0)$ (mT)	25 ± 4	17 ± 2	52 ± 5	64 ± 4	20 ± 4	calculated from (6.37)
$\xi_\perp(0)$ (Å)	7.9	11	$5 - 9$	$5 - 12$	$9 - 14$	[348, 789, 868, 103, 208, 452, 863]
$\xi_\parallel(0)$ (Å)	144	410	$53 - 74$	$28 - 64$	$87 - 143$	[348, 789, 868, 103, 208, 863, 864]
$\lambda_\perp(0)$ (μm)	800	70	$40 - 200$	$38 - 133$		[869, 868, 870, 871]
$\lambda_\parallel(0)$ (μm)	$1 - 2$	0.3	$0.51 - 2$	$0.65 - 1.5$	$0.15 - 0.72$	[869, 872, 873, 874, 871, 875, 868, 452, 876]
$\kappa_\parallel = \lambda_\parallel / \xi_\parallel$	$75 - 150$	10	$100 - 200$	$200 - 300$	$20 - 50$	[348, 869, 859, 103, 452, 863]

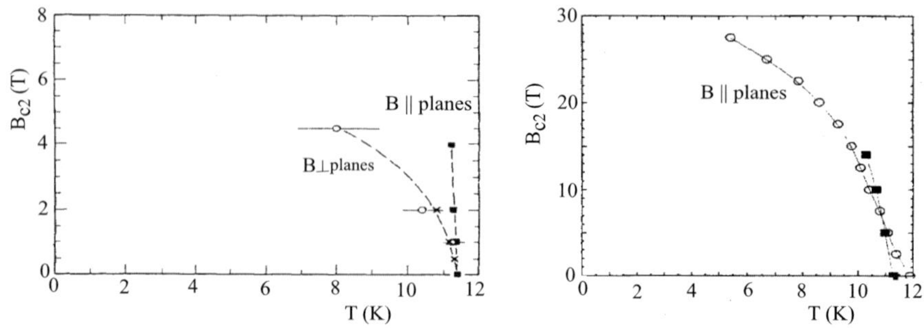

Fig. 6.10. Upper critical fields of κ-(ET)$_2$Cu[N(CN)$_2$]Br. Left panel: anisotropy of B_{c_2} as a function of temperature as determined from dc-magnetization measurements [789]. Right panel: $B_{c_2}(T)$ for fields aligned parallel to the conducting planes as determined from resistivity measurements using different criteria (closed and open symbols), after [856].

crystals with an in-plane mean free path of typically $\ell \sim 2000\,\text{Å}$ [249], the transition can be broadened due to internal strain fields as a consequence of the extraordinarily large pressure dependence of T_c. In addition, the quasi-2D nature of the electronic structure may give rise to pronounced fluctuation effects [277].

According to recent magnetoresistivity studies, the in-plane Fermi surface of λ-(BETS)$_2$GaCl$_4$ strongly resembles that of κ-(ET)$_2$Cu(NCS)$_2$ with the effective masses being almost identical for both compounds [452]. However, the interplane-transfer integral of $t_\perp \approx 0.21\,\text{meV}$ for the BETS salt is about a factor of 5 larger [452], indicating that λ-(BETS)$_2$GaCl$_4$ is more three-dimensional. Figure 6.11 compares the magnetic field-temperature phase diagrams of the above compounds for fields aligned perpendicular to the planes. A dimensional crossover has been suggested to account for the unusual temperature dependence of $B_{c2}^{\perp}$(T) observed for the λ-(BETS)$_2$GaCl$_4$ compound (left panel): B_{c_2}(T) shows a 3D-like linear behavior close to T_c which turns into a power-law dependence characteristic for a 2D superconductor with weakly coupled layers below some crossover temperature [452, 277]. In contrast, B_{c_2}(T) for the κ-(ET)$_2$Cu(NCS)$_2$ salt follows $B_{c_2}(T) \propto (T - T_c)^{2/3}$ over the whole temperature range. The various crossover and transition lines in the mixed state below the $B_{c_2}(T)$ curves, as indicated in Fig. 6.11, will be discussed in Sect. 6.5 below. The superconducting phase diagram of a λ-(BETS)$_2$GaCl$_4$ sample with $T_c = 4.8\,\text{K}$ has more recently been determined from microwave loss [864]. In contrast to Fig. 6.11, the authors find a more normal behavior for $B_{c_2}(T)$, i.e., without a crossover behavior, and a larger initial slope $\mathrm{d}B_{c_2}/\mathrm{d}T = -0.82\,\text{T/K}$ which gives $B_{c_2}(0) \sim 3\,\text{T}$ (from $B_{c_2}(0) = -0.69 T_c (\mathrm{d}B_{c_2}/\mathrm{d}T)|_{T=T_c}$ [847, 877]) and 87 Å for the GL coherence length (6.25) in the a^*c-plane [864].

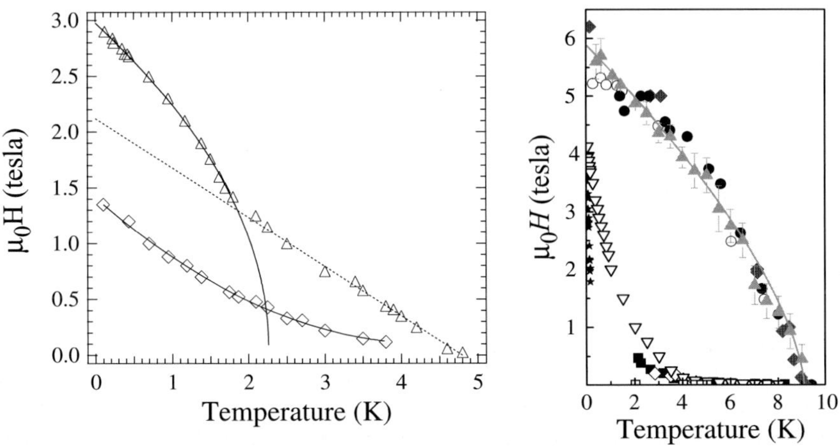

Fig. 6.11. Left panel: Upper critical field B_{c_2}(T) (triangles) of λ-$(BETS)_2GaCl_4$. Upper solid curve is $B_{c_2} \propto (T^* - T)^{1/2}$; the dashed curve $B_{c_2} \propto (T_c - T)$ (see text). Diamonds and the lower solid curve indicate the flux-lattice melting line (see also Sect. 6.5). Right panel: $B_{c_2}(T)$ for B ⊥ to the planes of κ-$(ET)_2Cu(NCS)_2$ as determined from different experimental techniques, see [452]. The solid curve is $B_{c_2} \propto (T_c - T)^{2/3}$ (see text). Also shown are transition and crossover lines in the mixed state, cf. the discussion in Sect. 6.5 below. Fields are aligned perpendicular to the planes. After [452].

Important information on the spin state of the Cooper pairs can be gained by comparing the experimentally determined $B_{c_2}(T)$ for $T \to 0$ with the Pauli-limiting field as defined in (6.32). Using this formula, which neglects any orbital effects, and assuming a weak-coupling BCS relation for the gap, i.e. $\Delta_0 = 1.76\, k_B T_c$, B_P amounts to ~ 18 T and ~ 21 T for κ-$(ET)_2Cu(NCS)_2$ and κ-$(ET)_2Cu[N(CN)_2]Br$, respectively. Apparently, these numbers are significantly smaller than the $B_{c_2}^{\|}(0)$ values found experimentally and listed in Table 6.3. On the other hand, clear evidence for a spin-singlet pairing state has been inferred from Knight shift measurements on κ-$(ET)_2Cu[N(CN)_2]Br$ yielding a vanishingly small spin susceptibility at low temperatures [862]. These deviations might find a natural explanation by recalling that (6.32) is valid only in the weak-coupling limit. For a strong-coupling superconductor, on the other hand, the density of states is renormalized leading to a Pauli field which is enhanced by a factor of $(1 + \lambda)^{1/2}$ [845, 797], where λ denotes the interaction parameter. Clear evidence for a strong-coupling type of superconductivity has been found in specific heat experiments on the κ-$(ET)_2Cu[N(CN)_2]Br$ and κ-$(ET)_2Cu(NCS)_2$ salts [456, 457, 458, 801], see Sect. 6.6.

As an alternative mechanism to account for a $B_{c_2}^{\|}(0)$ value in excess of the Pauli field, a transition into a Fulde-Ferrell-Larkin-Ovchinnikov (FFLO) state [735, 736] has been proposed [866, 878] and discussed controversially, see e.g. [879, 277]. In such a scenario, a superconductor with suitable materials parameter adopts a new state at sufficiently high fields where the order parameter is spatially modulated. A more detailed discussion on this issue will be given in Sect. 6.5 below.

Besides T_c and the upper critical fields, Table 6.3 contains further superconducting-state parameters such as the lower and the thermodynamic critical fields B_{c_1} and $B_{c_{th}}$, respectively, as well as the GL coherence lengths, ξ_{GL}, and the London penetration depths λ_L. B_{c_1} is usually determined by measuring the magnetization as a function of field under isothermal conditions where B_{c_1} corresponds to the field, above which magnetic flux starts to penetrate the sample. Due to the smallness of B_{c_1}, the plate-like shapes of the crystals and the peculiar pinning properties of these layered superconductors, an accurate determination of B_{c_1} is difficult. A more reliable way to determine $B_{c_1}^{\perp}(0)$ has been proposed by J. Hagel et al. based on a model for thermally activated flux creep yielding $B_{c_1}^{\perp}(0) = (3\pm0.5)\,\mathrm{mT}$ for κ-$(ET)_2Cu[N(CN)_2]Br$ [860]. The values for the thermodynamic critical fields, $B_{c_{th}}$, in Table 6.3 are estimated from specific heat results using

$$B_{c_{th}}(0) = T_c \cdot \sqrt{\mu_0\,\gamma/(2\,V_{\mathrm{mol}})}, \tag{6.37}$$

where γ is the Sommerfeld coefficient of the electronic specific heat. These values roughly agree with those calculated from $B_{c_{th}} = B_{c_2}/(\kappa\sqrt{2}) = B_{c_1}\sqrt{2}\kappa/\ln\kappa$, where $\kappa = \lambda/\xi$ is the GL parameter. The large numbers of κ reflect the extreme type-II character of these superconductors.

Also listed in Table 6.3 are values of the magnetic penetration depth λ, the characteristic length over which magnetic fields are attenuated in the superconductor. While the absolute values for λ, derived from various experimental techniques, are in fair agreement within a factor $4-5$, no consensus has yet been achieved concerning its temperature dependence, see Sect. 6.6 below.

6.5 The Superconducting Mixed State

6.5.1 Overview

In 1957 A.A. Abrikosov derived a mean-field version of the B-T phase diagram of type II superconductors characterized by $\kappa > 1/\sqrt{2}$ [880]. His solution of the Ginzburg-Landau (GL) equations corresponds to a regular array of vortices of current of radius $\sim\lambda$ surrounding a core region of radius $\sim\xi$ where the order parameter vanishes. For $B_{c_1} < B < B_{c_2}$ the magnetic field penetrates the bulk of the superconductor forming a triangular vortex lattice with nearest neighbor distance $a_\triangle = (2/\sqrt{3})^{1/2}\cdot(\phi_0/B)^{1/2}$. Each unit cell of the array carries total flux equal to $\phi_0 = h/2e = 2.068\times10^{-11}\,\mathrm{Tcm}^2$. This mean-field B-T phase diagram, comprising a Meissner-Ochsenfeld phase characterized by complete flux repulsion at low magnetic fields $B < B_{c_1}$ separated from the Shubnikov phase (or mixed state) at higher fields $B_{c_1} < B < B_{c_2}$, has proven to accurately describe the phenomenological behavior of all the conventional low-temperature superconductors. The fundamentals of type-II superconductors are discussed, e.g., in [587, 365]. The physics of vortices as a new state of matter has been comprehensively reviewed in [881, 787, 882] and more recently in [883, 790].

For the present low-dimensional systems, however, a mean-field description is no longer appropriate since they are strongly affected by thermal fluctuations, the strength of which is governed by the Ginzburg number $G_i \propto [T_c/(B_{c_{th}}^2 V_{\mathrm{coh}})]^2$ for layered systems (6.23), see Sect. 6.3.1. Compared to 3D materials, the layered crystal structure of the present quasi-2D compounds reduces the coherence volume V_{coh} by the anisotropy parameter γ: $V_{\mathrm{coh}} = \xi^3/\gamma$ with $\xi \equiv \xi_{\parallel}$ the (in-plane) coherence length and $\gamma = (m_{\perp}^*/m_{\parallel}^*)^{1/2}$, where $m_{\perp}^*$ and $m_{\parallel}^*$ denote the effective masses for the superconducting carriers moving perpendicular and parallel to the conducting planes, respectively. As has been discussed in Sect. 6.3.1, see (6.22), in the London and Ginzburg-Landau theories, γ is directly related to the anisotropies in the magnetic penetration depths $\gamma = \lambda_{\perp}/\lambda_{\parallel} = \xi_{\parallel}/\xi_{\perp}$, and is strongly enhanced for the present $(ET)_2X$ compounds (p. 152). An important consequence from the enhanced fluctuation effects ($G_i \propto \gamma^2 \equiv \Gamma$) is that the vortex matter is melted over large portions of the B-T phase diagram. In addition to thermal fluctuations, vortex matter is also strongly affected by quenched disorder which can be quantified by the dimensionless critical current density ratio j_c/j_0 which is a measure of the strength of the pinning force density [790], where $j_0 \propto B_c/\lambda$ denotes the depairing current density. Vortex lines are pinned by the disorder potential transforming the vortex solid into a vortex glass. At low fields, the glassy state is a quasi-ordered *Bragg glass* showing no topological disorder and exhibiting a superconducting response in all directions. At higher fields, a disorder-driven melting transition into a more disordered *glassy* or *frozen liquid state* occurs, see Fig. 6.12. In the present materials, pinning is usually weak due to the small coherence length and the absence of generic extended pinning sites.[12] If the pinning is weak enough, the melting transition of the vortex matter is only weakly influenced by quenched disorder and a generic melting transition is observed.

The above-mentioned characteristics of strongly anisotropic or layered materials lead to new phase transitions in the vortex system and a wealth of novel phenomena, among them a substantial broadening of the resistive

[12] Intrinsic pinning sites like molecular defects, dislocations, grain boundaries, or twinning planes (individual defects) could serve as strong pinning centers. A different phase coexisting with the actual crystal phase (e.g., α-phase in a κ-phase crystal) would be another possible pinning site. Weak (collective) pinning is due to point-like disorder. In [832] also, domain structures of ethylene-ordered and ethylene-disordered regions are discussed as possible pinning centers, see Sect. 3.3. Furthermore, the plate-like shape of many quasi-2D organic superconductors produces strong surface and geometrical barriers.

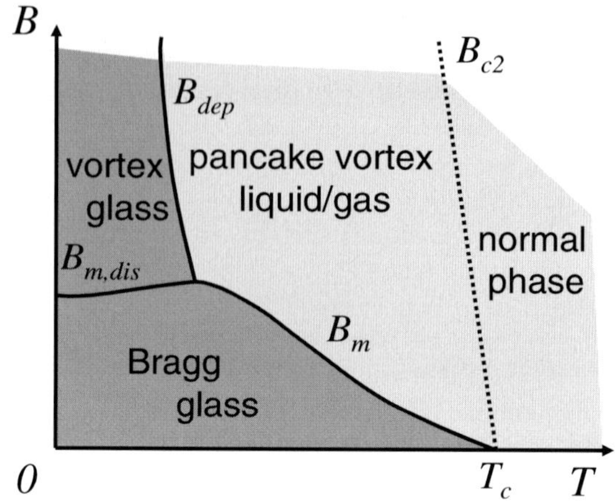

Fig. 6.12. Conceptual phase diagram for a strongly fluctuating quasi-2D superconductor in the presence of disorder (the Meissner-Ochsenfeld phase is not shown), after [790]. The (first-order) melting line B_m at high temperatures smoothly crosses over into a (first-order) disorder-induced melting/decoupling transition ($B_{m,dis}$). The depinning line B_{dep} denotes a (second-order) phase transition into a more disordered glassy or frozen liquid state different from the low-field Bragg glass and the high-temperature (viscous) vortex liquid. The irreversibility line B_{irr} (see text) defined as $j_c = 0$ is sensitive to the experimental conditions, while for ideal j_c measurements it matches the melting line.

transition, a dimensional crossover of the vortex structure, an irreversibility line in the B-T phase diagram possibly associated with a vortex glass transition, a first-order melting transition of the vortex lattice observed in samples with a low degree of disorder, a layer decoupling transition, flux creep phenomena, a rapid decrease of the critical current density with increasing temperature, and the observation of quantum effects for the temperature approaching zero. Depending on the interplay between quenched disorder, thermal fluctuations (which reduce the pinning strength by smoothing the disorder potential resulting in thermal depinning), and the actual dimensionality of the system, the above phenomena lead to a variety of possible B-T phase diagrams which have been studied extensively both theoretically and — in the high-T_c cuprates — experimentally, see, e.g., [790] and references therein. In this section we discuss the experimental observations for quasi-2D organic charge-transfer salts.

6.5.2 Basic Concepts of Vortex Matter in Quasi-Two-Dimensional Systems

For quasi-2D systems, the theoretical description of the mixed state is based on the *continuous* anisotropic Ginzburg-Landau model and — in the extreme limit of layered materials consisting of stacks of only electromagnetically coupled superconducting planes — the *discrete* Lawrence-Doniach model [781]. A finite coupling of the layers by the Josephson current density then establishes a bulk superconducting response.

Josephson and Pancake Vortices

The layered crystal structure modifies the structure of the flux lines compared to the Abrikosov vortices in the isotropic 3D case. For fields $B > B_{c_1}$ aligned parallel to the layers, in the limiting case of a quasi-2D superconductor characterized by a cross-plane coherence length $\xi_\perp$ being smaller than the interlayer distance s, the vortex cores slip into the insulating layers where the superconducting order parameter is small. Such *Josephson vortices* [884] lack the normal core of the Abrikosov vortices and have an elliptically-deformed cross section with screening currents extending a distance $\lambda \equiv \lambda_\parallel$ along the interlayer axis and $\lambda_\perp = \gamma\lambda_\parallel$ along the in-plane axis perpendicular to the magnetic field direction.[13] Since the Josephson screening currents across the insulating layers are very weak, the material is almost transparent for fields parallel to the planes corresponding to a large value for the upper critical field $B_{c_2}^{\parallel}$, cf. Table 6.3.

When a layered superconductor is exposed to a magnetic field $B > B_{c_1}$ aligned perpendicular to the planes, the confinement of the screening currents to the superconducting layers results in a segmentation of the flux lines into two-dimensional objects, the so-called *pancake vortices* [885]. The coupling between vortex segments of adjacent layers is provided by their electromagnetic interaction and the Josephson coupling. The latter interaction drives tunneling currents when two vortex segments are displaced relative to each other. As a result of both effects, the vortex pancakes tend to align thereby forming extended stacks.

Dimensional Crossover For increasing magnetic fields applied perpendicular to the planes of a layered superconductor, a *magnetically driven* crossover occurs from 3D vortex lines to 2D vortex points, i.e. from a system where the lateral position of pancake vortices in successive planes is strongly correlated so as to define a continuous vortex line to a system where pancake vortices in successive planes move essentially independently [886, 887]. This crossover has consequences for the melting transition of the vortex lattice (see below). Considering the relative strength of the restoring forces exerted on a given pancake vortex by adjacent pancake vortices in the same plane and by pancake vortices in the plane above and below the given one, allows for the estimation of the crossover field $B_{cr} \approx \phi_0/(s\gamma)^2$. As will be discussed below, the physical significance is that when $B \gg B_{cr}$, the interaction between

[13] Note that $\lambda_\perp$ and $\lambda_\parallel$ describe the screening by supercurrents flowing perpendicular and parallel to the conducting planes, respectively, and *not* the screening of a magnetic field along those directions.

adjacent pancake vortices in the same layer is stronger than the interaction between those in adjacent layers but on the same vortex line. This causes the thermal fluctuations to have a quasi-2D character in the high-field regime $B \gg B_{cr}$.

Vortex-Lattice Melting In analogy to crystalline solids, vortex matter might undergo a melting transition where the long range crystalline order would eventually be destroyed discontinuously by a first-order transition as the temperature is raised. The thermal melting transition of the vortex-lattice has been studied theoretically for continuous anisotropic, electromagnetically- and Josephson-coupled superconductors. Likewise, disorder induced melting has been investigated. An elementary analysis of vortex-lattice melting, based on nearest-neighbor interactions, employs the Lindemann criterion stating that the vortex crystal melts when the mean-squared thermal displacement of vortex lines becomes comparable to the magnetic lattice spacing $a_\triangle = (2/\sqrt{3})^{1/2} \cdot (\phi_0/B)^{1/2}$. Comparison with experiments on high-T_c cuprates show that once the materials' anisotropy parameter γ and the superfluid density $n_s(T) = m^*/(4\pi e^2 \lambda(T)^2)$ — the latter is experimentally determined via measurements of the magnetic penetration depth $\lambda(T)$ — are known, the location of the melting line $T_m(B)$ can be accurately predicted. The predicted $T_m(B)$ in the 3D fluctuation regime below the dimensional crossover field B_{cr} is different from that in the quasi-2D regime above B_{cr}.

Due to the first-order nature of the melting line, a jump ΔB in the magnetic induction occurs at T_m which is related via the Clausius-Clapeyron formalism $\Delta S = -1/4\pi \cdot \mathrm{d}B_m/\mathrm{d}T \cdot \Delta B$ to a discontinuity ΔS in the entropy. Since the vortex liquid is denser than the vortex solid, the vortex crystal melts like ice. It is interesting to note that due to strong thermal fluctuations, the second-order superconducting-normal metal phase transition at B_{c_2} can be smeared out to a mere crossover in layered materials so that the true (symmetry breaking) superconducting transition is shifted to the vortex melting transition.

Decoupling Transition An additional thermodynamic transition in layered systems is the loss of interlayer coherence due to strong thermal fluctuations of pancake vortices within individual layers [888], i.e. the 3D-2D decoupling transition, which is *thermally driven* and also can be investigated by the Lindemann approach [790]. It is often difficult to distinguish between the melting and the decoupling transition as they both can be experimentally related to the so-called irreversibility line in the B-T phase diagram, B_{irr}, which separates the B-T plane into an extended range $B_{irr} < B < B_{c_2}$ where the magnetization is entirely reversible from a magnetically irreversible state at $B < B_{irr}$. In addition, the temperature dependencies of the melting and decoupling transitions depend on the actual coupling between the superconducting layers: while for stronger coupling strengths the melting and decoupling transitions are theoretically expected to be separated, for weak interlayer coupling characterized by $\lambda < s\gamma$ the decoupling and melting lines merge at intermediate temperatures.

6.5.3 Experimental Findings for κ-(BEDT-TTF)$_2$X

The degree of flexibility of the vortex lines — which together with the degree of intrinsic and extrinsic disorder determines the various transitions in the vortex matter — depends on the nature and strength of interlayer coupling, temperature, and applied magnetic field. The novel aspects of the vortex phase diagram in quasi-2D systems have been investigated experimentally in the high-T_c cuprates where the high transition temperatures allow for an easy access to phenomena close to T_c. The present organic materials allow for an investigation of the vortex matter in the whole B-T plane, even when the field is aligned parallel to the planes. Below we will discuss some of the anomalous mixed-state properties such as the vortex-lattice melting and layer decoupling transition, and the irreversibility line found for fields aligned perpendicular to the planes as well as the lock-in transition and the possible realization of an anomalous high-field state for fields parallel to the planes.

Vortex Phase Diagram for $B \perp$ Planes

3D Flux-Line Lattice and Dimensional Crossover

First evidence for the existence of a 3D vortex-line lattice in organic superconductors at low fields and temperatures came from muon spin rotation (μSR) measurements on the κ-(ET)$_2$Cu(NCS)$_2$ salt [889, 890]. A recent review on μSR studies of molecular conductors and superconductors is given in [891]. Upon increasing the field to above some dimensional crossover field, $B_{cr} \equiv B_{\rm 2D}$, the vortex lattices of adjacent layers then become effectively decoupled: the dimensional crossover to a quasi-2D lattice occurs at $B_{cr,45^\circ} \sim 7\,$mT (magnetic field applied at an 45° angle with respect to the superconducting planes). Theory predicts that the crossover field is related to the anisotropy parameter γ and the interlayer distance s by $B_{\rm 2D} = \phi_0/(s\gamma)^2$ [787], which for the κ-(ET)$_2$Cu(NCS)$_2$ salt results in $B_{\rm 2D} = 7 - 30\,$mT [889, 892]. In a more recent μSR study where the field has been applied perpendicular to the planes, the internal field distribution is found to closely match that expected for a 3D Abrikosov flux line lattice (FLL) at temperatures below 5 K and magnetic fields below 5 mT [890]. Figure 6.13 shows the low-field vortex phase diagram for κ-(ET)$_2$Cu(NCS)$_2$ derived from μSR studies. A crossover from a 3D FLL to decoupled layers occurs at $5 - 8\,$mT. The skewness parameter β describes the μSR line shape and takes on a finite value for a static 3D flux line lattice that is characteristic for the lattice geometry. Above $4 - 5\,$K, β falls to zero, reflecting a thermally induced breakup of electromagnetically coupled pancake stacks, i.e. a breakup of vortex lines and interlayer order between the pancakes. The value of β below about 4 mT

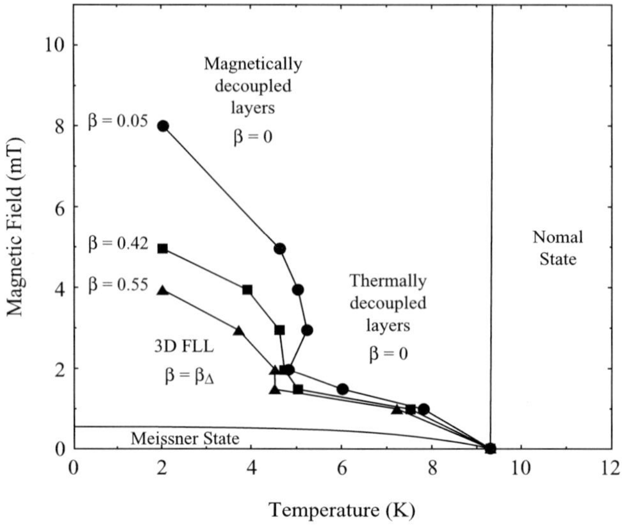

Fig. 6.13. Low-field vortex phase diagram for κ-(ET)$_2$Cu(NCS)$_2$ derived from μSR studies. β takes on a finite value for a static 3D FLL that is characteristic for the lattice geometry. A reduction of β reflects a reduction of pancake vortex correlations along the field direction. After [890].

is consistent with the value of 0.6 calculated for a triangular Abrikosov vortex structure. A plateau at $4 - 5\,\mathrm{K}$ around the square lattice value ($\beta = 0.47$) suggests the possibility of an intermediate vortex state before decoupling occurs, although it is difficult to distinguish between a possible change in the vortex lattice geometry from the onset of the dimensional crossover.

Using a decoration technique with Fe nanoparticle clusters, a slightly distorted triangular vortex lattice could be imaged for low magnetic fields up to 23 Oe applied perpendicular to the planes [893, 894].

Melting and Decoupling Transitions

A striking property common to the present quasi-2D superconductors is an extended vortex-liquid phase above B_{irr}. Here the magnetization behaves entirely reversible upon increasing and decreasing the magnetic field, indicating that in this range flux pinning is ineffective. The abrupt onset of magnetic hysteresis at $B \leq B_{irr}$ indicates a drastic increase in the pinning capability. The temperature dependence of the irreversibility field has been studied in detail for κ-(ET)$_2$Cu(NCS)$_2$ and κ-(ET)$_2$Cu[N(CN)$_2$]Br using a variety of techniques including, e.g., ac-susceptibility [860], dc-magnetization [789], magnetic torque [892, 670] or Josephson-plasma-resonance experiments [895]. In [892], $B_{irr}(T)$ has been found to follow a power-law behavior $B_{irr} \propto (1 - T/T_c)^{\alpha}$, with $\alpha = 1.9$ in the 3D regime below B_{2D}. Following an analysis similar to the one given for the high-T_c compound $Bi_2Sr_2CaCu_2O_8$ [896], this has been explained by 3D vortex lattice melting in a Lindemann approach, where $B_m \simeq \phi_0^5 c_L^4/[\pi^4(k_B T_c)^2\gamma^2\lambda_{\parallel}^4] \cdot (1 - T/T_c)^{\alpha}$ with $\alpha \leq 2$ is expected with c_L the Lindemann number in the order of ~ 0.1 [892]. In the same study, an anomalous second peak in magnetization curves is reported, similar to what has been observed for various high-T_c cuprates. The peak at $B_p \approx 150\,\mathrm{Oe}$ for κ-$(ET)_2Cu(NCS)_2$ indicates enhanced pinning as the field increases due to a dimensional crossover in the vortex system. The latter results in a redistribution of pancake vortices at more suitable pinning centers and has been associated with the crossover field $B_{2\mathrm{D}}$ [892, 897, 832]. This behavior is also reflected in the temperature dependence of the irreversibility line. In the temperature regime above $B_{2\mathrm{D}}$, quasi-2D vortex fluctuations become dominant and an exponential temperature dependence of the irreversibility line $B_{irr} \propto \exp(a/T)$ is observed, which, in turn, can be explained by vortex-lattice melting in the quasi-2D fluctuation regime, where $B_m \simeq \phi_0/r_j^2 \cdot \exp[\phi_0^2 c_L^2 s/(8\phi\lambda_{\|}^2)k_B T]$ is expected with the Josephson length $r_j = \gamma s$, see upper panel of Fig. 6.14. Further deviations from this behavior at higher fields and lower temperatures are believed to be related to quantum effects on the vortex-lattice-melting transition, see below. By studying the field dependence of the critical current density j_c, different pinning regimes have been found above and below $B_{2\mathrm{D}}$ in agreement with collective pinning theory in the presence of thermal fluctuations [892].

The lower panel of Fig. 6.14 shows a similar behavior also for the compound κ-$(ET)_2Cu[N(CN)_2]Br$ [860]. At $T/T_c \approx 0.75$, a crossover from an exponential behavior $B_{irr} \propto \exp(-AT/T_c)$ at low temperatures to a power-law behavior close to T_c, $B_{irr} \propto (1 - T/T_c)^{\alpha}$ with $1.3 < \alpha < 2$ is found. This is again interpreted as a crossover from 2D to 3D vortex pinning behavior, i.e. the breakdown of the proximity coupling between the superconducting layers. This is corroborated by measurements of the angular dependence of B_{irr} which shows a 2D behavior at low temperatures and a more 3D behavior close to T_c.

According to measurements of the interlayer Josephson-plasma resonance [895], a long-range quasi-2D order among vortices within the individual layers characterizes the state above $B_{2\mathrm{D}}$ and persists up to the irreversibility line. In this region of the B-T plane, the pancake vortices of adjacent layers become effectively decoupled leading to a pinned quasi-2D vortex solid in each layer with no correlations between the locations of vortices among the layers [898]. A somewhat different point of view is taken in [897, 899], where $B_{2\mathrm{D}}$ marks the crossover from a 3D flux-line lattice below $B_{2\mathrm{D}}$ to a state with less strong interlayer coupling on a long-range scale above, where a coupling between the layers is, to some extent, still present, see text to Figs. 6.16 and 6.17.

Besides the studies of the magnetization behavior, the above described issues of the mixed state of quasi-2D organic superconductors have been addressed also by employing electrical-transport measurements, see Fig. 6.15.[14]

[14] Flux-flow effects cause a substantial broadening of the superconducting transition in resistivity measurements, see, e.g., [776] for an early study on κ-$(ET)_2Cu[N(CN)_2]Br$ (Sect. 6.3.1, Fig. 6.2). In particular, a comparison with magnetization data show that resistivity measurements are ineffective for determining the onset of superconductivity [776].

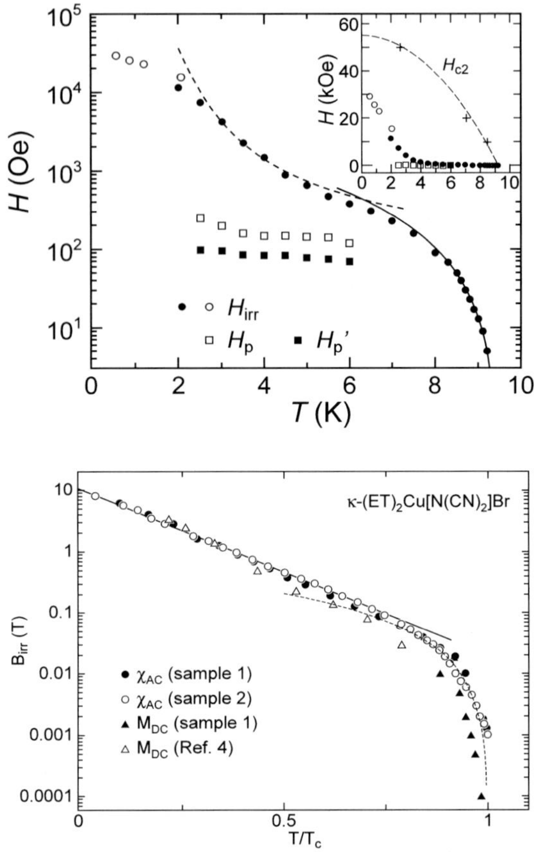

Fig. 6.14. Irreversibility field on a logarithmic scale as a function of temperature for κ-(ET)$_2$Cu(NCS)$_2$ (linear temperature scale, top) and κ-(ET)$_2$Cu[N(CN)$_2$]Br (reduced temperature scale, bottom), after [892] and [860], respectively. The vortex lattice is melted in the field and temperature region above the irreversibility lines in each case (see text). In both cases the data can be described by an exponential dependence $B_{irr} \propto \exp(a/T)$ in the high-field/low-temperature (quasi-2D) regime and a power law $B_{irr} \propto (1-T/T_c)^{\alpha}$ in the low-field/high-temperature (3D) regime marking a dimensional crossover in the vortex system. For κ-(ET)$_2$Cu(NCS)$_2$ (upper panel) the second peak in magnetization B_p (or the peak in the magnetic field derivative B'_p) mark the almost temperature-independent crossover field. The inset also shows the upper critical field curve.

In [795], it is argued that the line of approximately vanishing resistivity at different magnetic fields can be identified as the vortex-melting line. In the thermally activated flux-flow picture, the temperature- and field-dependent activation energy can then be deduced from the in-plane resistivity in the region where the vortex lattice is melted [901]. It is worth noting that the so-defined melting line deviates from the irreversibility line determined from magnetization measurements and — in contrast to these results — can be described by a 3D Lindemann-criterion scenario in the whole field region up to 3 T [900]. The vortex-liquid phase is described as a rather viscous liquid with activation barriers arising from disentanglement of vortex lines [902]. However, a 3D to 2D crossover behavior has been suggested based on the change of the temperature dependence of the characteristic pinning energy (activation energy) U_0: at low fields, the authors find $U_0(H) \propto 1/\sqrt{H}$ being a signature of the 3D nature of vortices [901, 900, 902]. At a crossover field $B_{3D \rightarrow 2D}$ in the order of 1 T, a logarithmic decrease of U_0 associated with a two-dimensional system of vortices is observed. This 3D-2D crossover (see Fig. 6.15) occurs when the intraplane interaction between pancake vortices dominates the inter-plane interaction.[15] In addition, a second type of crossover occurs when the Josephson coupling between the planes becomes dominated by quasiparticle tunneling between the superconducting planes: upon increasing temperature, the motion of pancake vortices becomes so intense that the Cooper pair tunneling between the layers is strongly suppressed. This has been argued to result in the distinct peak structure which

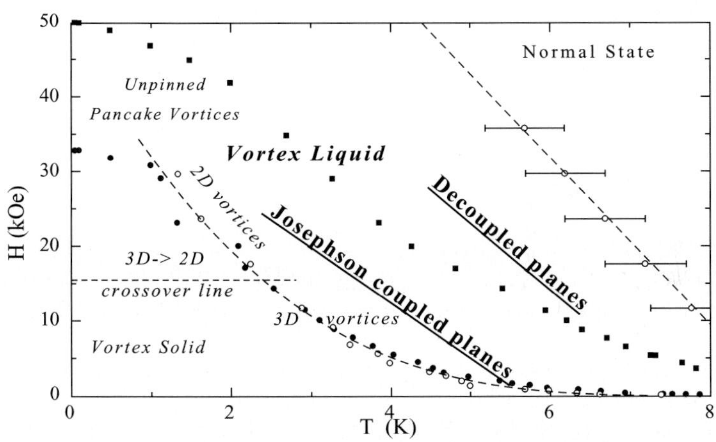

Fig. 6.15. *B-T* phase diagram of κ-$(ET)_2Cu(NCS)_2$ as determined from electrical transport measurements (see text), after [900]. Open and closed circles correspond to the melting line, the dashed line is a fit using a 3D Lindemann criterion image. Note that in contrast to Fig. 6.14 the description is valid in the whole field range up to about 30 kOe. Deviations at low temperatures are attributed to quantum effects, cf. Fig 6.14. The underlined text corresponds to electrical properties, the other text corresponds to the nature of vortices and their pinning properties.

[15] In κ-$(ET)_2Cu[N(CN)_2]Br$ also, at about 1 T a crossover in the shape of magnetoresistance accompanied by an abrupt change in slope of $B_{c2}^{\perp}$ has been observed [776]. Also from NMR experiments on κ-$(ET)_2Cu[N(CN)_2]Br$ and κ-$(ET)_2Cu(NCS)_2$ a 3D-2D crossover at similar fields has been inferred [903, 904].

is observed in the out-of-plane resistivity, see e.g. [905]. The peak position has been suggested to correspond to another "decoupling line" in the vortex-liquid region of the B-T phase diagram (black squares in Fig. 6.15).

To lift the apparent controversy in terms and concepts, simultaneous magnetization and transport measurements on the same sample of κ-(ET)$_2$-Cu(NCS)$_2$ are highly desirable. Such an experiment has been performed for the compound κ-(ET)$_2$Cu[N(CN)$_2$]Br [906]. In this study, high-resolution resistivity measurements reveal sharp drops in the in-plane resistivity with decreasing temperature at certain fields *before* the resistivity becomes unmeasurably small. Remarkably, these anomalies occur at the same temperatures and fields $T_m(B)$ as steps in the magnetization indicating the existence of a first-order phase transition due to either vortex-lattice melting or simultaneous melting and decoupling [906]. Here, the step of $M(T)$ occurs in the reversible region, i.e. above the irreversibility line. Again, comparing the $T_m(B)$ determined experimentally from resistivity and magnetization measurements, it is difficult to distinguish between different theoretical models suggesting the existence of either the 3D melting or the simultaneous melting and decoupling.

The first indications for a first-order phase transition in organic compounds associated with a melting and/or a decoupling of the quasi-2D vor-

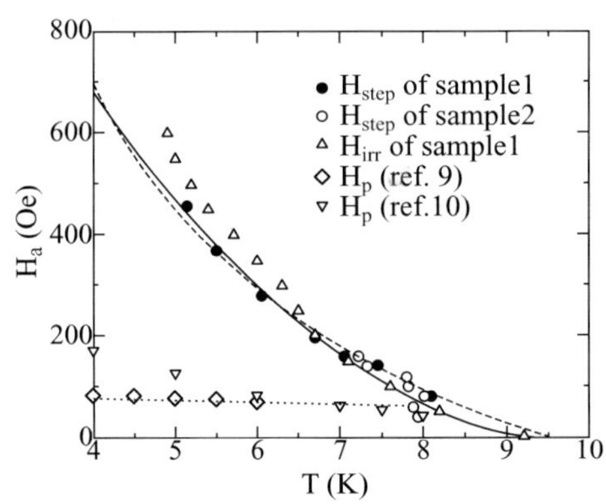

Fig. 6.16. First-order transition line for κ-(ET)$_2$Cu(NCS)$_2$(open and closed circles) determined by a step in the local induction. Solid and dashed curves represent theoretical predictions for the melting and decoupling transitions, respectively, see text. Open diamonds indicate the anomalous second peak in magnetization associated with the dimensional-crossover field B_{2D}, see Fig. 6.14. After [897].

tex lattice near the irreversibility line — similar to the findings in some high-T_c cuprates, see, e.g., [907, 908, 909, 910] — have been reported for κ-(ET)$_2$Cu[N(CN)$_2$]Br [911, 899] and κ-(ET)$_2$Cu(NCS)$_2$ [897, 912]. In magnetization studies using SQUID and micro Hall-probe magnetometry — the latter technique probing the local magnetic induction — step-like changes in the equilibrium and local magnetization, respectively, as a function of temperature at constant fields, have been observed. The results are interpreted as signatures of a first-order phase transition in the vortex matter occurring along with or close to the irreversibility line. Below the dimensional-crossover line B_{cr}, a first-order melting of the 3D FLL occurs, similar to that in the low-field state of the highly 2D material $Bi_2Sr_2CaCu_2O_8$. Above B_{2D}, the vortex state becomes a 2D pancake state with a weaker interlayer-coupling state on a long-range scale. However, the coupling between layers must exist to some extent. The transition from this state to the liquid phase may occur as the first-order decoupling transition. According to [897], see Fig. 6.16, the first-order transition line for κ-$(ET)_2Cu(NCS)_2$ can be fitted equally well by a melting or a decoupling transition. From the Clausius-Clapeyron relation, an entropy change of about $0.15\,k_B$ per vortex per layer has been deduced.

For κ-$(ET)_2Cu[N(CN)_2]Br$, a melting line which practically coincides with the irreversibility line has been reported in [911], based on step-like magne-

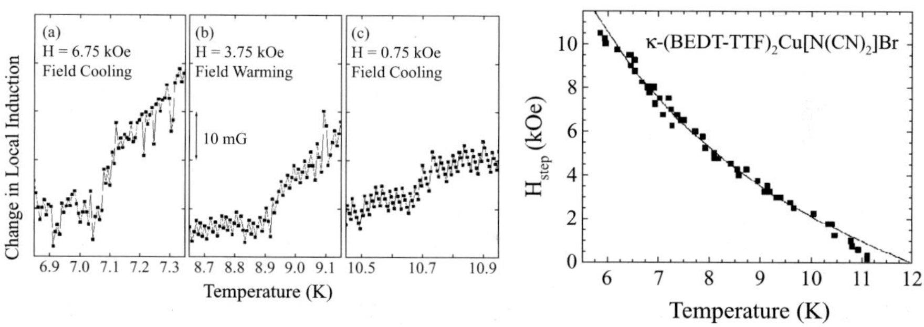

Fig. 6.17. Left: Changes in local magnetic induction ΔB measured by micro-Hall probes for κ-$(ET)_2Cu[N(CN)_2]Br$. The field is applied perpendicular to the planes and a linear background has been subtracted from the data. Right: First-order phase transition line determined by ΔB. The solid line is a fit to the prediction for a decoupling scenario. The fit for a melting transition works equally well. After [899].

tization changes using a SQUID magnetometer. For large cooling rates, the magnetization steps vanish at the higher temperatures indicating a different degree of quenched disorder and hence changes in the pinning behavior. In a micro Hall-probe experiment, corresponding discontinuities ΔB have been observed in the local magnetic induction [899], see Fig. 6.17. In this study, as for the resistivity and magnetization experiments in [906], the step in magnetization occurs in the reversible region above $B_{irr}(T)$. From the step height ΔB, the entropy change per vortex per layer $\Delta S \sim 0.1\,k_B$ is estimated. Again, the data are equally well described by a melting scenario (anisotropic 3D case) $B_m = B_{m0} \cdot (1 - T/T_c)^\alpha$ with $\alpha \leq 2$ and a decoupling scenario $B_d = B_{d0} \cdot (T_c - T)/T$, where $B_{m0} = \phi_0^5 c_L^4 / [\pi^4 (k_B T_c)^2 \gamma^2 \lambda_\parallel^4]$ and $B_{d0} = \alpha_d \phi_0^3 / [16\pi^2 \lambda_\perp^2 (k_B T_c) s]$ with c_L the Lindemann number in the order of ~ 0.1, $\alpha_d = (\pi e)^{-1} \simeq 0.12$ a constant, and s the interlayer spacing. Remarkably this has been confirmed by complementary studies of the

Josephson-plasma resonance which measures the interlayer phase coherence. The latter results reveal independent and direct evidence for the loss of phase coherence between adjacent layers on crossing the first-order transition line with increasing field, indicating that the melting and the decoupling occur simultaneously [899].

Vortex Dynamics Probed by NMR

NMR is a powerful method to study the intrinsic kinetics of the vortex matter which is due to the NMR time scale of 10^{-8} s not affected by pinning [903]. The weak coupling of the ^{1}H-nuclei, which are part of the ethylene end-groups with the π-electronic system, allows the protons to be sensitive probes to other sources of fluctuating magnetic fields such as fluxoids [913]. A pronounced peak in the ^{1}H-NMR spin lattice relaxation rate, T_1^{-1}, has been observed below T_c in κ-(ET)$_2$Cu(NCS)$_2$ [799, 914] and κ-(ET)$_2$Cu[N(CN)$_2$]Br [913, 903, 915]. From the magnitude and position of the peak, it soon became clear that it was not the usual coherence peak observed in s-wave superconductors but originated from the motion of the fluxoid system, as first concluded by S. De Soto et al. from the large anisotropy of the effect [913], see also Fig. 6.21.

Figure 6.18 shows the peak in the ^{1}H-NMR relaxation rate measured in κ-(ET)$_2$Cu[N(CN)$_2$]Br at different magnetic fields [903]. The peak position lies very close to the irreversibility transition line related to vortex-lattice melting. The theoretical analysis by H. Mayaffre et al. revealed that only the 2D regime can provide the observed magnitude of the relaxation peak. In this model, the enhanced relaxation near T_m is explained by Brownian motion of plastic deformations, interstitials, or vacancies which disappear at the melting transition [903].

For κ-(ET)$_2$Cu(NCS)$_2$, K. Kanoda et al. have been able to separate the contributions to the relaxation rate originating from quasiparticles and vortex

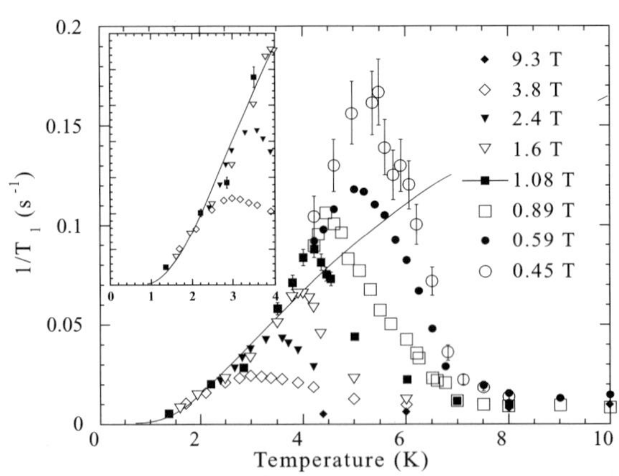

Fig. 6.18. ^{1}H-NMR relaxation rate for different magnetic fields applied perpendicular to the conducting layers in κ-(ET)$_2$Cu[N(CN)$_2$]Br. After [903].

dynamics by comparative NMR studies on different sites, i.e. the central double-bonded carbon (^{13}C), ethylene carbon (^{13}C), and ethylene hydrogen (^{1}H) sites [915], which have different hyperfine couplings to the conduction holes. A huge enhancement of $(T_1T)^{-1}$ with peak formation at $T = 2\,\mathrm{K}$ and $B_e = 1.5\,\mathrm{T}$ is found at the ethylene ^{13}C- and ^{1}H-sites while the central ^{13}C-site shows only a small peak. From these data a qualitative change in vortex dynamics such as melting is concluded, not a simple crossover behavior .

By analyzing anomalous NMR line shapes, a metastable solid vortex state, created by the radiofrequency pulses used in a NMR experiment, has been identified [904]. This state corresponds to a distorted flux-line lattice. For the temperature range 1.50 – 4.2 K the effects of the radiofrequency pulses are strongest around 0.8 T. As a possible explanation, this field has been attributed to a 3D-2D crossover field (see p. 179) or a transition from a weakly disordered vortex solid to a glassy one in which the susceptibility of the vortices to the pulses could be maximum at the transition [904].

Quantum Effects

The irreversibility line in the *B*-*T* phase diagram of κ-(ET)$_2$Cu(NCS)$_2$, deduced from torque-magnetometry measurements in fields perpendicular to the planes [670], has been found to lie well below the upper critical field at the lowest temperatures. For $B_{irr} \leq B \leq B_{c_2}$, a vortex-liquid state exists even at $T \simeq 0$. Thus, quantum fluctuations of the vortices, as opposed to thermally driven motions, are responsible for the vortex-liquid state in this region of the phase diagram [670]. A quantitative analysis in comparison with various theoretical models for quantum melting using the material parameters for κ-(ET)$_2$Cu(NCS)$_2$ yielded a value of about 4 T for the quantum melting transition field, in excellent agreement with the experimental obser-

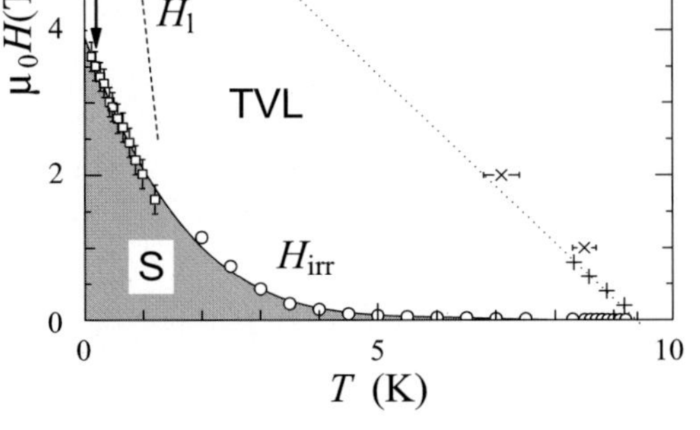

Fig. 6.19. Vortex phase diagram showing the quantum (QVL) and thermal vortex liquid (TVL) states in κ-(ET)$_2$Cu(NCS)$_2$ as suggested in [672, 916, 917]. The broken line B_l separates the vortex liquid in the TVL state at higher T and a non-ohmic low-resistance state at lower T. The low-T and low-H vortex solid region (S) is characterized by zero resistance and irreversible magnetization. After [916].

vation. The crossover from quantum to thermal fluctuations manifests itself in the temperature dependence of $B_{irr}(T)$. Below $T \sim 1\,\mathrm{K}$, where quantum fluctuations are predominant, $B_{irr}(T)$ varies linearly with temperature [670], whereas in the thermal fluctuation regime an exponential behavior has been observed above the dimensional-crossover field $B_{2\mathrm{D}}$ in contrast to a power-law behavior in the 3D vortex-line lattice region below $B_{2\mathrm{D}}$ [789, 892], see above. In addition, transport measurements revealed a transition line separating the thermal vortex-liquid state (TVL) at higher T from the quantum-vortex liquid (QVL) at lower T, see Fig. 6.19 and also Sect. 5.2.4. The QVL region below about 1 K is characterized by a non-linear low-resistance state: upon decreasing the temperature from the TVL region, a characteristic drop in the in-plane resistivity occurs. This novel low-temperature "quantum vortex slush" phase (in analogy to the vortex-slush phase between vortex-liquid and glass phase of high-T_c cuprates) is suggested to be characterized by a short-range order of vortices and the transition is argued to be a hidden first-order melting transition obscured by strong quantum fluctuations [672].

Similarly, changes in the stiffness of the vortex system have been inferred from magnetic-torque measurements in the very low-temperature regime $T_c/10^3$ [898]. The results are interpreted as a (possibly first-order) quantum melting transition between a quasi-2D vortex-lattice phase and a quantum-liquid phase.

Figure 6.20 shows in a semi-logarithmic plot the B-T phase diagram for κ-$(ET)_2Cu(NCS)_2$, as proposed in [898] summarizing many of the results described above. From Josephson-plasma resonance measurements [895], it has been inferred that in spite of the layer decoupling, a long-range quasi-2D order among vortices within the individual layers persists over most of the region in the phase diagram between $B_{2\mathrm{D}}$ and B_{irr} (shaded region). Note that here the crossover field $B_{2\mathrm{D}}$ separating the 3D flux-line lattice from the quasi-2D vortex solid is temperature dependent in contrast to the results shown in Figs. 6.14 and 6.16. Right below B_{irr} a melting transition of the quasi-2D vortex lattice occurs. This transition is driven by quantum fluctuations as temperature approaches zero and is thermally driven at higher temperatures. A slightly different scenario suggests that in the regime of thermal fluctuations close to T_c, a decoupling of the layers takes place, simultaneously with the melting transition, i.e. a transition from a coupled line lattice to a decoupled pancake liquid, see Fig. 6.17 and discussion above. Other researchers in turn suggest that the solid state below the melting line is still a (disordered) 3D solid formed of entangled flux lines which is ordered only at very low magnetic fields below the crossover line $B_{2\mathrm{D}}$ [902]. This crossover line is viewed as a transition from a weak-field Bragg glass phase to a disordered state at larger fields. In this picture, the real 3D-2D crossover line is at larger fields, see Fig. 6.15.

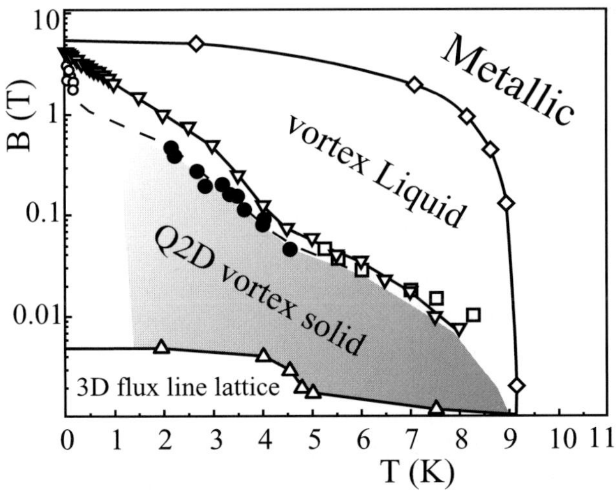

Fig. 6.20. Mixed-state B-T phase diagram of κ-$(ET)_2Cu(NCS)_2$ for $B \perp$ planes in a semi-logarithmic representation. Open diamonds correspond to the upper critical field [789]. B_{irr} as determined from magnetic-torque measurements [670] is indicated by down triangles, the dimensional-crossover field B_{2D} [889, 890] by up triangles, see also Fig. 6.13. Open squares denote a first-order decoupling and/or melting transition [897], see also Fig. 6.16. The shaded region represents long-range quasi-2D order among vortices within individual layers. Closed and open circles refer to thermal melting or depinning [895] and quantum melting transition, respectively, from a quasi-2D vortex lattice to a liquid phase. After [898].

B || Planes

An interesting situation arises when the magnetic field is aligned parallel to the planes enabling the vortices to slip in between the superconducting layers where the order parameter is small. For small tilt angles of the field with respect to the exact alignment, the vortex lattice can gain energy by remaining in this parallel "lock-in" configuration implying a "transverse Meissner effect", with the component of B perpendicular to the planes being zero [365]. Such a lock-in transition was predicted by M. Tachiki and S. Takahashi for the high-T_c cuprates [918] and later observed in resistivity measurements on $YBa_2Cu_3O_{7-\delta}$ by W. Kwok et al. [919]. This new state remains stable until the perpendicular field component, $B_\perp$, exceeds a threshold field at which the energy required to expel $B_\perp$ exceeds that associated with the creation of normal cores in the layers [920, 921]. One expects that the applied field first penetrates the plane when its perpendicular component exceeds a threshold value which is of the order of $B_{c1}^{\parallel}$. Upon increasing field, vortices then develop a kinked structure consisting of pancakes linked by Josephson vortex segments. Evidence for coreless Josephson vortices parallel to the superconducting layers and a lock-in state have been reported, e.g., from ac-susceptibility

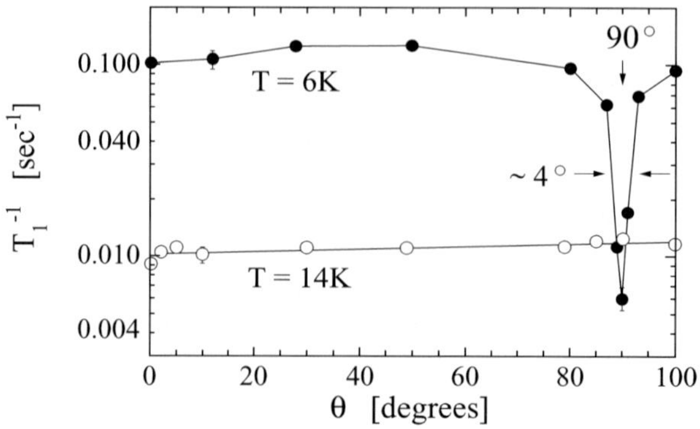

Fig. 6.21. Orientation dependence of T_1^{-1} for κ-$(ET)_2Cu[N(CN)_2]Br$ in the superconducting and normal state. After [913].

measurements for κ-$(ET)_2Cu(NCS)_2$ [802, 870], from torque magnetometry on κ-$(ET)_2Cu(NCS)_2$ and κ-$(ET)_2Cu[N(CN)_2]Br$ [804, 922], and rf susceptibility on κ-$(ET)_2Cu[N(CN)_2]Br$ [923], see also [103]. Figure 6.21 shows the angular dependence of the ^{1}H-NMR relaxation rate T_1^{-1} at a temperature below T_c nearly corresponding to the peak in T_1^{-1} (see Fig. 6.18) and in the normal state where fluxoids are not present [913]. The decrease of T_1^{-1} by more than an order of magnitude in a narrow window of only $\sim 3°$ from the orientation corresponding to the field being parallel to the superconducting layers is a signature of the lock-in transition with the fluxoids effectively locked when $\Theta = 90°$. The necessary condition $\xi_\perp < s$ is fulfilled for the present systems. In the lock-in state the fluxoids are pinned by the crystal structure itself (*intrinsic pinning*), as opposed to the normal pinning mechanisms.

For $\gamma s > \lambda_\|$ it is predicted [921] that the lock-in threshold field B_J is equal to $B_{c1}^\perp$, and marks the transition to a "combined" lattice of independent perpendicular Abrikosov vortices and parallel Josephson vortices. In [802] the threshold field was found to follow the expected dependence $B_J \approx B \sin\Theta$ and has the same temperature dependence as $B_{c1}^\perp$ but is about a factor of two smaller. Accurate measurements of $B_{c1}^\perp$, however, are difficult to make due to the complications of demagnetization effects, field penetration at corners, and the partial exclusion of flux above $B_{c1}^\perp$ due to pinning.

The lock-in transition has also been discussed for λ-$(BETS)_2Fe_xGa_{1-x}Cl_4$ (Sect. 7.2.6), see e.g. [924].

Further interest in the behavior of the present quasi-2D organic superconductors in fields precisely aligned parallel to the planes arose from the proposal that these systems are possible candidates for a Fulde-Ferrell-Larkin-Ovchinnikov (FFLO) state [926]. Under suitable conditions, a spin-singlet superconductor can reduce the pair-breaking effect of a magnetic field by adopting a spatially modulated order-parameter along the field direction [735, 736]. The wavelength of the modulation is of the order of the coherence length which results in a periodic array of nodal planes perpendicular to the vortices [735, 736]. In the case of an anisotropic superconductor, calculations show that the FFLO state might lead to an enhancement of the upper critical field to between 1.5 and 2.5 times the Pauli paramagnetic limit [926, 927].

By studying the magnetic behavior and the resistivity of κ-(ET)$_2$Cu(NCS)$_2$ in high magnetic fields employing a tuned circuit differential susceptometer, changes in the rigidity of the vortex arrangement at certain fields B_L *within* the superconducting state have been found for fields precisely aligned parallel to the planes [866, 878]. These effects have been interpreted as the manifestation of a phase transition from the superconducting mixed state into a FFLO state.[16] (A possible FFLO state in κ-(ET)$_2$Cu[N(CN)$_2$]Cl under pres-

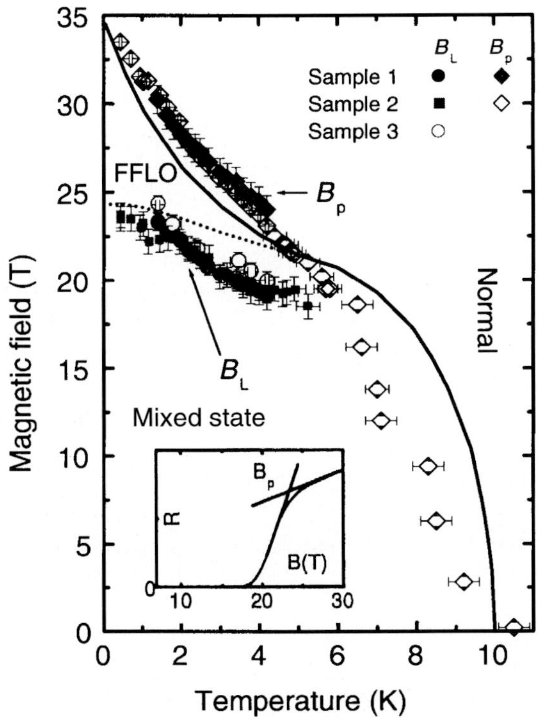

Fig. 6.22. Magnetic field-temperature phase diagram of κ-(ET)$_2$Cu(NCS)$_2$ for B aligned parallel to the planes compared with the theoretical FFLO phase diagram discussed in [925]. B_P as defined in the inset serves as a rough estimate for B_{c2}. B_L indicates changes in the rigidity of the vortex arrangement within the superconducting state. After [866, 277].

[16] According to [735, 736], see also [928, 929], the stabilization of a FFLO state requires: (i) a large electronic mean free path $\ell > \xi_0$, i.e. clean limit, (ii) a Pauli limiting that predominates the orbital pair-breaking effect, (iii) a Zeeman energy that overcompensates the loss of superconducting condensation energy and (iv) a short coherence length (or a large GL parameter $\kappa = \lambda/\xi$). As already discussed above, the criteria (i) and (iv) are met for the present κ-(ET)$_2$X salts. Condition (ii) is fulfilled since $B_P = \frac{1}{2}B_c(\pi\chi_{\mathrm{Spin}})^{-1/2} \approx 17\,\mathrm{T}$ is considerably smaller than the orbital critical field as derived from (6.25) taking the values from Table 6.3. To check for condition (iii) one has to compare the Zeeman energy with the condensation energy. Using $\chi_{\mathrm{Spin}} = 4 \cdot 10^{-4}$ emu/mol for κ-(ET)$_2$Cu(NCS)$_2$ [224] and the upper critical field of 35 T yields a Zeeman-energy density of $E_Z =$

sure is discussed in [798].) Figure 6.22 shows the temperature dependence of B_L and B_P, the latter serves as a rough estimate of B_{c_2} [866, 878], see inset of Fig. 6.22. Comparing the results with theoretical calculations, derived for a generic quasi-2D metal [925] (solid and dotted lines in Fig. 6.22) using the parameter $B_{c_2}(0) = 35\,\mathrm{T}$, yielded a fairly good agreement with the predictions of the FFLO model. In particular, the temperature T^* below which the new state is stabilized was found to meet the theoretical predictions of $T^* = 0.56\,T_c$. However, recent magnetic torque measurements failed to detect any indication for such a transition [898]. Also, in studies of the critical field under pressure using the tunnel diode oscillator technique, no evidence for a FFLO phase has been seen and the data suggest that κ-$(ET)_2Cu(NCS)_2$ is always Pauli limited [930].

6.6 Superconducting Order-Parameter

Conventional vs Unconventional Pair States

More insight into the nature of the superconducting state can be gained by considering the symmetry breaking at the phase transition [767, 931]. At the normal-to-superconducting phase transition — as for any other second-order phase transition — the occurrence of long-range order is accompanied by a lowering in symmetry. Since the symmetry breaking across the transition is continuous, the symmetries above and below the transition temperature are related to each other. In the language of group theory this means that the symmetry group, describing the superconducting state, must be a subgroup of the full symmetry group G describing the normal state at $T > T_c$:

$$G = X \times R \times U(1) \times T, \tag{6.38}$$

where X denotes the translational symmetry of the crystal lattice, R the spin-rotation symmetry, $U(1)$ the one-dimensional global gauge symmetry, and T the time-reversal-symmetry operations.

For *conventional* superconductors, the only symmetry which is broken below T_c is the global gauge symmetry $U(1)$ giving rise to all the well-known manifestations of the macroscopic phase-coherent quantum phenomena such as the Meissner effect, the flux quantization or the Josephson effects.

On the other hand, a superconductor is called an *unconventional* superconductor, if one or more symmetries, in addition to $U(1)$, are spontaneously broken below T_c. The symmetry of the resulting superconducting pair state is reflected in the symmetry of the order parameter or, equivalently, of the gap

$\frac{1}{2}\chi_{\mathrm{Spin}}B_{c2}^2 = 5\,\mathrm{mJ/cm^3}$ $(= 5 \cdot 10^4\,\mathrm{erg/cm^3})$, which exceeds the condensation-energy density of $E_c = B_c^2/(8\pi) = \frac{1}{4} \cdot \gamma/V_{\mathrm{mol}} \cdot T_c^2 = 1\,\mathrm{mJ/cm^3}$ $(= 10^4\,\mathrm{erg/cm^3})$ calculated by employing the experimentally determined Sommerfeld coefficient $\gamma = (23 \pm 1)\,\mathrm{mJ/mol\,K^2}$ for κ-$(ET)_2Cu(NCS)_2$ [456, 457].

function. Although the knowledge of the symmetry is of crucial importance in discriminating between *conventional* and *unconventional* pairing schemes, it provides only indirect hints to the kind of interaction which is responsible for the broken symmetry.

Symmetry Classifications of the Pair State

The Pauli principle requires the pair wave function to be antisymmetric under particle exchange. As a consequence, spin singlet pairs with a total spin $S = 0$ can only be in a pair state with an angular momentum $L = 0, 2, 4, \ldots$. This contrasts with spin triplet ($S = 1$) pair states, where the angular momentum must be odd-numbered, i.e. $L = 1, 3, \ldots$. Provided the crystal structure possesses a center of inversion symmetry, the superconductors can be classified by the parity of the pairing state: while the triplet state has an order parameter (gap function) with odd parity, singlet superconductors are characterized by an even-parity order parameter, i.e. $\Delta(\boldsymbol{k}) = \Delta(-\boldsymbol{k})$. For superconductors where the spin-orbit interaction is small, such as for the present molecular materials, singlet and triplet states are well-defined and can be used to classify the pairing state.

Further classifications of the pairing state are possible by including the symmetry of the underlying lattice. According to the Landau theory of second-order phase transitions, the corresponding order parameter must transform according to one of the irreducible representations of the symmetry group of the high-temperature phase [767]. Thus, a decomposition of the irreducible representations of the normal-state symmetry allows the categorization of possible forms of the order parameter, see e.g. [932, 933, 934] and [935] for a recent review on the pairing symmetry in the cuprate superconductors. As a result, the gap function can be expressed as a linear combination of the basis functions χ^j_μ of the irreducible representation Γ^j

$$\Delta(\boldsymbol{k}) = \sum_{\mu=1}^{l_j} \eta_\mu \chi^j_\mu(\boldsymbol{k}), \tag{6.39}$$

where l_j is the dimensionality of Γ^j, and η_μ is a complex number which is invariant under all symmetry operations of G in (6.38). For the basis functions of the irreducible representations for various symmetry groups and various crystal structures, see [932, 933, 934]. For the present molecular superconductors, the crystal structures usually have rather low symmetry. In a first approximation, the systems may be considered as layered materials with the molecules arranged in a square-lattice configuration with point-group symmetry C_{4v}, consisting of the following symmetry operations: mirror reflections with respect to the principal crystallographic axes and the diagonals as well as a 4-fold and 2-fold rotation about the tetragonal axis. In Fig. 6.23 we show a schematic $\boldsymbol{k}$-space representation for the possible spin-singlet states with point-group symmetry C_{4v}.

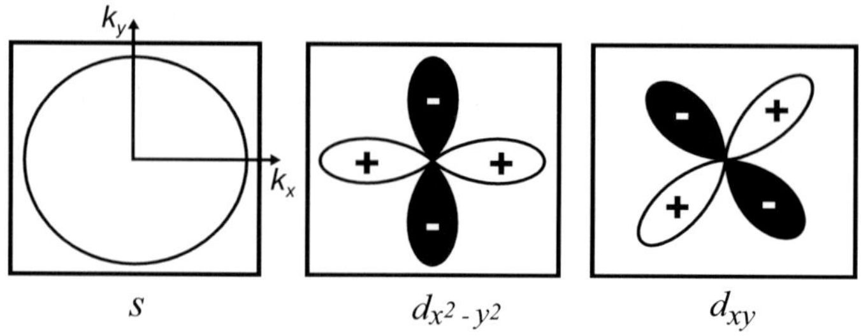

Fig. 6.23. $\boldsymbol{k}$-space representation of allowed symmetry basis functions for the C_{4v} point-group symmetry. The corresponding order parameter basis functions (group-theoretical notations) are from left to right: *constant* (A_{1g}), $x^2 - y^2$ (B_{1g}), and xy (B_{2g}).

Simple superconductors like Al or Zn are well described by the above-mentioned BCS model [364] which envisages pairing between quasiparticles with total spin $S = 0$ (spin singlet) and relative angular momentum $L = 0$ (s-wave). At $T = 0$, all electron pairs are in the same quantum mechanical state described by a single macroscopic wave function. In the simplest approximation, the gap is assumed to be isotropic in $\boldsymbol{k}$-space. Its magnitude at zero temperature, $\Delta(0)$, is related to the condensation energy, E_{cond}, i.e. the difference in the free energy between the normal and the superconducting state, via $E_{\text{cond}} = \frac{1}{2} \cdot N(E_F) \cdot \Delta(0)^2$, where $N(E_F)$ is the density of states at the Fermi level. For these conventional superconductors the gap function in $\boldsymbol{k}$-space, $\Delta(\boldsymbol{k})$, usually assumes a non-zero value over the entire Fermi surface. As a consequence, all properties of the superconductor, which are related to the excitation of electronic quasiparticles, i.e. excitations across the finite energy gap, exhibit an exponentially-weak temperature dependence at low temperatures. As an example, we mention the electronic quasiparticle contribution to the specific heat in the superconducting state C_{es}, which for a conventional superconductor at $T \ll T_c$ varies as

$$C_{es}(T) \propto \exp\left(-\frac{\Delta}{k_B T}\right) . \qquad (6.40)$$

A possible weak $\boldsymbol{k}$-dependence of the gap in a conventional superconductor may have its origin in anisotropies of the Fermi surface, i.e. a $\boldsymbol{k}$-dependent strength of the pairing interaction, see e.g. [936].

In contrast to the conventional superconductors where the quasiparticle density of states $N(E)$ is zero between $E_F - \Delta$ and $E_F + \Delta$, most of the unconventional pair states are characterized by nodes in the gap function at certain regions of the Fermi surface.[17] For the axial- and the polar states for example, the gap function is zero at points or along lines on the Fermi surface, respectively. This is accompanied by a density of states which vanishes in power-laws of $(E - E_F)$ as $E \to E_F$. In the presence of impurity scattering, however, the regions where the gap function vanishes can be widened significantly [934].

As a consequence of the gapless regions, quasiparticle excitations are possible even for $T \to 0$ and manifest themselves in power-law temperature dependencies in all the quantities which depend on the number of excited quasiparticles. Thus, for an unconventional superconductor a quasiparticle contribution to the specific heat of

$$C_{es}(T) \propto T^n \tag{6.41}$$

is expected, where the exponent n depends on the topology of the gap nodes.

A careful measurement of the temperature dependencies of physical properties which depend on the quasiparticle density of states such as specific heat, NMR relaxation rate or penetration depth should, in principle, enable the identification of an unconventional superconductor. However, as will be discussed below, the numerous experiments aiming at an identification of the pairing state for the present molecular superconductors give rather inconsistent results. Besides the above-mentioned differences in the density of states which manifest themselves only at temperatures far below T_c, unconventional superconductors are distinct in lacking the so-called coherence effects close to T_c. The latter are a consequence of the special Cooper pair configuration considered by BCS which involves quasiparticle states of opposite momenta and opposite spins. The best-known coherence effect is the so-called coherence peak in the temperature dependence of the NMR relaxation rate directly below T_c [938]. The observation of these coherence effects [938] — besides the isotope effect [939] — in the classical superconductors was considered to be a convincing confirmation for the validity of the BCS theory.

Experiments Probing the Gap Anisotropy in κ*-(ET)*$_2$*X*

For measuring the gap anisotropy for the present molecular superconductors, numerous studies employing a variety of different techniques have been performed. Most of these investigations have focused on the determination of the temperature dependence of quantities which provide information on the density of states such as the specific heat, the NMR relaxation rate or the magnetic penetration depth. More recent developments involve orientational-dependent studies aiming at a direct determination of the angular-dependence of the gap. These techniques include mm-wave transmission [940], STM spectroscopy [941] and thermal conductivity [942] studies.

[17] A well-known exception is the unconventional wave function of the superfluid B phase of liquid ^{3}He which has a non-zero gap function at the entire Fermi surface [937].

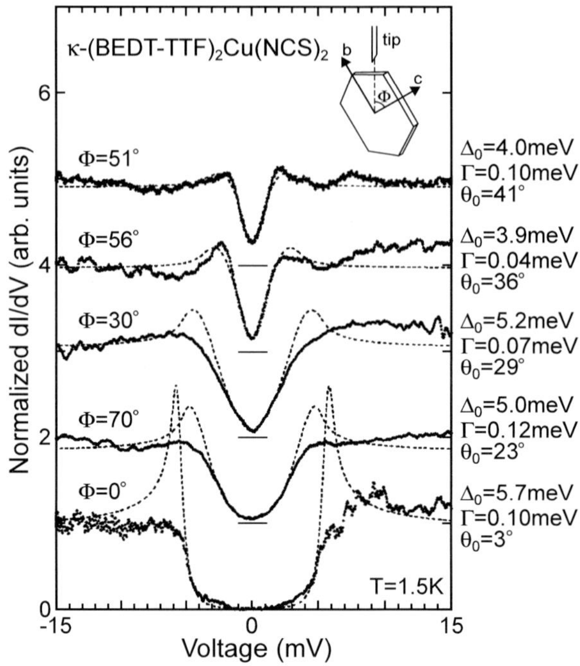

Fig. 6.24. dI/dV-V curves taken at 1.5 K on the lateral surfaces of κ-$(ET)_2Cu(NCS)_2$ single crystals. Data have been taken along various tunneling directions at different angles ϕ as defined in the inset. The dashed line represents the calculated curve based on the d-wave gap model with a $\boldsymbol{k}$-dependent tunneling. After [941].

A ***mm-Wave Magneto-Optical Technique*** was used to determine the angular dependence of the high-frequency conductivity of κ-$(ET)_2Cu(NCS)_2$ [940, 943]. The results have been interpreted to support an anisotropic gap with "X shape", i.e. with nodes along the b- and c-direction [940], consistent with a $d_{x^2-y^2}$-wave symmetry of the order parameter as theoretically suggested by J. Schmalian [944].[18] However, these results have been critically commented upon by other groups [569, 945].

The superconducting gap structure of the same compound has been investigated using ***STM Spectroscopy*** by T. Arai et al. [941], see also [946] for a recent review. The tunneling curves observed on the bc-plane (parallel to the conducting layers) in the low-energy region could be well fitted by a d-wave gap model. The corresponding $2\Delta_0/(k_BT_c)$ ratio was found to be 6.7 which is smaller than a previously reported value of 9 [947] but substantially larger than the BCS value of 3.53. In addition, the in-plane gap anisotropy was investigated, see Fig. 6.24. The dI/dV-V curves observed on the lateral surfaces were found to be also consistent with a d-wave gap. For this configuration a

[18] In the literature different conventions for the definition of the magnetic Brillouin zone are used. The assignment of the nodal directions in accordance with $d_{x^2-y^2}$ is apparently mistaken in [940]. From STM studies a $d_{x^2-y^2}$-wave gap with line nodes along the directions $\pi/4$ from the k_b- and k_c-axes has been inferred. However, an identical nodal structure found in thermal conductivity (see below) has been assigned to a d_{xy}-wave gap.

very large $2\Delta_0/k_B T_c$ ratio of $8.7 \sim 12.9$ has been obtained. The analysis of the angular dependence revealed that the direction of the line nodes of the gap is $\pi/4$ from the k_b- and k_c-axes, i.e. the gap has $d_{x^2-y^2}$ symmetry [941]. It has been noted that these orientations of the gap nodes are at variance with those inferred from the above mm-wave-transmission experiments (see above).

The ***Thermal Conductivity*** has been used as another directional-dependent probe. When compared to STM measurements, for example, this quantity has the advantage that it is free of surface effects. The implications of the symmetry of gap zeroes on the thermal conductivity in the vortex state have been theoretically investigated by various authors, see e.g. the work by H. Won and K. Maki [948]. Measurements have been performed for the κ-$(ET)_2Cu(NCS)_2$ salt in a magnetic field rotating within the 2D superconducting plane. Figure 6.25 shows the angular variation of the thermal

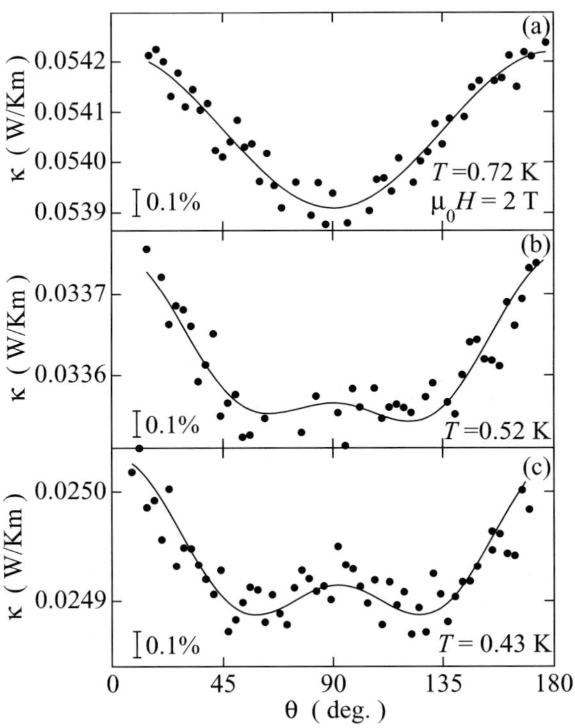

Fig. 6.25. Angular variation of $\kappa(B, \Theta)$ at 2 T for different temperatures where Θ denotes the angle between a rotating magnetic field in respect to the heat current flowing along the b-axis of the κ-$(ET)_2Cu(NCS)_2$ crystal. The solid lines represent the results of a fit using the function $\kappa(B, \Theta) = C_0 + C_{2\Theta} \cos 2\Theta + C_{4\Theta} \cos 4\Theta$, where C_0, $C_{2\Theta}$ and $C_{4\Theta}$ are constants. After [942].

conductivity κ at a fixed field of $B = 2\,\mathrm{T}$. Θ denotes the angle between the heat current flowing along the crystallographic b direction and the magnetic field, i.e. $\Theta = 0°$ for $B \parallel b$. The salient result of this study is the occurrence of a $\kappa(\theta)$ contribution with a four-fold symmetry, $\kappa_{4\theta}$, at low temperatures $T \leq 0.52\,K$ that adds to a predominant term with two-fold symmetry. While the latter has been interpreted as being mainly phononic in origin, it is argued that the former is of a purely electronic nature and reflects the nodal gap structure [942]. Their analysis revealed that the gap zeroes are oriented along the directions rotated by 45° relative to the b- and c-axes. It has been pointed out in [942] that this nodal structure is inconsistent with the theories based on antiferromagnetic spin fluctuation where the nodes are expected to be along the b- and c-directions. Based on this observation, K. Izawa et al. proposed a d_{xy} symmetry (referring to the magnetic Brillouin zone, see inset of Fig. 1 in [942]) which has been theoretically suggested for a charge-fluctuation scenario [949, 950].

NMR Measurements

A more indirect information on the symmetry of the superconducting order parameter is provided by temperature-dependent measurements of quantities which depend on the quasiparticle excitation spectrum. In this context NMR experiments, i.e. measurements of the Knight shift K_S and the spin-lattice relaxation rate T_1^{-1} are of particular interest. While the latter quantity measures the relaxation of nuclear spins via their interaction with the electronic quasiparticles, the former probes — via its proportionality to the spin susceptibility χ_P of the quasiparticles — the density of states at the Fermi

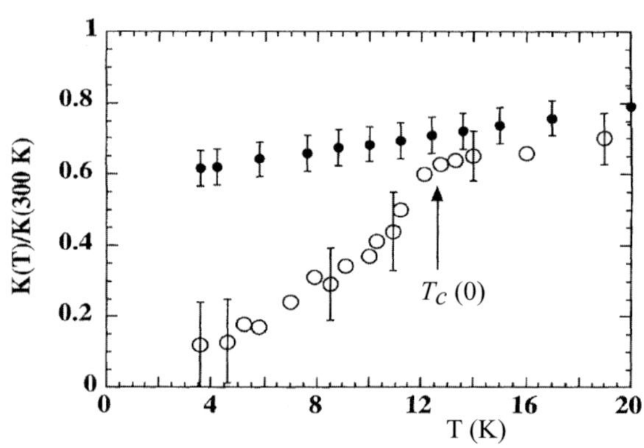

Fig. 6.26. Temperature dependence of the normalized Knight shift of ^{13}C-NMR signals in the superconducting state of κ-$(ET)_2Cu[N(CN)_2]Br$ at a field of 7.8 T perpendicular (black circles) and parallel to the conducting planes (open circles). After [862].

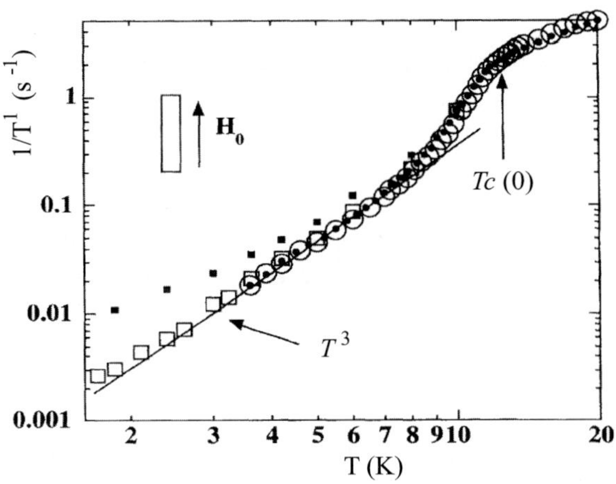

Fig. 6.27. Spin-lattice-relaxation rate $(T_1)^{-1}$ at fields of 5.6 T (open circles), 7.8 T (black circles) and 7 T (open squares) applied parallel to the conducting planes of κ-(ET)$_2$Cu[N(CN)$_2$]Br. Black squares correspond to a field of 7 T with a small misalignment. After [862].

energy. For conventional superconductors, the temperature dependence of the Knight shift in the superconducting state, $K_s(T)$ is given by the Yosida function [951] and simply reflects the T-dependence of the Pauli-type magnetic susceptibility in the superconducting state, $\chi_s(T)$, given by

$$\chi_s = -4\mu_B^2 \int \frac{N(E_F) \cdot |E|}{(E^2 - \Delta^2)} \left(\frac{\partial f}{\partial E} \right) dE, \tag{6.42}$$

where f is the Fermi-Dirac distribution function. While for conventional superconductors with a nodeless gap function Δ and a spin singlet $S = 0$ pair state, χ_s shows an exponential decrease for $T \to 0$, gap functions with nodes result in a non-exponential decay of $K_s(T)$ at low temperatures. In contrast, the Knight shift for spin triplet (S $=$ 1) pairing remains unaffected upon cooling through T_c, see, e.g., [952]. The ^{13}C spin-lattice relaxation and Knight shift of κ-(ET)$_2$Cu[N(CN)$_2$]Br have been investigated by various groups [478, 862, 953] with similar results. In these experiments, single crystalline material was used where both ^{12}C atoms in the central double bond of the ET molecule had been replaced by ^{13}C. For the investigation of electronic properties, these nuclei are superior since their coupling to the π-electron system is much stronger than that of the protons in the ethylene endgroups of the ET molecules. The salient results of these studies are: (i) Knight-shift measurements performed in fields aligned parallel to the conducting planes reveal a spin susceptibility that tends to zero at low temperatures, cf. Fig. 6.26. Since for this field configuration contributions from the pancake vortices are not expected, the above results have been taken as evidence for the spin-singlet character of the pairing state.[19] (ii) The spin-lattice relaxation rate, T_1^{-1}, measured in the same parallel field configuration lacks the Hebel-Slichter coherence peak and shows a power-law T^n behavior at low T, with n being close to 3, see Fig. 6.27. For the experimental conditions chosen, the authors ascribed the dominant source of relaxation to the quasiparticle excitations in the superconducting state. Consequently, the power-law temperature dependence in T_1^{-1} has been interpreted as indicating an anisotropic pairing with line nodes in the gap function [478, 862, 953].

Thermal Conductivity

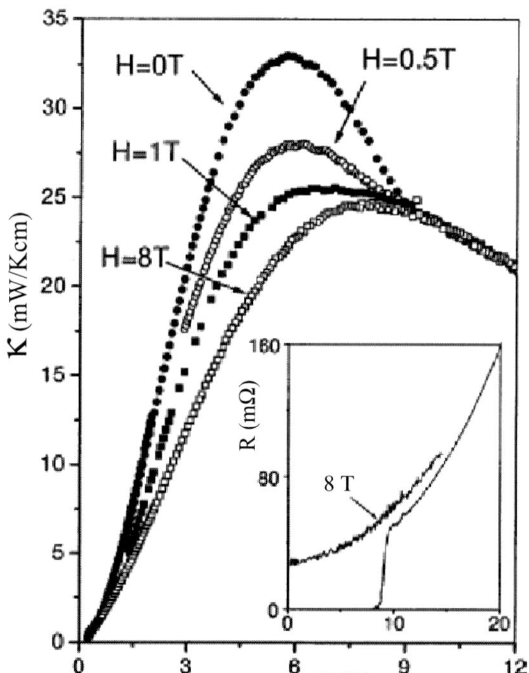

Fig. 6.28. Temperature dependence of the thermal conductivity of κ-(ET)$_2$Cu(NCS)$_2$ for B applied perpendicular to the planes. The inset shows the temperature dependence of the resistance, after [955].

Investigations of the thermal conductivity, κ, on quasi-2D organic superconductors at low temperatures have been performed first on κ-(ET)$_2$Cu(NCS)$_2$ [955] and more recently also on κ-(ET)$_2$Cu[N(CN)$_2$]Br [457]. As shown in

[19] This is corroborated by the Pauli paramagnetic-limited upper critical field, e.g., in κ-(ET)$_2$Cu[N(CN)$_2$]Br [954].

Fig. 6.28 the onset of superconductivity is associated with a sudden increase of $\kappa(T)$ which can be suppressed by a moderate magnetic field. The enhancement of $\kappa(T)$ at the onset of superconductivity has been attributed to a strengthening of the phonon heat transport by reducing the scattering of the phonons by electrons due to the opening of a gap below T_c. The authors' argument is based on a quantitative analysis of the data employing the Wiedemann-Franz law. It showed that just above T_c the electronic contribution amounts to only 5 % of the total thermal conductivity. A lattice-dominated thermal conductivity above T_c is also consistent with the absence of a magnetic field dependence of κ in this temperature range [955]. As for the question of the gap symmetry, the data at low temperature have been interpreted as indicating an excitation spectrum with gap zeroes: an extrapolation of the data for κ-(ET)$_2$Cu(NCS)$_2$ to $T \rightarrow 0$ revealed a finite in-T linear term which has been attributed to a residual electronic contribution [955]. The latter is expected for an unconventional superconductor due to impurity scattering of residual quasiparticles [956, 957]. For unconventional superconductors with certain gap topologies — including the one associated with a $d_{x^2-y^2}$ symmetry — theory even predicts a universal transport behavior, i.e., a quasiparticle heat transport independent from the scattering rate at small impurity concentrations, for $T \rightarrow 0$ [958]. This universal limit is the consequence of the compensation between the growth of the normal fluid density (the normal quasiparticles) with increasing impurity concentration and the concomitant reduction of their mean free path.

On the other hand, recent thermal conductivity measurements on the κ-(ET)$_2$Cu[N(CN)$_2$]Br salt showed that down to the lowest temperatures the phonon scattering length is strongly influenced by quasiparticle scattering [457] which renders the analysis of the data on κ-(ET)$_2$Cu(NCS)$_2$[955] as being questionable.

Magnetic Penetration Depth

The quantity, which has been studied most intensively for the κ-(ET)$_2$X superconductors with respect to the order-parameter symmetry, is the magnetic penetration depth. It defines the length scale over which the magnetic field changes in the superconductor, i.e., the depth it penetrates into the superconductor.

According to the London theory, the penetration depth λ_L in the limit $T \rightarrow 0$ is directly related to the density of superconducting electrons n_s via $\lambda_L(0) = (m^*/\mu_0 n_s e^2)^{1/2}$ (6.13). Employing a two-fluid model with $n_e = n_s(T) + n_n(T)$ and n_e the number of conduction electrons, the temperature dependence of $\lambda_L(T)$ provides information on the normal-conducting component $n_n(T)$, i.e. the quasiparticle excitation spectrum. Since $n_s(T \rightarrow 0) = n_e$, the low-temperature value $\lambda(T \rightarrow 0)$ is a measure of the pair condensate, i.e. $\lambda_L^2(0) \propto m^*/n_s(0)$. For a conventional weak-coupling superconductor, the BCS theory predicts a mean-field temperature dependence of λ_L around T_c

and an exponentially small variation at low temperatures $T \ll T_c$ [774]:

$$\lambda_L(T) \simeq \lambda(0)\left[1+\left(\frac{2\pi\Delta}{k_BT}\right)^{\frac{1}{2}}\exp\left(-\frac{\Delta}{k_BT}\right)\right]. \tag{6.43}$$

This holds true also for an anisotropic gap function without nodes, where for $k_BT \ll \Delta_{\mathrm{min}}$ the exponential low-temperature behavior is governed by the minimum value of the gap Δ_{min}. In contrast, an energy gap, which vanishes along lines or at points at the Fermi surface, will result in a power-law dependence of $\lambda_L(T) \propto T^n$ for $T \ll T_c$.

For the present materials, the magnetic penetration depth has been determined by a variety of different techniques including ac-susceptibility [959, 870, 871, 829], muon-spin relaxation [872, 960, 961, 889], magnetization [873, 874, 828, 841], surface impedance [962, 963, 964, 875, 868], tunnel diode oscillator [965] and decoration experiments [893, 894]. The results of these studies, however, were quite inconsistent regarding the temperature dependence. Likewise, the range of absolute values of $\lambda_L(T \to 0)$ is within a factor of 4 – 5 (see Table 6.3 in Sect. 6.4). In [841] it has been pointed out that the marked differences in the absolute value of the in-plane penetration depth, inferred from these studies, can be linked to the strength of the magnetic field at which the experiment had been performed, see Fig. 6.29: for the measurements with external fields below the lower critical field $B < B_{c_1}$ (surface impedance [962, 963, 964, 875, 868], ac inductance [965] and ac-susceptibility [959, 870, 871, 829]), i.e. in the shielding state, an in-plane penetration depth $\lambda_{\|}(0)$ ranging from 1.4 – 2 μm has been reported for κ-$(ET)_2Cu(NCS)_2$. On the other hand, measurements performed in the superconducting mixed state (μSR [872, 960, 889], magnetization [873, 874, 828, 841] and decoration [893, 894]), i.e. for fields $B_{c_1} < B < B_{c_2}$, yield considerably lower values for $\lambda_{\|}(0)$ with comparable values extracted from the different experiments. Apparently, measurements performed in the shielding configuration, Fig. 6.29 (a), where the shielding currents flow at the sample surface with $\lambda_{\|}(0)$ mea-

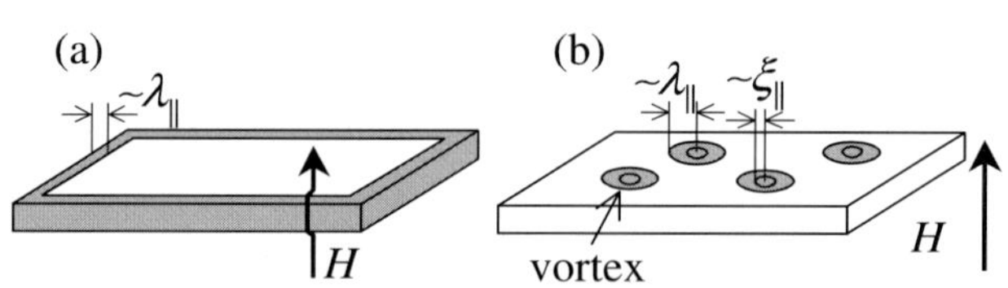

Fig. 6.29. Schematic view of the in-plane magnetic penetration depth in (a) the shielding state for fields $B < B_{c_1}$ and (b) the mixed state for $B > B_{c_1}$. Bold arrows indicate the magnetic field perpendicular to the conducting planes. The grey-shaded regions indicate the area where the magnetic fields penetrate the sample. After [841].

suring the field penetration from the sample edge, results in a penetration depth which is much larger compared to values extracted from mixed-state experiments, cf. Fig. 6.29. It has been suggested in [841] that this may originate in surface states which are different from those of the bulk. In addition, damages or degradation of the crystals at the surface may also result in an enlarged value for $\lambda_{\|}(0)$ [841].

Not surprisingly, the above-mentioned differing experimental results have led to quite different conclusions as to the symmetry of the superconducting order parameter. Interestingly enough, the inconsistencies do not only involve results from different experimental techniques. Contradictory conclusions have also been drawn on the basis of seemingly identical experiments performed by different groups. This is the case for surface impedance studies where the penetration depth can be extracted from the complex conductivity. The latter is derived from the frequency shifts and variations of the quality factor of the resonator caused by the sample. While the surface impedance studies using a microwave perturbation technique on κ-(ET)$_2$Cu(NCS)$_2$ and κ-(ET)$_2$Cu[N(CN)$_2$]Br by O. Klein et al. [963] and M. Dressel et al. [868] were found to be in good agreement with the BCS predictions, other studies by D. Achkir et al. [875] on κ-(ET)$_2$Cu(NCS)$_2$ revealed an in-T linear behavior at low temperatures indicative of an order parameter with zeroes on the Fermi surface. Deviations from an exponential temperature dependence of $\lambda_L(T)$ for the above two κ-(ET)$_2$X compounds have been observed also in an experiment by A. Carrington et al. using an rf tunnel diode oscillator [965]. In contrast to the above measurements by D. Achkir et al., however, their data of the in-plane penetration depth rather follow a $T^{\frac{3}{2}}$ power law (see Fig. 6.30). As has been argued by the authors, the data would still be consistent with a quasi-linear variation of the superfluid density as expected for a d-wave superconductor with impurities or a small residual gap [965]. Alternatively, the authors point out that the exponent 3/2 may arise naturally in a model proposed for short-coherence-length superconductors exhibiting a pseudo-gap [966, 967, 968].

An inconsistency exists also for μSR experiments performed by different groups. Here λ_L can be determined by measuring the field inhomogeneities in the mixed state, i.e. the spatial variation of the local induction of the vortex lattice. This technique was first applied to κ-(ET)$_2$Cu(NCS)$_2$by D.R. Harshman et al. [872] who could fit their data by a BCS temperature dependence. Subsequently, L.P. Le et al. [961] carried out similar experiments on the same system as well as on the κ-(ET)$_2$Cu[N(CN)$_2$]Br salt and found an in-T linear variation for the in-plane penetration depth at low temperatures $\lambda_{\|}(T) \approx 1 + \alpha \cdot (T/T_c)$.

A power-law temperature dependence $\lambda_{\|}(T)/\lambda_{\|}(0) - 1 \propto (T/T_c)^2$ consistent with d-wave superconductivity has been observed by ac-susceptibility measurements performed by different groups [959, 871, 829]. In this context it has been pointed out by A.F. Hebard et al. that an additional, extrinsic

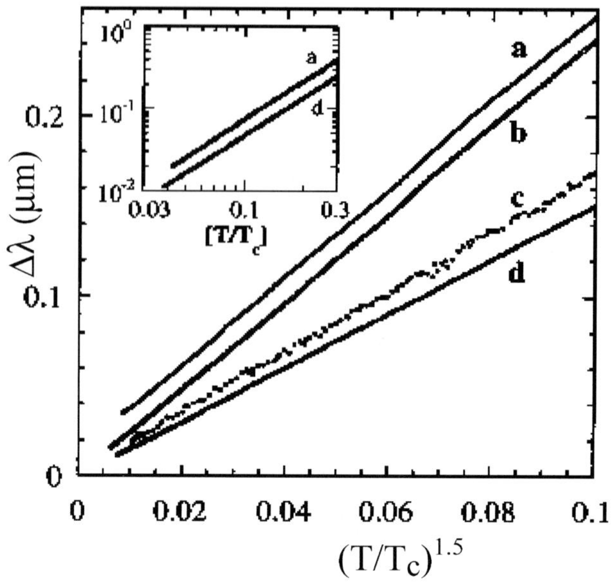

Fig. 6.30. Changes of the in-plane penetration depth $\Delta\lambda_{\|}(T)$ of κ-$(ET)_2Cu[N(CN)_2]Br$ [two samples (a),(b)] and κ-$(ET)_2Cu(NCS)_2$ [(c),(d)] plotted versus $(T/T_c)^{\frac{3}{2}}$. The data have been offset. After [965].

contribution to λ arising from vortex-core pinning at defects gives a low-temperature T^2 power-law behavior of $\lambda(T)$ which can easily be mistaken as evidence for nonconventional pairing [969]. This has been corroborated by ac-screening measurements on κ-$(ET)_2Cu(NCS)_2$ which yield the same quadratic low-temperature power-law dependence of $\lambda(T)$ independently of whether the pinned vortices are created by an external field or created as vortex-antivortex pairs at a defect. Thus, the observed $(T/T_c)^2$ component most likely reflects the extrinsic contributions to screening [969].

An alternative method to determine the penetration depth is to make use of the reversible mixed-state magnetization, a peculiarity of these strongly anisotropic superconductors with short coherence lengths.

According to the London model, the field dependence of the magnetization is given by

$$\frac{\mathrm{d}M}{\mathrm{d}(\ln B)} = \frac{\phi_0}{32\pi^2\lambda_{\mathrm{eff}}^2}, \tag{6.44}$$

where $\lambda_{\mathrm{eff}}^2 = \lambda_{\|}^2$ for B perpendicular and $\lambda_{\mathrm{eff}}^2 = \lambda_{\|}\lambda_{\perp}$ for B parallel to the planes [971]. For superconductors in the clean, local limit, i.e. for $\lambda_{\|} > \ell_{\|} > \xi_{\|}$, with $\ell_{\|}$ the in-plane mean free path, $\lambda_{\|}$ is given by

$$\lambda_{\|}(0) = \lambda_L(0)\sqrt{1 + \xi_{\|}/\ell_{\|}}, \tag{6.45}$$

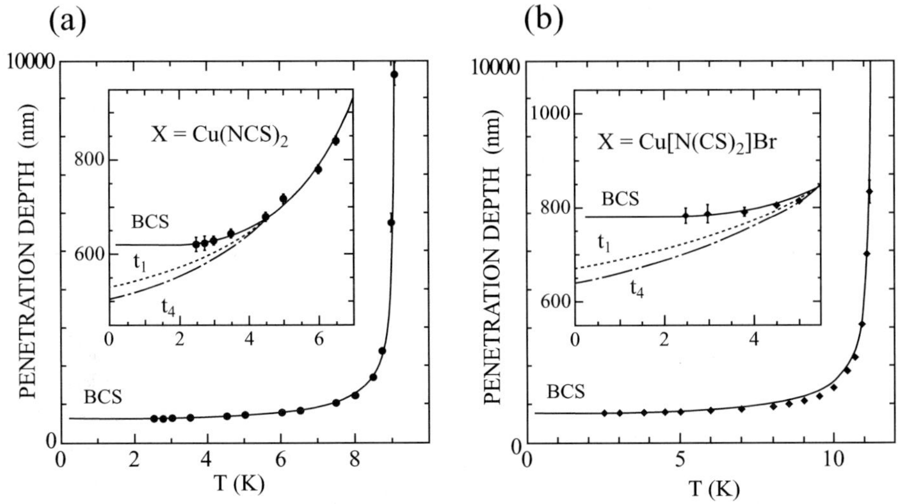

Fig. 6.31. In-plane penetration depth for (a) single crystalline κ-(ET)$_2$Cu(NCS)$_2$ and (b) κ-(ET)$_2$Cu[N(CN)$_2$]Br. The solid lines represent BCS fits. The model calculations labeled t_1 and t_4 represent those anisotropic states proposed by Y. Hasegawa et al. [970] and used by L.P. Le et al. to explain their μSR data [961] which have the weakest and strongest T dependencies, respectively. After [873, 874].

where $\lambda_L(0)$ is the London penetration depth for a clean sample as defined in (6.13) and $\xi_{\parallel}$ the in-plane Ginzburg-Landau coherence length. While for the κ-(ET)$_2$Cu(NCS)$_2$ salt the mean free path $\ell_{\parallel} = 100 - 240\,$nm, see, e.g. [841] and references cited therein, greatly exceeds the in-plane coherence length $\xi_{\parallel} = 3 - 7\,$nm, cf. Table 6.3, a much shorter $\ell_{\parallel} = 26 - 38\,$nm [448, 449, 164] exceeding $\xi_{\parallel} = 2.4 - 3.7\,$nm by a factor 10 has been reported for the κ-(ET)$_2$Cu[N(CN)$_2$]Br salt.

In 3D superconductors, vortex pinning usually gives rise to an inhomogeneous distribution of the vortices in the mixed state and thus to an irreversible behavior of the magnetization upon increasing and decreasing the field. This may cause substantial uncertainties in determining the penetration depth from magnetization data. On the contrary, for quasi-2D superconductors with short coherence length, the magnetization is entirely reversible over an extended field range, i.e. $B_{c_1} < B_{irr} < B < B_{c2}$ with B_{irr} being the temperature-dependent irreversibility line (see Sect. 6.5). For κ-(ET)$_2$Cu(NCS)$_2$and κ-(ET)$_2$Cu[N(CN)$_2$]Br a reversible magnetization has been observed over an extended range in the B-T plane which thus allows for a determination of the in-plane penetration depth [873, 874, 841]. From the slopes $\mathrm{d}M/\mathrm{d}(\ln B)$ of the isotherms taken at different temperatures and using (6.44) the in-plane penetration depths $\lambda_{\parallel}(T)$ have been determined, see Fig. 6.31. The solid line represents a BCS fit [774] to the data. For both systems, the data reveal only a weak variation with temperature at low T consistent with an exponential temperature dependence as expected for a finite gap.

Specific Heat

The above controversial results on the magnetic penetration depth indicate an extraordinarily high sensitivity of this quantity to extrinsic effects such as disorder, pinning-related phenomena and surface properties. A quantity which is less sensitive to such problems but provides fundamental information on the gap structure is the specific heat. In case this integral thermodynamic quantity were to find an electronic quasiparticle contribution in the superconducting state, C_{es}, far below T_c that varies exponentially weakly with the temperature, the existence of gap zeroes on the Fermi surface could be definitely ruled out. On the other hand, the observation of a non-exponential temperature dependence does not necessarily prove the existence of gap zeroes as this result might originate in extraneous contributions such as impurity phases, normal-conducting regions or pair-breaking effects. First specific heat measurements were focussing on the determination of the discontinuity at T_c which provides information on the coupling strength. From the results of B. Andraka et al. [820] and J.E. Graebner et al. [801] on κ-$(ET)_2Cu(NCS)_2$ yielding a ratio of $\Delta C/\gamma T_c > 2$ a strong-coupling behavior has been inferred for this salt. In a series of subsequent experiments, the temperature dependence of the electronic contribution, C_{es}, at lower temperatures was at the

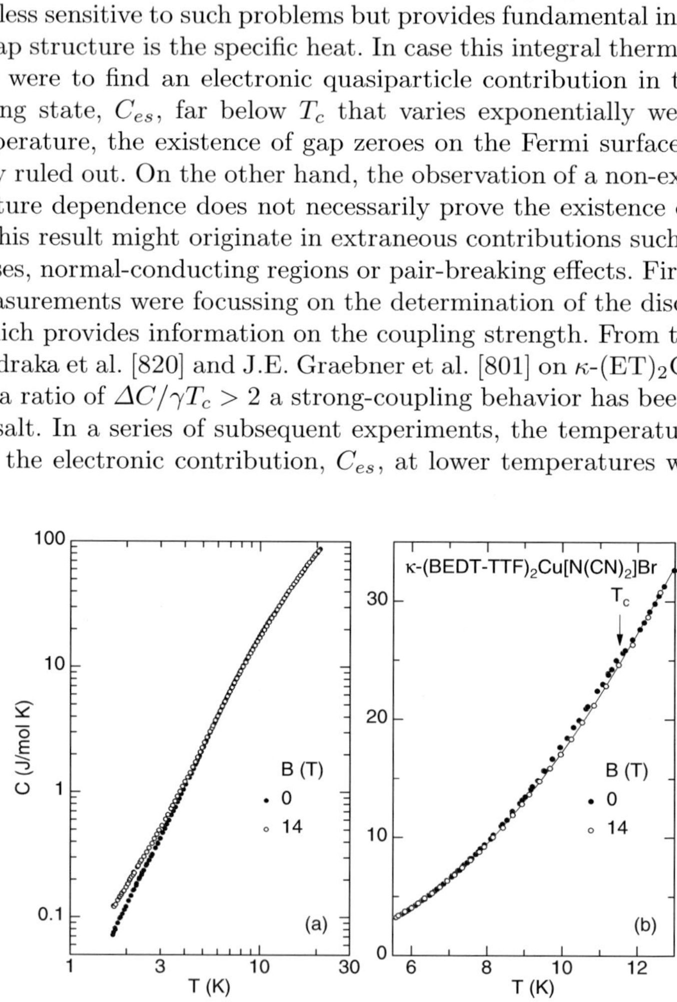

Fig. 6.32. Temperature dependence of the specific heat of κ-$(ET)_2Cu[N(CN)_2]Br$ in the superconducting ($B = 0$) and normal state ($B = 14\,T$) over an extended temperature range (left panel) and in the vicinity of $T_c = 11.5\,K$ (right panel). The solid line is a polynomial fit to the 14 T data. After [458].

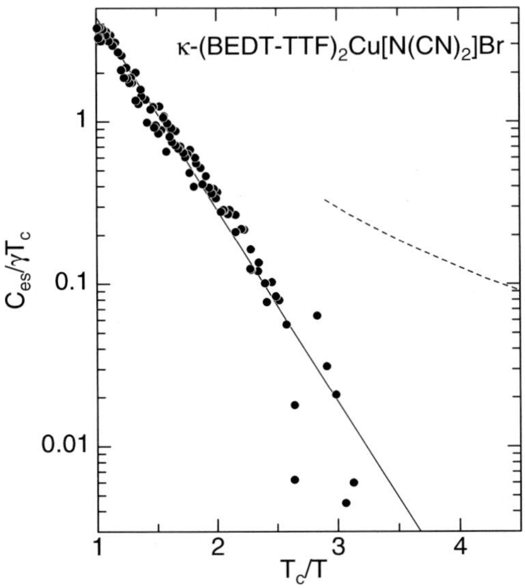

Fig. 6.33. A semi-logarithmic plot of $C_{es}/(\gamma T_c)$ vs T_c/T as determined from the data shown in Fig. 6.32. The solid line indicates the exponential variation of C_{es}. The dashed line corresponds to C_{es} as determined in [972] which has been critically commented in [458] (see text). After [458].

focus of the investigations. From experiments on κ-(ET)$_2$Cu[N(CN)$_2$]Br, Y. Nakazawa and K. Kanoda [972] reported a quadratic temperature dependence of C_{es} at low temperatures, which was taken as indication for line nodes in the gap. However, more recent high-resolution specific heat measurements on the same compound revealed an exponentially weak low-T electronic contribution to the specific heat implying a finite energy gap [458]. Moreover it has been shown in the latter study that the T^2 dependence of the C_{es} data in [972] most likely originates in their incorrect determination of the phonon background.[20] Figure 6.32 shows the results of specific heat measurements performed by H. Elsinger et al. [458]. The phonon contribution, which predominates the specific heat near T_c, has been determined from measurements in an overcritical field. This procedure, which is usually applied, is valid as long as there are no magnetic contributions to the specific heat which might change with the field. From the absence of any measurable field dependence in the data above T_c, this assumption appears justified. Figure 6.33 shows the normalized electronic contribution to the specific heat $C_{es}/\gamma T_c$ as a function of T_c/T in a semi-logarithmic plot. From the exponential decrease of C_{es} with

[20] In [972] the lattice specific heat of a κ-(ET)$_2$Cu[N(CN)$_2$]Br crystal was estimated by measuring a second, deuterated sample which was quench cooled to suppress superconductivity. The data analysis in [972] is based on the assumption that by this procedure the lattice specific heat remains unaffected and is identical to that of the hydrogenated superconducting compound. In [458] it has been shown, however, that the so-derived phonon background differs substantially from that determined in an overcritical field. The reason for this might be related to the glass-like transition at T_g observed in this system [155], see Sect. 3.3.

temperature, a finite gap everywhere on the Fermi surface has been inferred [458].

A similar behavior has been observed also for κ-$(ET)_2Cu(NCS)_2$ as well as other $(ET)_2X$ superconductors [456, 457]. Figure 6.34(a) shows the specific heat difference between the superconducting and normal state $\Delta C(T) = C(T, B = 0) - C(T, B = 8\,\mathrm{T} > B_{c_2})$ used to analyze the specific heat data. The advantage of discussing $\Delta C(T)$ instead of $C(T)$ is that the unknown phonon contribution drops out. As Fig. 6.34 demonstrates, $\Delta C(T)$ deviates markedly from the weak-coupling BCS behavior in both the jump height at T_c as well as the overall temperature dependence. However, as was found also for κ-$(ET)_2Cu[N(CN)_2]Br$ [458], a much better description of the data is obtained by using the semi-empirical extension of the BCS formalism to strong-coupling superconductors, the so-called α-model [974], which was first used to describe conventional superconductors like In, Sn, Zn, and Pb/In alloys. Assuming that the electronic excitations above the superconducting ground state are described by non-interacting quasiparticles, the entropy in the superconducting state is given by

$$S_{\mathrm{es}} = -2k_B \sum [f_{\boldsymbol{k}} \ln f_{\boldsymbol{k}} + (1 - f_{\boldsymbol{k}}) \ln(1 - f_{\boldsymbol{k}})], \qquad (6.46)$$

where $f_{\boldsymbol{k}} \equiv [\exp(\beta E_{\boldsymbol{k}}) + 1]^{-1}$ with $\beta \equiv (k_B T)^{-1}$. For the quasiparticle excitation spectrum $E_{\boldsymbol{k}}^2 = (\epsilon_{\boldsymbol{k}}^2 + \Delta^2)$ with the single-particle energy $\epsilon_{\boldsymbol{k}}$ and the temperature dependence of the energy gap, $\Delta(T)/\Delta(0)$, the BCS relations are used, the latter being tabulated in [774]. The model contains a single free parameter $\alpha \equiv \Delta(0)/k_B T_c$ which scales the BCS energy gap $\Delta(T) = (\alpha/\alpha_{\mathrm{BCS}}) \cdot \Delta_{\mathrm{BCS}}(T)$ with $\alpha_{\mathrm{BCS}} = 1.764$. Although there is some sample dependence as shown by the results from different groups, Fig. 6.34 clearly demonstrates that the strong-coupling BCS model with $\alpha = 2.4 - 2.8$ provides an excellent description of the data over the entire temperature range investigated [456, 457]. From similar experiments on various $(ET)_2X$ superconductors with T_c ranging from 1.1 K to 11.5 K, a monotonic increase of the normalized specific heat discontinuity $\Delta C/\gamma T_c \propto \alpha$ with T_c has been inferred [975, 457] indicating an increasing coupling strength with T_c, see Table 6.4. Besides the coupling parameter α, which is directly determined from specific heat measurements, the electron-phonon coupling constant λ_{ep} can be indirectly determined based on the Eliashberg theory for strong-coupling superconductors which provides a phenomenological relation between $\Delta C/\gamma T_c$ and $T_c/\bar{\omega}$ where $\bar{\omega}$ denotes the mean phonon energy [976]. For a coupling not too strong $T_c/\bar{\omega}$ is in turn linked to λ_{ep} via (6.20), the McMillan relation, see [775, 977].[21] The observation that α and λ_{ep} increase similarly with T_c as $\Delta C/\gamma T_c$ increases with $T_c/\bar{\omega}$ indicates that the characteristic phonon energy

[21] The described determination of λ_{ep} is only a rough estimate, especially since (6.20) is not valid for large coupling constants $\lambda_{ep} > 1.5$. Therefore, in [458] for κ-$(ET)_2Cu[N(CN)_2]Br$ an empirical relation between $T_c/\bar{\omega}$ and λ_{ep} based on tunneling experiments on BCS superconductors, see [977], has been used.

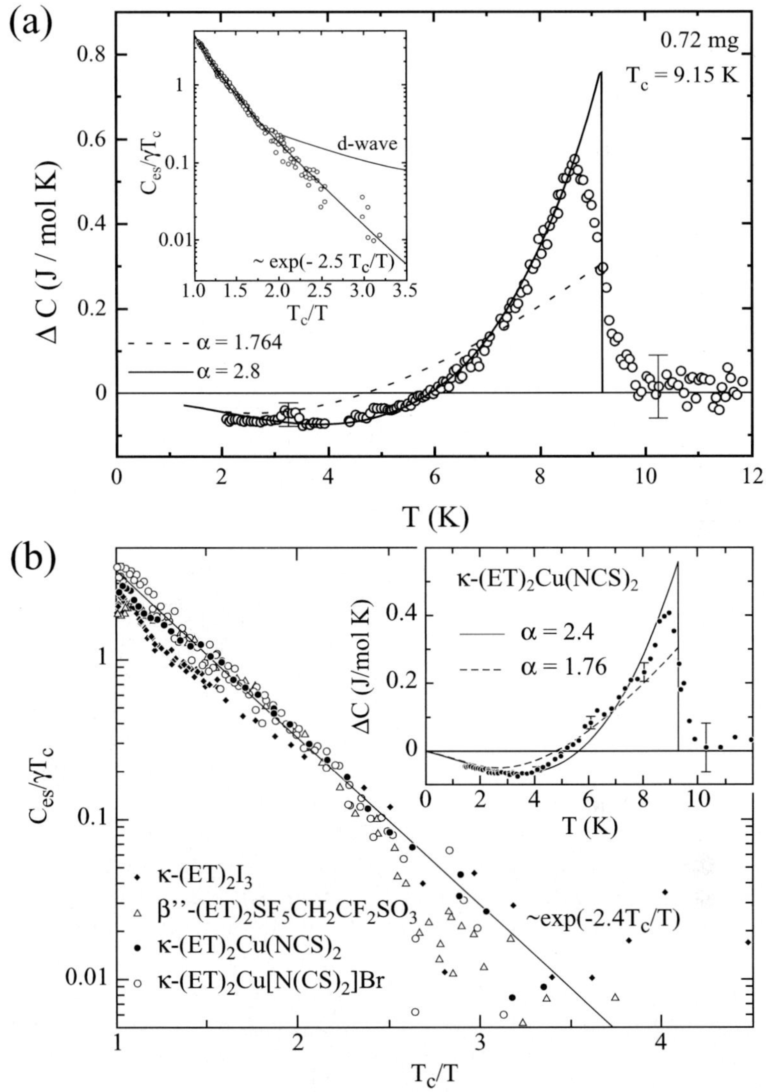

Fig. 6.34. (a) ΔC of a small κ-(ET)$_2$Cu(NCS)$_2$ single crystal with $m = 0.72$ mg (main panel). The dotted and solid lines represent the BCS curves for weak and strong coupling, respectively. The inset shows $C_{es}/\gamma T_c$ vs T_c/T in a semi-logarithmic representation. Here the solid line represents the strong-coupling BCS behavior while the dotted line indicates a T^2 behavior as expected for a d-wave order parameter [973]. After [456, 115]. (b) $C_{es}/\gamma T_c$ vs T_c/T for different (ET)$_2$X superconductors. The solid line shows the exponential vanishing of C_{es} towards low T (fit to the κ-(ET)$_2$Cu(NCS)$_2$ data). Open circles correspond to κ-(ET)$_2$Cu[N(CN)$_2$]Br. The inset shows ΔC measured on a large single crystal with $m = 3.07$ mg. After [457].

Table 6.4. Parameters determined from specific heat measurements (for κ-$(ET)_2Cu(NCS)_2$ results from three different groups are given). For the ET salts, note the monotonic increase of the electron-phonon coupling constant λ_{ep} with T_c.

	T_c (K)	γ ($\frac{\text{mJ}}{\text{molK}^2}$)	$\Delta C/\gamma T_c$	α	λ_{ep}	Ref.
α-$(ET)_2NH_4Hg(NCS)_4$	1.1	$25-26$	1.43	1.76	0.4	[112]
κ-$(ET)_2I_3$	3.5	18.9 ± 1.5	1.6 ± 0.2	1.94	0.8	[455]
κ-$(ET)_2Cu(NCS)_2$	9.1	25 ± 3	$2-2.8$		1.5	[820]
κ-$(ET)_2Cu(NCS)_2$	9.3	23 ± 1	2.05	2.4	1.65	[457]
κ-$(ET)_2Cu(NCS)_2$	9.1	23 ± 1	3.6	2.8		[456]
κ-$(ET)_2Cu[N(CN)_2]Br$	11.5	25 ± 2	3.3	2.7	2.5	[458]
β″-$(ET)_2SF_5CH_2CF_2SO_3$	4.5	18.7 ± 1	2.1	2.15	1.1	[459]
λ-$(BETS)_2GaCl_4$	4.8	14.4 ± 3.4	1.37 ± 0.32	1.76		[460]

$\bar{\omega}$ is of similar order for all the $(ET)_2X$ superconductors discussed here. In [458] it has been pointed out that using the BCS relation [977]

$$\lambda_{ep} = \frac{N(E_F)\langle I^2 \rangle}{M\bar{\omega}^2} \tag{6.47}$$

with $\langle I^2 \rangle$ the electron-phonon matrix element averaged over the Fermi surface, M the atomic mass, and $\bar{\omega}$ the average phonon frequency, gives $\lambda_{ep} \propto N(E_F)\langle I^2 \rangle$. T_c is mainly controlled by $\langle I^2 \rangle$ since according to Table 6.4 there is no simple relation between T_c and $\gamma \propto N(E_F)$.

Similar to what has been found for κ-$(ET)_2Cu[N(CN)_2]Br$ (Fig. 6.33), the data for κ-$(ET)_2Cu(NCS)_2$ (Fig. 6.34) are fully consistent with an exponentially small C_{es} at low temperatures, i.e. an energy gap without zeroes at the Fermi surface. The same behavior has been previously observed for other $(ET)_2X$ superconductors, see Fig. 6.34(b) [455, 459, 457]. The above findings of an exponentially weak specific heat at low temperatures are clearly incompatible with the existence of gap zeroes as inferred by various of the abovementioned experiments. It remains to be shown by future studies whether or not this controversy can be removed by considering carefully the influence of magnetic fields even when applied parallel to the planes and other extraneous effects such as the cooling-rate-dependent disorder associated with the ethylene groups.

On the Pairing State in $(TMTSF)_2X$

As for the ET systems, the nature of superconductivity in the Bechgaard salts $(TMTSF)_2X$ is still far from being understood and continues to be a controversial issue. Early suggestions of a spin-triplet ($S = 1$) state were based on the proximity of superconductivity to a SDW state in the pressure-temperature phase diagram (cf. Fig. 4.31 in Sect. 4.6). This situation resembles the theoretical expectation for an interacting 1D electron gas [978] where a SDW phase lies next to triplet superconductivity. Further arguments for a spin-triplet p-wave superconductivity were derived from the observation that T_c is extremely sensitive to the introduction of non-magnetic defects [979, 980] and substitutional impurities [981, 982], see also [983, 984].

Among the $(TMTSF)_2X$ superconductors, the X = PF_6 [33] and ClO_4 [36] salts are the most extensively studied materials, see e.g. [146, 86, 985, 90] for a review, although the number of experimental investigations of the superconducting properties is much less compared to that of the ET salts and their derivatives. The reason for this is related to the low T_c values of the $(TMTSF)_2X$ systems, requiring extensive low–temperature equipment and, in the case of X = PF_6, external pressure of $p \geq 5.8$ kbar to stabilize the superconducting state. For the ambient-pressure superconductor $(TMTSF)_2ClO_4$, it is the anion ordering which renders the experimental situation difficult (see Sect. 3.3). By slowly cooling through the anion-ordering temperature T_{AO} at around 24 K an ambient pressure superconducting state below $T_c = (1.2 \pm 0.2)$ K can be stabilized. In this case, the anions are believed to be well ordered. On the other hand, for samples that have been cooled rapidly across T_{AO}, an insulating state forms below $T_{MI} \simeq 6.1$ K.

The $(TMTSF)_2ClO_4$ salt is the only member of the $(TMTSF)_2X$ series where specific heat [851] or Meissner-effect measurements [849] have been performed. Figure 6.35 shows the specific heat of $(TMTSF)_2ClO_4$ in a temperature range close to T_c. For temperatures above T_c, the data reveal a behavior $C/T = \gamma + \beta T^2$ with a Sommerfeld coefficient of the electronic contribution to the specific heat of $\gamma = 10.5\,\mathrm{mJ\,mol^{-1}K^{-2}}$ and the βT^3 term accounting for the phonons. The discontinuity at T_c amounts to $\Delta C/\gamma T_c = 1.67$ which is slightly larger than the BCS ratio of 1.43 [851], consistent with a weak-coupling scenario.

On the other hand, deviations from a BCS-type of superconductivity have been observed in NMR measurements by M. Takigawa et al. [986]. These authors reported the absence of a Hebel-Schlichter peak and a T^3 dependence in the spin–lattice relaxation rate. This has led to the proposal of a d-wave order parameter with a gap that vanishes along lines on the Fermi surface [986]. These results are at variance with more recent thermal conductivity data on the same salt, showing a rapid decrease of the electronic contribution to the heat transport below T_c which indicates the absence of low-lying excitations [987]. The latter results provide strong evidence for a nodeless gap function. However, as pointed out by the authors, this result is not necessarily associated with an s-wave order parameter.

By enumerating possible gap functions for quasi-1D systems, Y. Hasegawa and H. Fukuyama [970] showed that besides an anisotropic spin-singlet d-wave, a spin-triplet p-wave state — in both cases the order parameters vanish

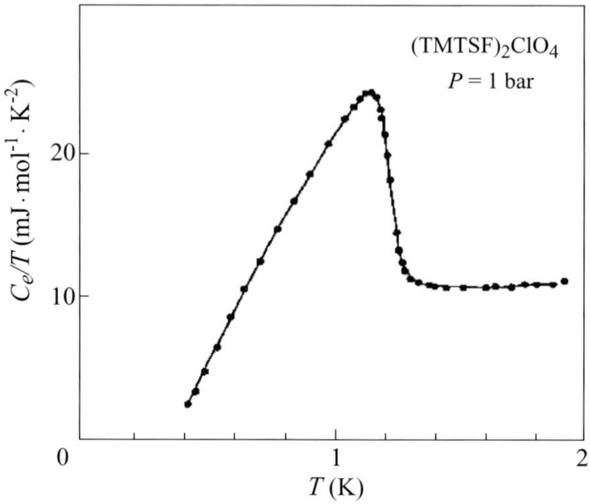

Fig. 6.35. Electronic contribution to the specific heat of $(TMTSF)_2ClO_4$ plotted as C/T vs T. After [851].

along lines on the FS — is also possible. The authors suggested the possibility of an antiferromagnetic spin-fluctuation pairing mechanism for the Bechgaard salts.

Arguments in favor of such a spin-fluctuation-mediated superconductivity with d-wave symmetry have been derived from a resistivity study under pressure [466, 214]. According to these results, a minimum in the resistivity $\rho_a(T)$ at $T_{\min}$ marks the onset of AF fluctuations before, at lower temperatures, an itinerant antiferromagnetic state (SDW) is stabilized. The width of the region of critical AF fluctuations in the T-p phase diagram (see Fig. 4.31 in Sect. 4.6) is enhanced when the SDW ground state is approached from the high-pressure side, where the system is close to the SDW/SC border, and largest where $T_c(p)$ reaches its optimum value. The correlation between the spin-fluctuation regime above the onset of superconductivity and the T_c value is taken as a strong argument for a pairing mechanism mediated by the exchange of these fluctuations [466, 214]. Further arguments supporting an unconventional gap symmetry has been presented by N. Joo et al. [843] based on their measurements on the influence of disorder on T_c in the solid solution $(TMTSF)_2(ClO_4)_{1-x}(ReO_4)_x$. They found a strong reduction of T_c with increasing disorder consistent with the Abrikosov-Gor'kov pair breaking function (6.29) discussed in Sect. 6.3.3. The non-magnetic character of the disorder has been inferred from EPR measurements showing no traces of localized spins [843].

In a recent series of papers on the upper critical fields, the discussion of a possible realization of a triplet-pairing state in the Bechgaard salts has again been raised [35, 988, 855]. I.J. Lee et al. examined the upper critical field $B_{c_2}(T)$ and its directional dependence in $(TMTSF)_2PF_6$ under pressure via resistivity measurements. More recently, a magnetic determination of B_{c_2}

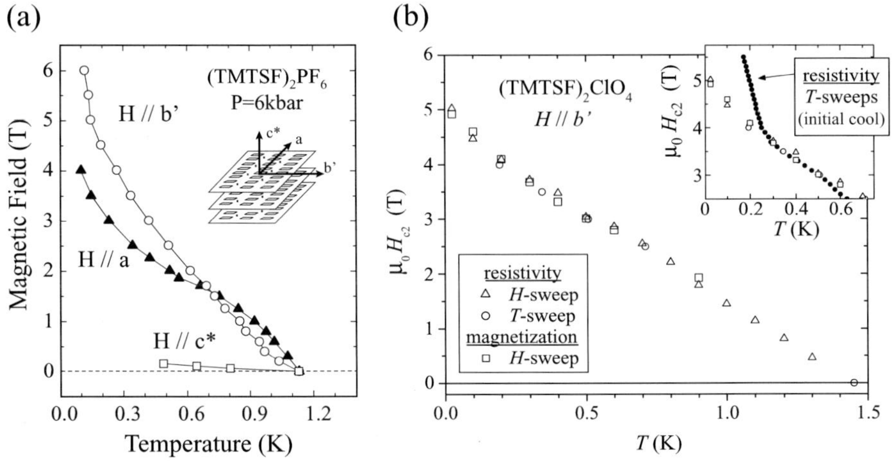

Fig. 6.36. B-T phase diagrams of (a) $(TMTSF)_2PF_6$ determined from resistivity measurements at a pressure of 6 kbar for magnetic fields aligned along three perpendicular axes as defined in the inset, after [35] and (b) $(TMTSF)_2ClO_4$ along the b'-axis, from resistivity and magnetization measurements. The inset shows the resistive B_{c2} from the same sample's initial cooldown, after [855].

under accurate alignment has been carried out also for the ambient-pressure superconductor $(TMTSF)_2ClO_4$ by J.I. Oh et al. [855]. The resulting magnetic field-temperature phase diagrams are depicted in Fig. 6.36. While the upper-critical-field curves near T_c were found to be consistent with earlier results on these compounds, which indicated a more conventional behavior (see also Fig. 6.9 in Sect. 6.4) [852, 853, 854], the extension of the measurements to lower temperatures revealed for some field directions a critical field of a factor of $2-4$ larger than the Pauli limiting field [845, 989] of $B_P = 1.84 \times T_c \simeq 2.2$ and 2.6 T for the X = PF_6 and ClO_4 salts, respectively.

Various scenarios have been discussed to explain the upturn in the upper critical field including a strong spin-orbit scattering [990, 991], a magnetic-field-induced dimensional crossover (FIDC) from 3D to 2D [992, 993], the formation of a spatially inhomogeneous Fulde-Ferrell-Larkin-Ovchinnikov (FFLO) state [735, 736], and the formation of slabs of superconductor sandwiched between insulating regions [994]. The insulator-superconductor slab model is more consistent with the data for $(TMTSF)_2PF_6$ at a pressure of 5.7 kbar since a cusp in the angular dependence of the upper critical field, as expected from the FIDC model, has not been observed [994]. Furthermore, no indications for a first-order phase transition into the FFLO state have been observed which points to the possibility of a spin-triplet superconducting state.

This interpretation has gained further support from recent ^{77}Se-NMR measurements of the Knight shift K_S in the superconducting state of (TM-

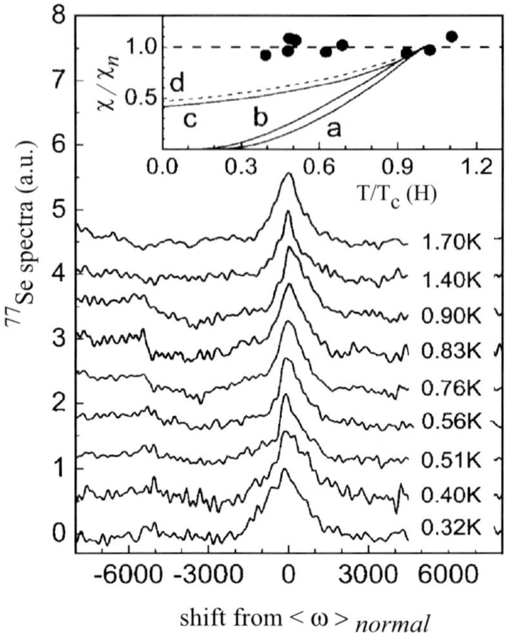

Fig. 6.37. ^{77}Se-NMR spectra of $(TMTSF)_2PF_6$ under 7 kbar at various temperatures above and below T_c which is 0.81 K at the applied field of 1.43 T aligned along the most conductive a-axis. The inset shows normalized χ_{Spin} data compared with theoretical calculations. Curves a and b are for a singlet state in applied fields B/B_{c2} near zero and 0.63, respectively, as calculated by P. Fulde and K. Maki [995] while curves c and d represent a scenario where the vortex cores induce normal regions of a fraction equal to B/B_{c2}, after [996].

$TSF)_2PF_6$ under 7 kbar of pressure [996]. Since K_S is proportional to the spin-susceptibility χ_{Spin}, which is different for spin singlet ($\chi_{\mathrm{Spin}} \rightarrow 0$ for $T \rightarrow 0$) and triplet (χ_{Spin} remains unchanged on cooling through T_c) superconductors, measurements of K_S provide a direct probe of the spin parity of the superconductor. Figure 6.37 shows that in contrast to the expectations for the various scenarios involving singlet superconductivity there is no significant change in K_S for $B \parallel a$ on cooling into the superconducting state. These observations, together with similar results of a previous study for $B \parallel b$ [997], thus argue for a spin-triplet pairing, see also [998] for a recent discussion. Besides K_S, the above studies also included measurements on the spin-lattice relaxation rate $1/T_1$ which exhibits a small peak near T_c. In light of the absence of such a peak in previous reports of zero-field proton-NMR measurements [986], its identification with a Hebel-Schlichter peak is unclear.

6.7 On the Pairing Mechanism

Understanding the mechanism that gives rise to the formation of Cooper pairs for the present molecular materials continues to pose a challenging problem. The excitonic mechanism proposed by W.A. Little [73, 999] for certain quasi-1D organic polymers was of great importance historically in giving the initial impetus to the development of the field. From today's perspective, however, it is fair to say that the search for materials with suitable chemical and physical properties for such a mechanism to work has so far been unsuccessful.

In the discussion of the transport and optical properties in Sect. 4.3.1, it became clear that there is a substantial coupling of the charge carriers to both intra- as well as intermolecular phonons. Consequently, some researchers in the field believe that the conventional electron-phonon coupling provides a substantial contribution to the pairing interaction. On the other hand, the fact that for the present materials — as for the cuprates and heavy-fermion systems — the superconducting region in the phase diagram lies next to a magnetically ordered state suggests that magnetic interactions are involved in the pairing mechanism.

Models Considering the Role of Phonons in the Pairing Mechanism

One example for a conventional electron–phonon pairing scenario has been discussed by K. Yamaji [1000]. In this model he considers an attractive interaction mediated by several high–frequency internal molecular vibrations, in addition to low-frequency intermolecular phonons.

A more general account for the complex role of phonons for superconductivity has been given by A. Girlando et al. [1001, 1002, 303]. By calculating the lattice phonons for the κ-$(ET)_2I_3$ and β-$(ET)_2I_3$ superconductors using the quasi-harmonic-lattice-dynamics method and evaluating the coupling to the electrons, λ, they succeeded in reproducing all available experimental data related to the phonon dynamics, for example the lattice specific heat. They showed that a lattice mode that couples particularly strongly to the electrons is one in which the relative displacement of the ET molecule is along the long axis of the molecule, i.e. perpendicular to the conducting planes. As an interesting side result of their study on κ-$(ET)_2I_3$, it is mentioned, that the coupling of acoustic phonons is very anisotropic and likely to cause gap anisotropies, though of a conventional type. In addition, it has been shown that the lattice phonons alone cannot account for the T_c values in these compounds. High–frequency intramolecular phonons modulating the on–site electronic energies have to be taken into account to reproduce the critical temperatures [1001].

Recently, a distinct kind of phonon–mediated pairing has been suggested for the $(ET)_2X$ salts [1003]. It is based on the idea that in a system in which Coulomb correlations are screened to be short range, i.e. Hubbard type, the electron-phonon scattering is dominated by forward processes. This results in an effective small-$\boldsymbol{q}$ pairing potential. Subsequent self-consistent solutions of the BCS gap equation using a band structure based on the effective-dimer approximation, lead to a gap structure where d- and anisotropic s-wave states are nearly degenerate. Furthermore it has been argued that the conflicting experimental results about the gap symmetry may originate in the decorrelation of superconductivity on various parts of the Fermi surface — a consequence of small-$\boldsymbol{q}$ dominated pairing — and the near degeneracy of s- and d-wave superconducting gap states [1003].

Models which Consider an Unconventional Pairing Interaction

On the other hand, the rich phase diagram and the anomalous properties of the metallic state of these materials may suggest that the key elements dominating the physical behavior are the layered structure and the strong interactions between the electrons [104]. As a consequence, some researchers even resign from considering any coupling to the phonons and instead consider mechanisms which are solely based on two-dimensional electronic interactions. Most of these proposals deal with a spin-fluctuation-mediated pairing mechanism. The latter is motivated by the close proximity of the superconducting region in the phase diagram to an antiferromagnetic insulating state which — in analogy to the high-T_c cuprates — suggests that both phenomena are closely connected to each other [189, 104, 944].

An approach in this direction has been proposed by H. Kino and H. Fukuyama [189] who studied the effects of on-site Coulomb interaction and dimer structure in a strictly two-dimensional system within the Hartree-Fock approximation, see also [1004]. In their picture, the antiferromagnetic insulating state of κ-(ET)$_2$X is a Mott insulator which is driven to a metal/superconductor by increasing the band width. The Mott-Hubbard scenario for the present organic superconductors implies a half-filled conduction band, together with strong electron correlations. Because of the approximate square-lattice configuration of the dimers, the authors expect a similar spin-fluctuation mediated superconductivity with probably $d_{x^2-y^2}$ symmetry as in the cuprates [189].

Such a possibility has been studied in detail by J. Schmalian [944]. Using a two-band Hubbard model to describe the antibonding orbitals on the ET dimer, he succeeded in creating a superconducting state with $T_c \simeq 10\,\mathrm{K}$ mediated by short-ranged antiferromagnetic spin fluctuations. It has been argued that despite the frustrating interactions and in-plane anisotropies which distinguish the organic materials from the high-T_c cuprates, the origin of superconductivity is very similar for both material classes [944].

A spin-fluctuation-based superconductivity similar to that of the cuprates has been claimed also by H. Kondo and T. Moriya [1005, 1006, 1007]. They investigated the properties of a half-filled Hubbard model in a fluctuation exchange approximation (FLEX) with a right-angle isosceles triangular lattice with transfer matrices $-\tau'$ and $-\tau$. They revealed an energy gap of $d_{x^2-y^2}$ symmetry which upon cooling grows much faster compared to that expected in the BCS model. In addition, they showed, that the appearance of d-wave superconductivity near an antiferromagnetic instability requires a suitable electronic structure, i.e. $\tau'/\tau > 0.3$ [1005, 1006, 1008]. A spin-fluctuation mediated d-wave superconducting state has been found also by several other approaches including FLEX, perturbation theory or quantum

Monte Carlo simulation applied to κ-$(ET)_2X$ [1009, 1010, 1011, 1012] or the quasi-1D $(TM)_2X$ salts [1011, 1013]. In [1014] and [1013] an explanation for the pseudo-gap behavior at elevated temperatures T^* has been proposed in terms of strong antiferromagnetic spin-fluctuations.

In [1015] it has been argued that a mean-field treatment of the above-mentioned Hubbard model, which has been considered a minimal model for the layered organic superconductors [1016, 189], will not correctly describe the materials as it neglects important spin correlations arising from superexchange. Instead, the authors considered the Hubbard-Heisenberg model which can be derived from the Hubbard model in the limit of large, but finite, on-site Coulomb repulsion U [1015]. Within a resonating-valence-bond (RVB) theory they found a first-order transition from a Mott insulator to a $d_{x^2-y^2}$ superconductor with a small superfluid stiffness $\rho_s \propto \lambda_L^{-2}$ (λ_L is the London penetration depth) and a pseudo-gap with $d_{x^2-y^2}$ symmetry.

A similar two-dimensional Hubbard model, including a spin-exchange coupling term, has been studied by J.Y. Gan et al. [1017] using the Gutzwiller variational method. They found that the ground state is a Gossamer superconductor at small on-site U and an antiferromagnetic Mott insulator at large U, separated by a first-order phase transition.

While the starting point of the above models is in the limit of strong correlations, i.e. near the Mott-insulating state, a somewhat different viewpoint is taken in the work by R. Louati et al. [1018]. These authors studied the effect of spin fluctuations in a two–dimensional model in the weak correlation regime by varying the bandwidth and the nesting properties of the Fermi surface. They argued that spin fluctuations are enhanced by the good nesting properties which may account for the anomalous NMR relaxation rate observed at temperatures T^* above T_c in the κ-$(ET)_2Cu[N(CN)_2]Br$ salt. Furthermore they found that spin fluctuations can induce a superconducting coupling with d-wave symmetry that lies next to a spin-density-wave instability [1018].

The above models address systems at half filling, which is realized in the dimerized κ-phase $(ET)_2X$ suggesting spin-fluctuation-mediated superconductivity with a $d_{x^2-y^2}$ symmetry. A different scenario has been proposed for the θ and β'' structures [950]. Here the ET molecules are not dimerized which results in a quarter-filled hole band. In this case, a nearby insulating phase is believed to be due to a charge ordering, driven by strong inter-site Coulomb correlations. Applying slave-boson theory to an extended Hubbard model (4.2) at quarter filling, superconductivity mediated by charge fluctuations has been found. This results in a d_{xy} symmetry of the superconducting state [950] as opposed to the $d_{x^2-y^2}$ symmetry for the above spin-fluctuation mechanism. The effect of charge fluctuations on the superconducting state has also been estimated by calculating pairing interaction within the random phase approximation [1019]. The authors found that the obtained SC state has d_{xy} symmetry in the vicinity of both spin-density-wave (SDW) phase

($U \geq V$) and charge-density-wave (CDW) phase ($U \leq V$) in a phase diagram of U and V at fixed temperatures.

Experimental Hints to the Pairing Mechanism

On the experimental side, the issue of identifying the mechanism that binds the electrons into Cooper pairs is even more difficult than the question of the pairing symmetry. Even under ideal experimental conditions, there is no decisive probe from which the relevant pairing interaction can be unequivocally determined. There are, however, some crucial experiments which may help to delineate the various possibilities. One of those key experiments is the isotope effect which addresses the question whether the lattice degrees of freedom are involved in the Cooper-pair formation.

Isotope Effect and Phonon Renormalization

Studying the effect of isotope substitutions on the superconducting transition temperature is one of the key experiments to illuminate the role of phonons in the pairing mechanism. Since a controlled exchange of isotopes of some or all the atoms of a material affect only the lattice properties, a finite isotope effect is a clear proof that phonons are involved, to some extent, in the pairing interaction. For materials consisting of only one atomic species, a T_c dependence of

$$T_c \propto M^{-\alpha} \tag{6.48}$$

is expected, where M is the isotopic mass, and $\alpha = 1/2$ in a first approximation [1020].

The effect of isotope substitution has been studied for various members of the $(ET)_2X$ superconductors. Isotopes have been substituted in the ET molecule by replacing ^1H by ^2D in the ethylene endgroups, by a partial exchange of ^{12}C by ^{13}C or ^{32}S by ^{34}S atoms in the inner skeleton of the molecule. In addition, systems have been studied where the acceptor molecule has been isotopically labelled. First experiments focused on the role of the electron-molecular-vibration (EMV) coupling by substituting ^{13}C for ^{12}C in the central double bond of the ET molecule. The large decrease of T_c of $\Delta T_c/T_c = -2.5\,\%$ found for the high-temperature variant (β_H) of β-$(ET)_2I_3$ by the Orsay group [1021] could not be reproduced by a subsequent study where no systematic decrease of T_c could be detected [1022]. Most intensive studies on the isotope effect have been carried out on the κ-$(ET)_2Cu(NCS)_2$ salt by the Argonne group [297, 1023]. Their investigations include isotope substitutions on both the ET molecules (all together seven differently labeled ET derivatives) as well as the anions. In each case, a batch of unlabeled samples has been synthesized under strictly parallel experimental conditions. These crystals serve as a reference for comparison. By sampling a large number of crystals, typically eight or more of both labeled and unlabeled material, a genuine mass-isotope effect on T_c has been found: upon replacing all eight

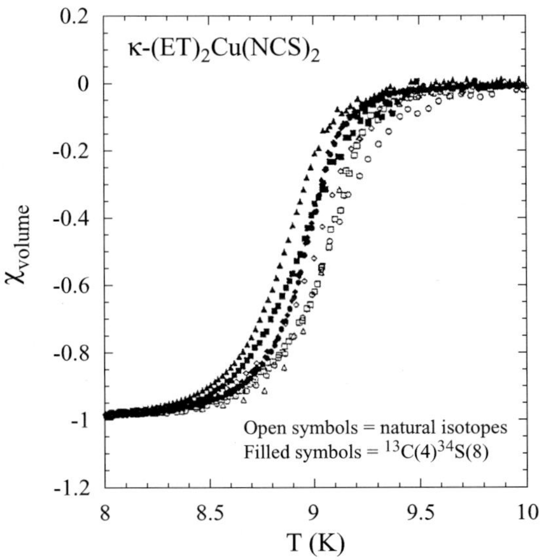

Fig. 6.38. Susceptibility curves of four crystals each of natural abundance (open symbols) and $^{13}C(4)^{34}S(8)$-labeled (filled symbols) κ-$(ET)_2Cu(NCS)_2$. After [1023].

sulfur atoms by ^{34}S and the outer-ring-carbon atoms of the $[(CH_2)_2]$ end-groups by ^{13}C, denoted by $^{13}C(4)^{34}S(8)$, which corresponds to a 5 % increase of the ET molecule's mass, a shift of $\Delta T_c = -(0.12 \pm 0.08)$ K has been observed, see Fig. 6.38. This value was obtained by measuring 19 crystals each of natural abundance and $^{13}C(4)^{34}S(8)$-labeled κ-$(ET)_2Cu(NCS)_2$. Figure 6.38 shows that each set of curves, despite being spread over $\Delta T \sim 0.1 - 0.15$ K shows a clean separation. Assuming that the whole ET molecule is the relevant mass entity M, this shift corresponds to $\alpha = 0.26 \pm 0.11$ in (6.48). This experiment provides clear evidence that the *inter*-molecular (lattice) phonon modes are strongly involved in the pairing mechanism. On the other hand, the same group demonstrated the absence of a comparable isotope effect on T_c for κ-$(ET)_2Cu(NCS)_2$ and κ-$(ET)_2Cu[N(CN)_2]Br$ upon a partial substitution of the central C=C and also a simultaneous replacement of both the central and ring C=C atoms. The same holds true for a substitution of the eight sulfur atoms, see [297, 1023] and references therein. These results indicate that the *intra*-molecular C=C and C−S bond-stretching vibrational modes of the ET molecule do not provide a substantial contribution to the Cooper pairing.

An *inverse* isotope effect on T_c has been observed for κ-$(ET)_2Cu(NCS)_2$ where T_c of deuterated samples κ-$(D_8$-$ET)_2Cu(NCS)_2$ was found to be higher than that of the hydrogenated salt, see [86]. This effect has been confirmed and quantified by the above-mentioned study where particular care has been taken to guarantee comparable quality of both the labelled and unlabelled crystals [297, 1023]. The physical reason for the inverse isotope effect is still unclear. Taking into account the results of thermal expansion and X-ray studies of the lattice parameters [348, 1024], it has been proposed that the inverse H-D isotope effect is not directly related to the pairing mechanism. Instead it has been attributed to a geometrical effect, i.e. changes in the chemical pressure: as the effectively shorter C–D bond [1025] corresponds to a higher chemical pressure perpendicular to the planes compared to that for the hydrogenated system, the negative values of $\partial T_c/\partial p_\perp$ (Table 6.2) result in a higher T_c for the deuterated compound [1026, 1027], see also [115]. A geometric H-D isotope effect has also been found for two other $(ET)_2X$ compounds κ_L-$(ET)_2Ag(CF_3)_4$(solvent) and β''-$(ET)_2SF_5CH_2CF_2SO_3$ having different packing motifs and anion structures. Although the T_c values are considerably different for these salts, the investigations reveal an almost identical "universal" shift of T_c of $\Delta T_c = +(0.26 \pm 0.06)$ K [1026, 1027].

An alternative explanation for the H-D isotope effect has been proposed recently by T. Biggs et al. based on their measurements of the Shubnikov-de Haas effect focussing on pressure-induced changes of the Fermi-surface topology of deuterated and protonated κ-$(ET)_2Cu(NCS)_2$ [462]. It has been suggested that the superconducting mechanism is most sensitively influenced by the exact topology of the Fermi surface. Since the latter has been found to change more drastically with pressure in the deuterated salt, this effect might also be responsible for the inverse isotope effect [462].

Further indications for a phononic contribution to the pairing interaction can be inferred from thermal conductivity measurements on κ-$(ET)_2Cu(NCS)_2$ and κ-$(ET)_2Cu[N(CN)_2]Br$ [955, 942, 457]. As discussed in Sect. 6.6, the significant enhancement of the thermal conductivity below T_c observed in these experiments has been assigned to an increase of the lattice thermal conductivity due to the condensation of the electrons into Cooper pairs. This contrasts with the situation for $(TMTSF)_2ClO_4$, where the lattice conductivity was found to remain unchanged by the superconducting transition [987].

A substantial contribution of the intermolecular phonons in the pairing interaction has been claimed also from the phonon renormalization measured by inelastic neutron scattering experiments [304]. As a consequence of the interaction of charge carriers with the phonon system, the opening of a gap in the electronic density of states below T_c induces changes in the phonon frequencies and linewidths. These effects were first observed in the classical superconductors Nb_3Sn and Nb [1028, 1029]. The results of these studies supported the generally accepted picture that superconductivity in these materials is phonon–mediated. Inelastic neutron-scattering experiments have been performed on single crystals of κ-$(ET)_2Cu(NCS)_2$ on both hydrogenated and deuterated crystals [1030, 304]. Due to the extraordinarily large incoherent cross section of the protons, the study of κ-$(H_8$-$ET)_2Cu(NCS)_2$ allows for a selective investigation of those vibrational modes that involve the hydrogen atoms at the ethylene endgroups. The analysis of measurements above and below T_c suggest a substantial coupling of these modes to the superconducting charge carriers [1030]. Figure 6.39 shows the results of inelastic

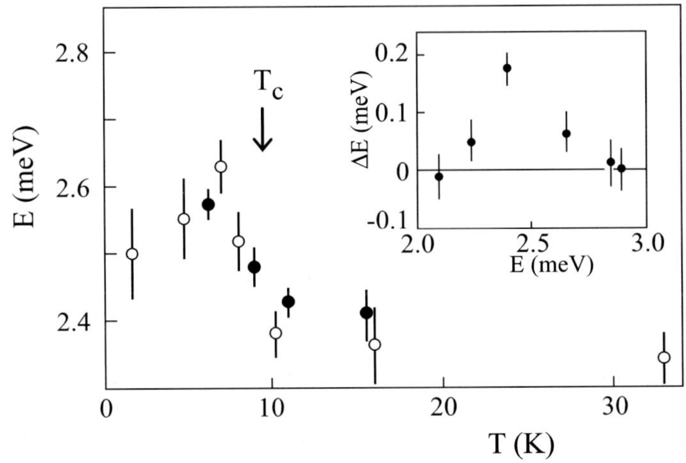

Fig. 6.39. Temperature dependence of the energy of the transverse acoustic phonon with wave vector $\boldsymbol{q} = (-0.225, 0, 0.45)$ derived from inelastic neutron scattering on two different single crystals (open and closed circles) of deuterated κ-$(ET)_2Cu(NCS)_2$. Inset: observed frequency shifts $[\Delta E = E(T < T_c) - E(T > T_c)]$ of transverse acoustic phonons in the $[-\zeta, 0, 2\,\zeta]$ direction, after [304].

neutron-scattering experiments on deuterated κ-$(ET)_2Cu(NCS)_2$ carried out by L. Pintschovius et al. [304]. The data reveal a sudden increase of the frequencies of transverse acoustic phonons upon cooling through T_c. This phonon hardening was found to be most pronounced for the wave vector $\boldsymbol{q} = (-0.225, 0, 0.45)$ and a phonon energy 2.4 meV. As discussed by R. Zeyher and G. Zwicknagl [1031], significant changes are expected only for those phonons whose energy $\hbar\omega$ is close to the gap value $2\,\Delta$ with a softening (hardening) for $\hbar\omega < 2\,\Delta$ ($\hbar\omega > 2\,\Delta$) [1031]. The above results thus imply $2\,\Delta \simeq 2.4$ meV, i.e. $2\,\Delta/k_BT_c \simeq 3.1$, which is close to the BCS ratio of 3.52. The salient feature of this study is that intermolecular modes strongly couple to the superconducting charge carriers and may thus provide a substantial contribution to the pairing interaction [304].

Likewise, a substantial hardening of phonon modes associated with the opening of the superconducting gap below T_c have been identified by a Raman scattering study on κ-$(ET)_2Cu[N(CN)_2]Br$ [88].

Universal Behavior

Other attempts to uncover the mechanism for superconductivity in the present class of molecular materials involves searching for universal trends. Here we mention the linear scaling between T_c and the superfluid stiffness $\rho_s = c^2/\lambda_L^2$ (λ_L is the London penetration depth), first identified by Y.J. Uemura et al. for the underdoped high-T_c cuprates [1032]. There ρ_s is directly proportional to the carrier density n, which is controlled by the doping level, and the linear T_c-ρ_s dependence has been attributed to a Bose-Einstein condensation of preformed pairs [1033] or to the effect of phase fluctuations of the superconducting order parameter [1034]. In contrast, for the molecular superconductors an approximate scaling of T_c with $\rho_s^{3/2}$ has been reported [858]. This scaling relation was found to hold for a wide range of molecular superconductors encompassing the quasi-1D $(TMTSF)_2ClO_4$, various examples of the quasi-2D $(ET)_2X$ superconductors, and also the three-dimensional fulleride superconductors [858]. Moreover, in an attempt to look for correlations between superconducting parameters and the normal-state conductivity σ_0, a marked difference to the cuprates has been observed: while for the molecular superconductors ρ_s decreases with increasing conductivity, the opposite trend has been reported for the cuprates [1035]. In [858] it has been argued that the simplicity of the scaling for the molecular superconductors suggests that this effect is controlled by one dominant parameter. One such possibility would be the ratio of the electron correlation energy U by the electronic band width W, cf. the proposed phase diagram for the κ-$(ET)_2X$ salts in Sect. 4.6. This interpretation would also be consistent with the effect of hydrostatic pressure which reduces U/W and which results in a depression of T_c and an increase of the normal state conductivity σ_0, see Sect. 4.3.1. However, as has been pointed out in [858], this contrasts to theoretical models – attempting to describe the transition from a Mott insulator to a superconductor – which predict that an enhanced T_c near the insulator state is accompanied by a depressed ρ_s [1015].

7 Phenomena of Interacting π- and d-Electrons

The interplay between magnetism and conductivity has continued to be an attractive subject in LDMM as it has been in inorganic magnetic materials with localized d- or f-electrons interacting with conduction electrons. Here we describe the phenomena of interacting π- and d-electrons in $(DCNQI)_2Cu$ with mixed-valent Cu ions and in quasi-2D $(BETS)_2FeX_4$ (X = Cl, Br) with localized d-spins.

7.1 $(DCNQI)_2Cu$

7.1.1 Crystal Structure and Electrical Conductivity

All the metal salts $(2\text{-R1,5-R2-DCNQI})_2M$ (M = Cu, Ag, Li) are isomorphous with space group $I4_1/a$, with a tetragonal unit cell [1036, 1037]; review articles are available in [1038, 1039, 1040, 40, 39]. The crystal structure of $(DMe\text{-}DCNQI)_2Cu$ ($R1=R2=CH_3$) is illustrated in Fig. 7.1. Planar DCNQI anions, uniformly stacked along the c-axis, form a columnar structure. These columns are interconnected with each other via Cu ions that are stacked like a string of pearls along the c-axis, and tetrahedrally coordinated with the local symmetry of D_{2d} with the N atoms of the terminal cyanoimino (=N–CN) groups. Thus the DCNQI molecule not only acts as an electron acceptor but also as a ligand for the metal ions. The bonding angle $\alpha = 122° - 127°$, which depends on the combination of the R1 and R2, is a measure for the degree of distortion from T_d symmetry, and therefore is a crucial parameter for the electronic states, see below. The overlap of DCNQI molecules within the stack is of the so-called ring-over-double bond type, resulting in a short inter-planar distance d of $3.13 - 3.24$ Å between DCNQIs. The tilting angle ϕ of the molecular plane from the chain axis is $33° - 38°$ depending on the ligands. This close stacking is responsible for the large transfer integral of π-orbitals along the c-axis. On the other hand, the Cu ions are separated from each other enough to prevent a direct metallic contact.

An extremely high conductivity in the range of $10^5\,\mathrm{S\,cm^{-1}}$ at low temperatures has been observed in $(DMe\text{-}DCNQI)_2Cu$, the first anion radical

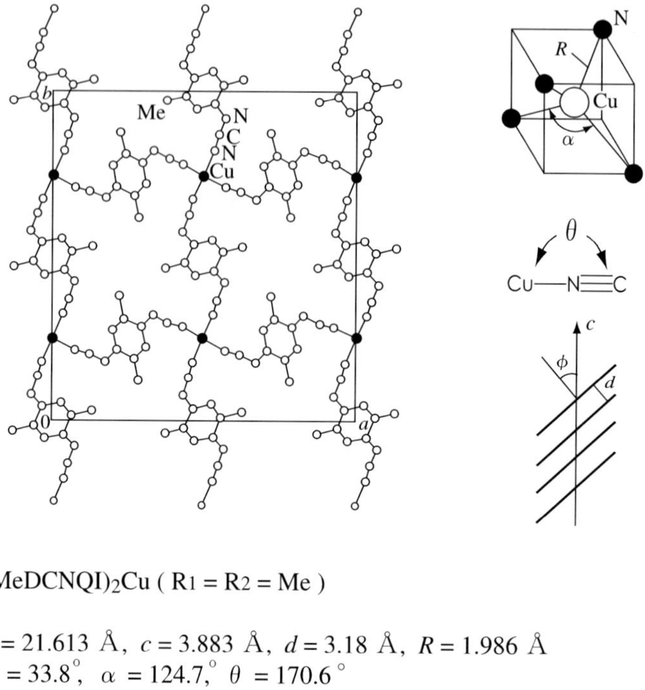

Fig. 7.1. Crystal structure projected along the c-axis, local molecule coordinations and structural parameters in $(DMeDCNQI)_2Cu$. After [40].

quasi-1D metal which is stable down to low temperatures without any metal-insulator (MI) transition.

Figure 7.2 shows the temperature dependence of the logarithmic conductivity of the various Cu salts. According to these conductivity behaviors a variety of salts can be classified into Groups I and II; DMe- and MeI-DCNQI salts belonging to Group I exhibit a metallic conductivity down to low temperatures, while the other salts with MeBr, MeCl and BrCl, belonging to Group II, undergo a first-order MI transition, respectively at 155 K, 210 K, and 213 K. The MI transition is sensitively controlled by applying hydrostatic pressure as well as chemical modifications such as alloying and isotope labeling. For example, the metallic state in the DMe salt exhibits not only an instability against forming an insulating state under small pressures $p > 100$ bar, but also a re-stabilization of the metallic state at slightly higher pressures [1041, 1042, 1043]. This so-called reentrant metallic transition, which is also observed in partially deuterated DMe salts and others, as e.g., the DMeO salt [1044], defines Group III.

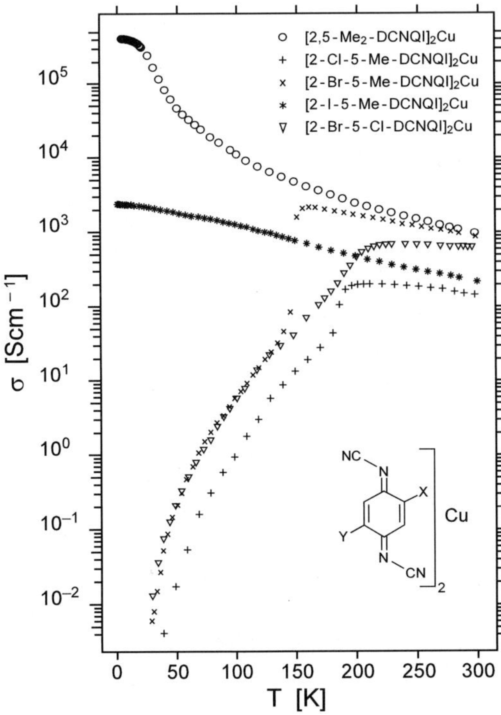

Fig. 7.2. Logarithmic electrical conductivity as a function of temperature in (2-R1, 5-R2-DCNQI)$_2$Cu. In the inset, R1 = X and R2 = Y. After [1038].

7.1.2 Reentrant Metal-Insulator Transition

The above-described three groups are mapped on the schematic pressure-temperature phase diagram for the Cu salts, as shown in Fig. 7.3(a) [40]. Here the horizontal axis measures not only physical but also chemical pressure controlled by substitution, alloying, and isotope labeling. For example, a selective deuteration R1 = R2= CH_2D by R. Kato and his coworkers [1045] clearly indicates the successive metal-to-insulator-to-metal phase transitions[1] as shown in Fig. 7.3(b). This deuteration for the methyl group induces an effective lattice contraction, in particular, a reduced inter-molecular spacing due to a steric effect caused by C–D bonds being shorter than C–H. This selective deuteration, with many different combinations of the sites and the number of deuterium atoms, made it possible for various measurements to be carried out on ambient-pressure samples with chemically-controlled electronic states.

[1] This reentrant transition was originally found in the methyl and fully deuterated (DMe-DCNQI-d_n)$_2$Cu (n = 3, 6, 8) by K. Sinzger et al. [1046].

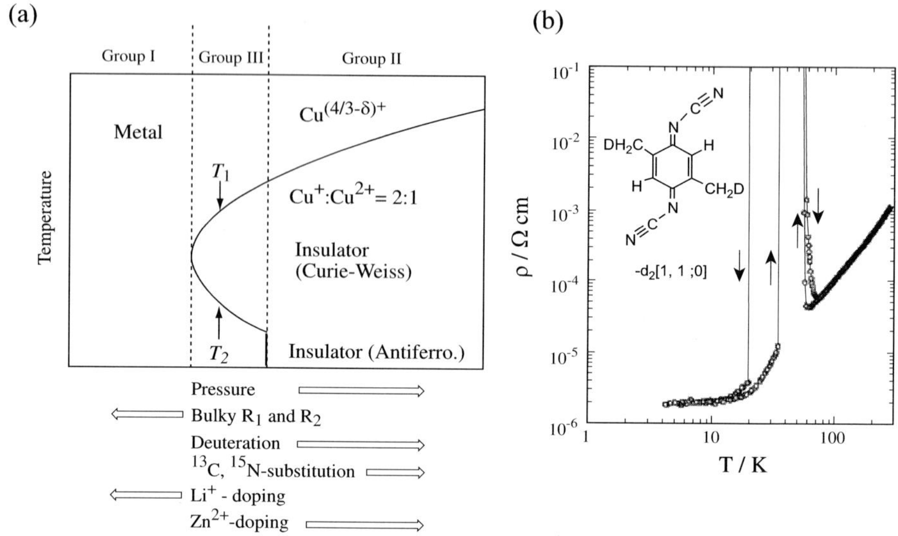

Fig. 7.3. (a) Schematic pressure-temperature phase diagram for $(DCNQI)_2Cu$ and (b) the reentrant metal-insulator transition observed in the selectively deuterated (DMe-DCNQI-d$_2$[1,1;0])$_2$Cu. After [40].

The electronic states and relevant physical properties are described within the p-T phase diagram. Satellite X-ray reflections[2] at $\boldsymbol{q} = (0, 0, c^*/3)$, corresponding to the three-fold superlattice formation ($a \times b \times 3c$), appear in the insulator phase of not only Group II [1037, 1047] and III compounds [1048, 1049] but also for Group I systems under high pressure [425]. This structural phase transition indicates that a commensurate $2k_F$-CDW is induced along the c-axis and the one-dimensional band in the high-temperature metallic phase is 1/3-filled. Therefore, the formal valences of DCNQI and Cu are $-2/3$ and $+4/3$, which is in contrast to that in the non-magnetic Ag salt (Ag^+), as described in Sect. 4.4.3. This unusual valence state of the Cu ions leads to a mixed-valent (or valence fluctuating) state in the metallic phase and a charge-ordered state of $-Cu^+Cu^+Cu^{2+}-$ in the insulating phase. These valence instabilities, responsible for the prominent magnetic behavior in both phases, were disclosed experimentally by X-ray photoemission spectroscopy (XPS) for the Cu $2p$ and/or N $1s$ core levels [1050] and were later studied by ^{1}H- [1051, 1052] and ^{63}Cu-NMR [1053].

Thus, the MI transition can be understood by the interplay between the π-electrons' Peierls instability and the $3d$-electrons' charge-ordering, which is attributed to the location of the $3d$-orbitals in the HOMO-LUMO

[2] Here, they emerge just at the first-order MI transition temperature without any diffuse scattering as seen in the metallic state of TTF-TCNQ (Sect. 4.4.1).

gap of about 1.5 eV. In particular, the $3d_{xy}$-orbital can be hybridized with the LUMO[3] so that an additional electron transfer of small quantity δ to $(DCNQI)^{-(2/3+\delta)}$ occurs, since it is separated from other d-orbitals due to the crystal-field effect which is sensitively tuned by the bonding angle α via a Jahn-Teller distortion. This MI transition has been discussed theoretically in terms of electron correlation effects [1056, 1057, 176].

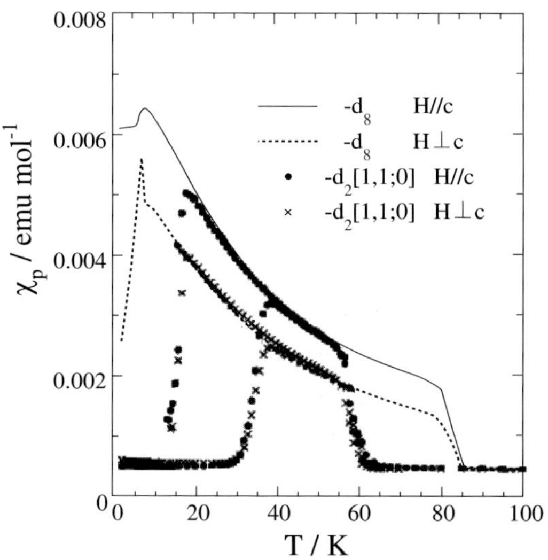

Fig. 7.4. Temperature dependence of the magnetic susceptibility for selectively deuterated (DMe-DCNQI-d_2[1,1;0]$)_2$Cu (filled circles and crosses for magnetic fields applied parallel and perpendicular to the chain axis, respectively) and the fully deuterated (DMe-DCNQI-$d_8)_2$Cu (solid and dotted lines). After [1058].

Figure 7.4 shows the magnetic susceptibility, demonstrating how the magnetic properties are directly related to the valence instability. The pristine (h_8) DMe salt (Group I) reveals a temperature-independent Pauli susceptibility (not shown). On the other hand, the fully deuterated (d_8) salt (Group II) exhibits a spontaneous appearance of the magnetization at $T_{MI} = 85$ K due to the spin 1/2 at the Cu^{2+} sites, which is followed by a Curie-Weiss law with a Weiss temperature $\Theta = -14$ K indicating the existence of an AFM exchange interaction [1059]. Upon further decreasing the temperature, AFM order appears at $T_N = 7$ K with a spin easy axis perpendicular to the c-

[3] The band structure calculated by the tight-binding approximation [1054] explains the ellipsoidal Fermi surface and its magnetic breakdown-like orbit with open sheets that has been observed in dHvA oscillations [1055]. The 3D Fermi surface was assigned due to the strong hybridization effect of the $3d_{xy}$-orbital with the LUMO. It has been confirmed fairly well by more elaborate calculations based on local-density approximation for (DMe-DCNQI$)_2$M (M = Cu, Li, Ag) [422], although a difference in the detailed dispersions of the hybridized bands just below E_F remains between these two calculations. In contrast to the 1D Li and Ag salts, the Cu salt is the first example with 3D electronic states at the Fermi level in coexistence with quasi-1D π-electronic states.

axis, which is the ground state of Group II materials, as shown in Fig. 7.3(a). For the selectively deuterated salt of Group III, however, the reentrant transition is seen in Fig. 7.4 [1058], where the susceptibility drops back to the temperature-independent Pauli paramagnetism. The corresponding hysteresis, as observed in the resistivity, Fig. 7.3(b), is seen also in the susceptibility (not shown in Fig. 7.4).

The reentrant transition in Group III compounds can be phenomenologically understood from a thermodynamic point of view [1060]. Taking into account both entropies $S_M = \gamma T$ for the hybridized (itinerant) electrons in the metallic (M) phase and $S_I = (1/3)N_A k_B \ln 2$ (= 1.9 J/mol K) for the localized spins at Cu^{2+} in the insulating (I) phase, the free-energy difference $\Delta F = F_M - F_I$ between these phases can be written as

$$\Delta F = E_M - \frac{1}{2}\gamma T^2 - (E_I - S_I T) = \Delta E_{MI} - \frac{1}{2}\gamma T^2 + S_I T, \qquad (7.1)$$

where $\Delta E_{MI} = E_M - E_I$ is the energy difference between the two phases. Here E_M and E_I stand for the energy gains due to the metallic band formation and the insulating charge ordering, respectively. ΔE_{MI} is assumed to be a parameter depending on the applied pressure, see Fig. 7.3(a). Since the phase transition is expected to occur at $\Delta F = 0$, the materials in Groups I – III are clearly classified as follows. When $\Delta E_{\mathrm{MI}} < -S_{\mathrm{I}}^2/2\gamma$, the metallic state is exclusively stable without any transitions as in Group I. Alternatively, two successive phase transitions in Group III are expected to show up when $-S_I^2/2\gamma < \Delta E_{MI} < 0$, while only one transition occurs in Group II when $\Delta E_{\mathrm{MI}} > 0$. This phenomenological theory, which can reproduce also the first-order nature of all the transitions with the latent heat $\Delta Q = (S_M - S_I)_{T_{MI}} \cdot T_{MI}$, explains the phase diagram at least semi-quantitatively. A quantitative comparison was made by specific heat measurements [1061]. The essential point is that the entropy of the spin degree of freedom in the charge-ordered insulating phase is almost independent of temperature, while the electronic entropy in the metallic phase is strongly temperature-dependent. It is notable that the vertical border line, Fig. 7.3(a), between the metallic and AFM insulating phase at low temperatures is due to a rapid decrease of the spin entropy in approaching the AFM order.

7.2 (BEDT-TSF)$_2$FeX$_4$ (X=Cl,Br)

The BETS molecule, which is a derivative of ET obtained by substituting the S atoms in the TTF skeleton by Se (Figs. 2.1 and 3.1), had been expected to be of great promise to bring forth new salts with stable metallic states because of its potential ability to enhance the intermolecular chalcogen-chalcogen contact due to the larger extension of the π-orbital wavefunction in Se. Employing the titanocene-involving synthetic route developed by R. Kato et al. [1062, 1063], a number of 2:1 BETS salts were synthesized with a variety of counter anions, in which quasi-2D metallic states at quarter filling remain stable. Novel phenomena induced by π-d interactions have been observed in the magnetic-anion salts with tetrahalometalates FeX_4(X=Cl, Br); review articles are available in [66, 67, 1064, 1065].

7.2.1 Crystal Structures and Transfer Integrals

The two different crystal morphologies called κ and λ are shown in Figs. 7.5 and 7.6 [121, 1066]. The κ-type salts (X=Cl, Br) have face-to-face coupled dimers in the most conducting ac-plane, similar to the κ-ET salts forming an orthorhombic structure with the space group $Pnma$. In the anion layer, the shortest Fe–Fe distance is about 5.88 Å with short contacts of 4.12 – 4.13 Å between chalcogen ions.

The λ-type[4] salt (X = Cl) is quite unique: The crystal symmetry is triclinic with the space group of $P\bar{1}$ having only one inversion center except for the translational symmetry operation. The BETS donors are stacked along the a-axis, forming a tetradic, zigzag column with two independent BETS molecules. As seen in Fig. 7.6, the tetrahedral $FeCl_4$ anions are located in between these columns with a *key-to-key hole* packing motif. The rather

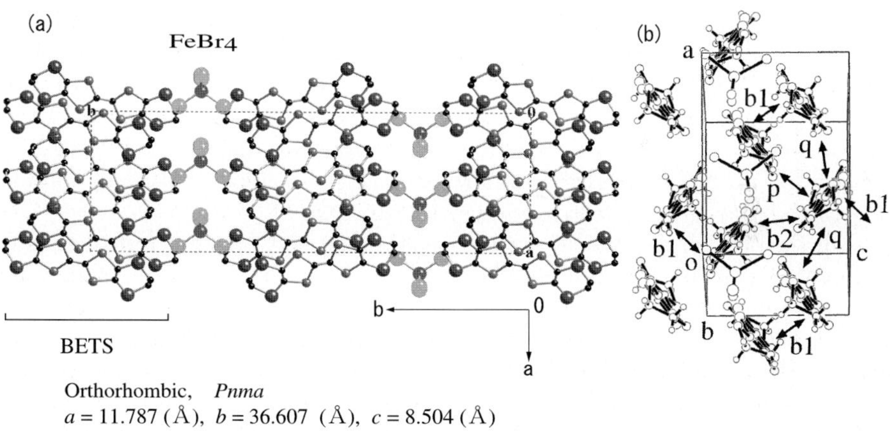

Fig. 7.5. Crystal structure of κ-(BETS)$_2$FeBr$_4$: (a) unit cell projected along the c-axis and (b) perspective view nearly along the interlayer b-axis (after [1067]). The tetrahedral $FeBr_4$ anions are located alternately along the b-direction. In the most conducting ac-planes (b) the BETS molecules form dimers overlapping with each other in a ring-over-double bond manner, and adjacent BETS dimers are arranged in a roughly orthogonal manner, characteristic of the κ-type packing motif. The overlap integrals (in units of 10^{-3}) [1067] are: b_1 = 29.2, b_2 = 18.0, p = 8.1 and q = −7.0.

[4] This structure is also seen in ET tetrahalometalates like (ET)$_2$MX$_4$ (M = Fe, Ga; X = Cl, Br) [1068, 1069, 1070, 1071].

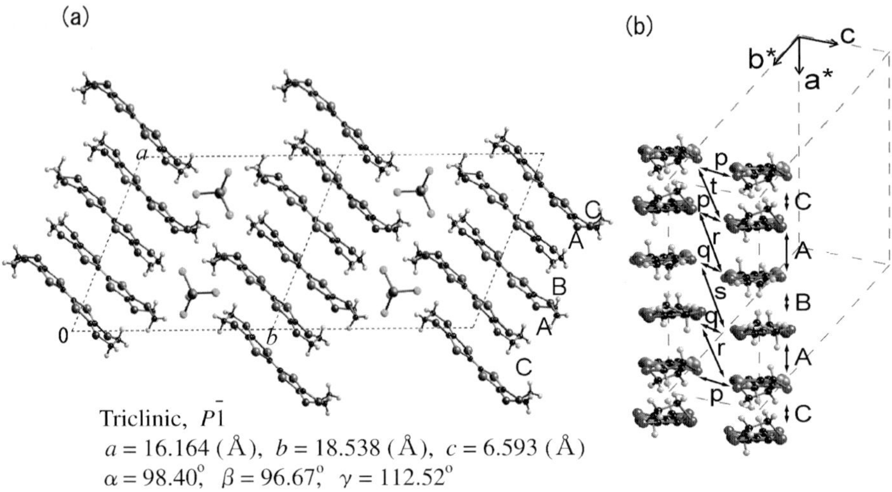

Fig. 7.6. (a) Crystal structure (double unit cells) of λ-$(BETS)_2FeCl_4$ projected along the c-axis and (b) the perspective view featuring a well-developed columnar structure in BETS stacks. Overlap integrals (in units of 10^{-3}) [1067] are: $A = 33.55$, $B = 18.27$, $C = 14.83$, $p = -2.80$, $q = -9.30$, $r = 13.00$, $s = 17.11$ and $t = 2.56$.

strong coupling along the c-axis between columns makes the ac-plane the most conductive one. The planes are bridged by the anions through several coupling routes between Se and Cl, shorter than the van der Waals radii. These characteristic couplings lead to sizable π-d exchange interactions via the Cl $3p$-orbitals, as will be described later. To note, the shortest Fe–Fe distance is about 6.59 Å which is somewhat longer than that of κ-type systems due to the above described characteristic packing.

For the overlap integrals S_{ij} of the π-orbitals [1067] (see the figure captions of Fig. 7.5), the κ-salt exhibits a strong dimerization in the ac-plane such that the intra dimer b_1 is larger by a factor of 1.6 than the inter dimer b_2, similar to the κ-ET salts. In the λ-salt having the tetradic sequence -A-B-A-C- along the column, the overlap integral A is almost twice as large as B and C. The reason is attributed to the peculiar face-to-face configuration of an ideal ring-over-double bond realized solely for A [1066]. For the intercolumn coupling due to the side-by-side configuration, the overlap integrals, in particular, along the routes of r and s are fairly large (comparable to B and C), resulting in a 2D conducting plane. It is important to note that these coupling routes provide a bridge across the dimers (A) on neighboring columns. (For a comparison to other calculations for the transfer integrals [1066, 1072], refer to the appendix in [1067].)

7.2.2 Conductivity and Magnetism

The κ- and λ-salts exhibit common features in transport and magnetic properties at high temperatures as shown in Figs. 7.7 and 7.8. One is a broad maximum in the electrical resistance around 70 K (κ) and 100 K (λ), similar to what is observed for the κ-ET salts as discussed in Sect. 4.3.1. Another is a paramagnetic metallic (PMM) state following a Curie-Weiss law,

$$\chi = \frac{C}{T-\Theta} = \frac{N_A(g\mu_B)^2 S_d(S_d+1)}{3k_B} \cdot \frac{1}{T-\Theta}, \tag{7.2}$$

where N_A is Avogadoro's constant. A localized spin $S_d = 5/2$ at Fe^{3+} and Weiss temperatures of $\Theta = -5.5$ K (κ) and -10 K (λ) have been obtained. These magnetic properties indicate that the crystal-field effect is strong enough to stabilize the high-spin state, and that there is an antiferromagnetic interaction among the spins. These high-temperature properties, independent of the crystal morphology, however, are followed by quite different properties at low temperatures.

κ-(BETS)$_2$FeBr$_4$

In κ-(BETS)$_2$FeBr$_4$, the susceptibility indicates AFM order at $T_N = 2.5$ K associated with a slight drop of the resistivity [1073]. The spin easy axis is

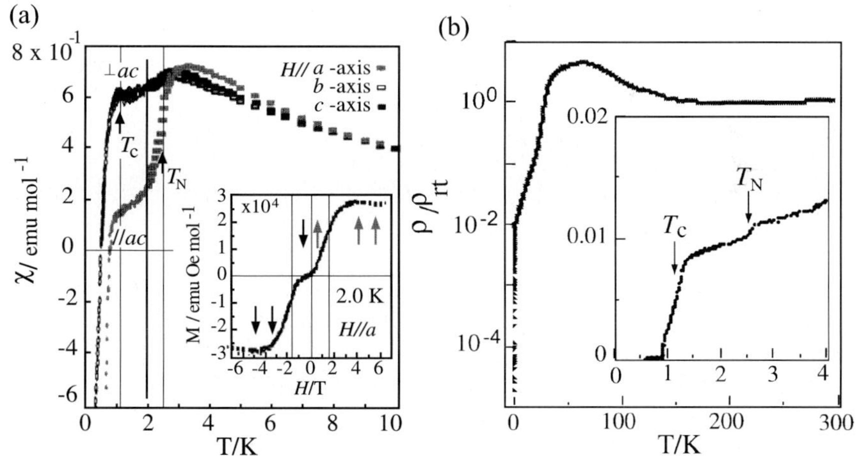

Fig. 7.7. (a) Temperature dependence of the magnetic susceptibility of κ-(BETS)$_2$FeBr$_4$ down to 60 mK for fields parallel and perpendicular to the *ac* conduction plane. Both antiferromagnetic and superconducting transitions are seen around $T_N = 2.5$ and $T_c = 1.1$ K. The inset shows the magnetization curve for the field applied parallel to the spin easy axis and the metamagnetic transition around 1.6 T. (b) Temperature dependence of the in-plane electrical resistivity ρ normalized to the room-temperature value. The inset shows the low-temperature resistivity behavior with anomalies at T_N and T_c. After [1073].

parallel to the a-axis and the metamagnetic transition (spin-flop) occurs at $B_{sf} = 1.6\,\mathrm{T}$. (In the FeCl4 salts with $T_N = 0.45\,\mathrm{K}$ and $T_c = 0.1\,\mathrm{K}$ [1074], $B_{sf} = 0.6\,\mathrm{T}$ [1075].) The specific heat exhibits a sharp λ-shaped peak at T_N and the entropy reaches $R\ln(2S_d + 1) = R\ln 6$, as expected for the high-spin $S = 5/2$ [1073]. This AFM metal undergoes a superconducting transition at $T_c = 1.1\,\mathrm{K}$. Muon spin rotation and relaxation measurements, carried out on these salts, are reported in [876, 1076].

λ-(BETS)$_2$FeCl$_4$

In λ-(BETS)$_2$FeCl$_4$, however, the PMM state turns into an AFM insulating state (AFMI) at $T_N(= T_{MI}) = 8.3\,\mathrm{K}$. (Note that the isostructural, non-magnetic GaCl$_4$ salt is a superconductor with $T_c = 4.8\,\mathrm{K}$ [121, 1078, 1066]). At T_{MI}, the localization of the π-electrons manifests itself in the occurrence of a quantum spin $s_\pi = 1/2$ which is proved by the magnetization drop just below T_{MI}, and thus the MI transition accompanies an s_π-S_d coupled AFM order [1079]. (Note that the ratio of the magnetization contributions from these spin systems, $s_\pi(s_\pi + 1)/S_d(S_d + 1)$ is about 0.086.) Indeed, the two-sublattice model [1080] for these coupled spins can successfully explain the AFM resonance [1080, 1081, 1082] that is attributed to the spin waves excited on the π-spin sublattice. The spin easy axis tilts from the c-axis by about $30° - 40°$ [67, 1083] and the metamagnetic transition occurs at $B_{sf} = 1.5\,\mathrm{T}$.

The conducting and magnetic properties in the λ-salts are sensitively dependent on pressure and anion alloying, see, e.g. [66, 67]. For example, the

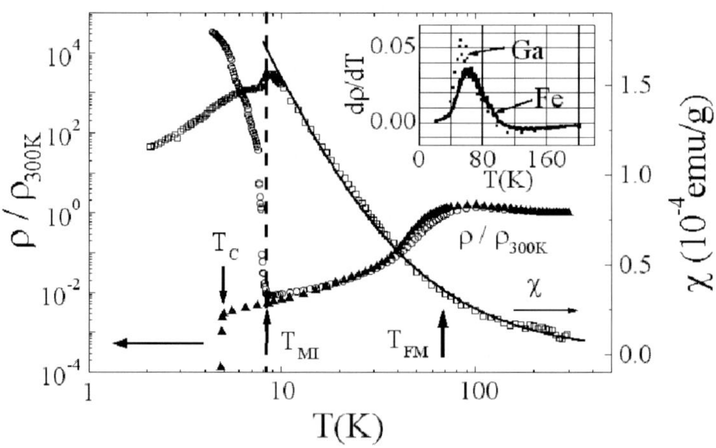

Fig. 7.8. Temperature dependence of the in-plane electrical resistivity normalized to its room-temperature value and magnetic susceptibility with the magnetic field parallel to the ac-plane for λ-(BETS)$_2$FeCl$_4$. The resistivity for the isostructural, superconducting GaCl$_4$ salt with $T_c = 4.8\,\mathrm{K}$ is also shown for comparison. The solid line for the susceptibility data shows the Curie-Weiss dependence and the inset the temperature derivative of ρ. After [1077].

isostructural mixed salts with $0.35 < \mathrm{x} < 0.5$ in $\mathrm{Fe_xGa_{1-x}Cl_4}$ undergo successive transitions such that the PMM state turns into a superconductor around 4 K, and then transforms to the AFMI ground state at lower temperatures [124].[5] These results clearly indicate that the AFMI state, which is brought about via magnetic interactions with localized d-spins, is energetically in close proximity to superconductivity as observed in the nonmagnetic $\mathrm{GaCl_4}$ salt, see Sect. 7.2.6.

7.2.3 Magnetic Interactions and Mean-Field Theories

Magnetic Exchange Interactions Based on Mori-Katsuhara Theory

Following the Mori and Katsuhara theory [1067] based on intermolecular overlap integrals, we first introduce the magnetic exchange interactions responsible for the AFM ordering transitions mentioned above. Hereafter we define the spin operators $\boldsymbol{S}_i$ and $\boldsymbol{s}_i$ for d-and π-spins at the i-th site, respectively.

Firstly, the exchange-interaction Hamiltonian between neighboring localized d-spins is given by

$$\mathcal{H}_{dd} = -2 \sum_{<i,j>} J_{dd}(i,j) \boldsymbol{S}_i \cdot \boldsymbol{S}_j, \tag{7.3}$$

where the coupling constant $J_{dd}(< 0) = -2t_{dd}^2/U_{dd}$. Here t_{dd} is the transfer integral (energy) between magnetic anions via $\mathrm{FeX_4 \cdots X_4Fe}$ and U_{dd} is the on-site Coulomb energy of d-orbitals assumed to be 1 eV. The numerical values of J_{dd}/k_B are very small: -0.64 K for λ-(BETS)$_2$FeCl$_4$ and -0.36 K for κ-(BETS)$_2$FeBr$_4$. The difference is simply attributed to the closer Cl–Cl contacts (3.56 – 3.75 Å) in the λ-salt compared to the Br–Br ones (4.14 Å) in the κ-salt.

Secondly, the so-called π-d interaction is expressed by the coupling constant $J_{\pi d}$ between π- and d-spins as

$$\mathcal{H}_{\pi d} = -2 \sum_{<i,j>} J_{\pi d}(i,j) \boldsymbol{S}_i \cdot \boldsymbol{s}_j, \tag{7.4}$$

where $J_{\pi d} = -2t_{\pi d}^2/\Delta_{\pi d}$. Here $t_{\pi d}$ is the transfer integral and $\Delta_{\pi d}$ is the energy difference between the donor HOMO and d-like molecular orbitals at $\mathrm{FeX_4}$. Since it is mediated via $3p$-orbitals of $\mathrm{X^-}$ located about 5 eV lower than the HOMO, this interaction is called a superexchange π-d interaction.

[5] To the authors' knowledge, among all known superconductors, this is the first system giving place to an insulating ground state at a lower temperature. An interplay between superconducting and insulating ground states is found in thin films of Bi and Pb [1084].

A strong intra-atomic exchange interaction (Hund coupling) induces a high-spin configuration ($S_d = 5/2$) in which the five d-orbitals (3-fold degenerate t_{2g} and 2-fold degenerate e_g) are occupied by unpaired electrons. To note, t_{2g} is located lower than the HOMO by 0.9 eV and the average magnitude of $\Delta_{\pi d}$ is assumed to be 1.0 eV. The total $J_{\pi d}$ is obtained by taking into account all these five d-orbitals, each coupled to neighboring seven (six) different π-orbitals for the λ-(κ-)type systems. An exclusively large $J_{\pi d} = -14.14$ K is obtained along the coupling channel Se(7)-Cl(2) (3.531 Å) for λ-$(BETS)_2FeCl_4$, while $J_{\pi d} = -1.22$ K along S(1)-Cl(3) (3.757 Å) for κ-$(BETS)_2FeBr_4$. These comparative results indicate that the λ-salt has a much stronger π-d exchange than the κ-salt. The reason is explained straightforwardly by the peculiar packing motif of BETS and $FeCl_4$ anions in the λ-salt, as described in Sect. 7.2.1.

With these exchange couplings, the magnetic structure can be discussed for both λ- and κ-salts with a dimer model in which one π-electron with $s_\pi = 1/2$ is located at the center of each dimer. For the present case, a third exchange interaction is introduced between π-spins:

$$\mathcal{H}_{\pi\pi} = -2 \sum_{<i,j>} J_{\pi\pi}(i,j) \boldsymbol{s}_i \cdot \boldsymbol{s}_j . \tag{7.5}$$

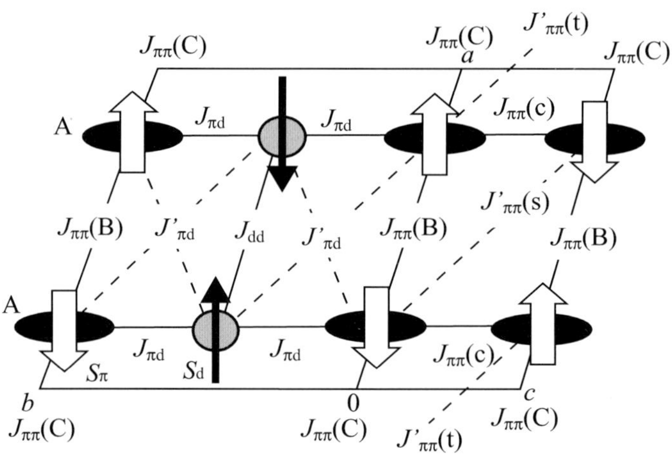

Fig. 7.9. The dimer model [1067] for the AFM spin arrangement in the ab-(left) and ac-planes (right), based on interdimer π-π interactions consisting of the intra-column $J_{\pi\pi}(C)$ and $J_{\pi\pi}(B)$ and the intercolumn $J_{\pi\pi}(c)$, $J_{\pi\pi}(t)$ and $J_{\pi\pi}(s)$. Open and closed arrows represent, respectively, $s_\pi = 1/2$ localized at the dimer A (Fig. 7.6) and $S_d = 5/2$ at Fe^{3+}. The solid and dotted lines connecting these spins indicate the cooperative and frustrated (denoted with prime) exchange couplings for $J_{\pi d}$ and $J_{\pi\pi}$. The symbols B and C in the parenthesis of $J_{\pi\pi}$ correspond to the transfer integrals as shown in Fig. 7.6.

Here $J_{\pi\pi}$ is given by $-2t_{\pi\pi}^2/U_{\pi\pi}$ where $t_{\pi\pi}$ is the interdimer transfer integral and $U_{\pi\pi}$ is the on-site Coulomb energy of the donor HOMO. In the strong dimerization limit, the effective $U_{\pi\pi}$ can be approximated to double the intradimer transfer integral[6] [1085], which is assumed to be 1 eV. Since the metallic state remains stable even below T_N in the κ-salt, we concentrate here on the insulating λ-salt with localized π-electrons below T_N. Figure 7.9 illustrates the AFM spin configurations expected from the dimer model. Three interdimer interactions work cooperatively giving $J_{\pi\pi} = J(B) + J(C) + 2J(c) = -622\,\mathrm{K}$, while the other two interactions give the frustrated coupling $J'_{\pi\pi} = J(t) + J(s) = -174\,\mathrm{K}$. The effective π-π exchange, $J_{\pi\pi} - J'_{\pi\pi}$, is then $-448\,\mathrm{K}$. (Note that for the κ-salt with a triangular dimer arrangement, frustration interactions are important and can be compared with κ-ET salts [1086].)

To evaluate T_N for both salts, mean-field approximations were used based on an itinerant model instead of the dimer model. With the Ginzburg-Landau free energy, T_N and Θ were discussed in terms of two independent contributions from the direct J_{dd} and indirect $J_{\pi d}$. As expected from above estimations for the exchange constants, T_N of λ-$(BETS)_2FeCl_4$ with the π-d-coupled AFM order is determined almost equally by both interactions, while that of κ-$(BETS)_2FeX_4$ predominantly by the direct interaction. For the latter AFM order of S_d spins, $J_{\pi d}$ plays a role in the interlayer magnetic coupling.

Kondo-Hubbard model

We proceed further to discuss the π-d interaction-induced phenomena with a so-called Kondo-Hubbard model [1081, 1087, 677, 1088]. The Hamiltonian,

$$\mathcal{H} = \sum_{i,j,\sigma}(t_{ij}c_{i\sigma}^{\dagger}c_{j\sigma} + h.c.) + U_{\pi\pi}\sum_{i} n_{i\uparrow}n_{i\downarrow} - 2J_{\pi d}\sum_{i,j}\boldsymbol{S}_i \cdot \boldsymbol{s}_j, \tag{7.6}$$

consists of three terms, the 2D tight-binding hopping, the on-site Coulomb repulsion $U_{\pi\pi}$, and the π-d exchange. Although the direct exchange interaction term between d-spins is absent here, it can be derived as a Ruderman-Kittel-Kasuya-Yoshida (RKKY) interaction, J_{RKKY}, mediated by a spin polarization of π-electrons, see, e.g. [1089]. The magnitude of J_{RKKY} is a rough measure for T_N, given by

$$J_{RKKY} \cong J_{\pi d}^2\widetilde{\chi}(\boldsymbol{q}) \sim k_{\mathrm{B}}T_N. \tag{7.7}$$

Here $\widetilde{\chi}(\boldsymbol{q})$ is the staggered susceptibility renormalized by $U_{\pi\pi}$,

$$\widetilde{\chi}(\boldsymbol{q}) = \frac{\chi(\boldsymbol{q})}{1 - U_{\pi\pi}\chi(\boldsymbol{q})}, \tag{7.8}$$

[6] The effective on-site Coulomb energy on the dimer is given by $U_{dimer} = 2t_d + (U/2)\left(1 - \sqrt{1 + (4t_d/U)^2}\right)$, where t_d and U are the intra-dimer transfer energy and on-site Coulomb energy on each site, respectively.

where $\chi(\boldsymbol{q})$ is the electronic response function (4.24). For both salts the band-structure calculations give a peak of $\chi(\boldsymbol{q})$ around $\boldsymbol{Q}_{AF} = (0, 0, \pi/c)$ approximately corresponding to the $2k_F$ vectors spanning two open FS sheets. The averaged values of $\chi(\boldsymbol{q})$ are 2.4 states/eV and 1.85 states/eV, respectively for the λ-(BETS)$_2$FeCl$_4$ and κ-(BETS)$_2$FeBr$_4$ [1067]. Therefore the AFM instability at $Q_{AF} = 2k_F$ is found to be more enhanced in λ-(BETS)$_2$FeCl$_4$ than in κ-(BETS)$_2$FeBr$_4$. Indeed, ^{77}Se- and ^{1}H-NMR measurements [1090, 1091] on λ-(BETS)$_2$GaCl$_4$ revealed the existence of strong AFM spin fluctuations in the metallic state.

The mean field calculations for λ-(BETS)$_2$FeCl$_4$ are summarized as follows. With the use of $J_{\pi d} = 8\,\mathrm{meV}$ (93 K) and $\widetilde{\chi}(\boldsymbol{Q}_{AF}) = 80$ states/eV, T_N was estimated to be 5 meV($\sim$ 60 K) [1081], while the AFMI transition was interpreted as a Mott transition due to the strong dimerization at the critical $U_{\pi\pi} \sim 0.27\,\mathrm{eV}$ [1087]. On the other hand, in order to explain the phase diagram for λ-(BETS)$_2$Fe$_x$Ga$_{1-x}$Cl$_4$ [124], the inherent superconductivity was discussed by adding an effective BCS Hamiltonian, see e.g. [365] to (7.6) [1088]. There, the π-d exchange interaction breaks up Copper pairs and hence could work as a strong pair breaker, reducing T_c as strongly as usually found in the so-called Kondo superconductor [836, 492, 1092, 1093]. The unusual insensitivity of T_c to x, however, was ascribed to a strong reduction of the effective magnetic-impurity scattering due to the electron correlation effect with $U_{\pi\pi} = 43\,\mathrm{meV}$ and $J_{\pi d} = 2.5\,\mathrm{meV}$.

These so-called π-d exchange interaction-induced phenomena have readdressed long-term issues of magnetism and conductivity/superconductivity, which so far have been extensively studied in s-d or s-f interacting systems containing transition-metal and rare-earth (RE) ions [1093, 252]. As will be described next, some phenomena specific to the present low-dimensional, magnetic molecular metals have been found both in the high-temperature PMM and low-temperature AFMI states.

7.2.4 Anomalous Metallic States

Microwave Conductivity and ^{1}H-NMR

High frequency conductivity, $\tilde{\sigma}(\omega) = \sigma_1 + \mathrm{i}\sigma_2$, measured by a microwave-cavity perturbation technique, suggested an anomalous metallic state in λ-(BETS)$_2$FeCl$_4$ [1094, 1095]. The resonance width along the c-axis shows a broad maximum around 100 K, followed by a shallow minimum around 30 K. While the maximum is in accordance with the broad peak of the dc resistivity ρ_{dc}, as shown in Fig. 7.8, the latter minimum indicates the existence of an anomalous microwave loss in the PMM state, which is absent in ρ_{dc} monotonously varying with temperature.

Figure 7.10 shows the conductivity σ_1 and the dielectric constant $\varepsilon_1 \equiv 1 - (4\pi/\omega)\sigma_2$, where $\omega = 2\pi f$, at $f = 44.5\,\mathrm{GHz}$ as a function of temperature.

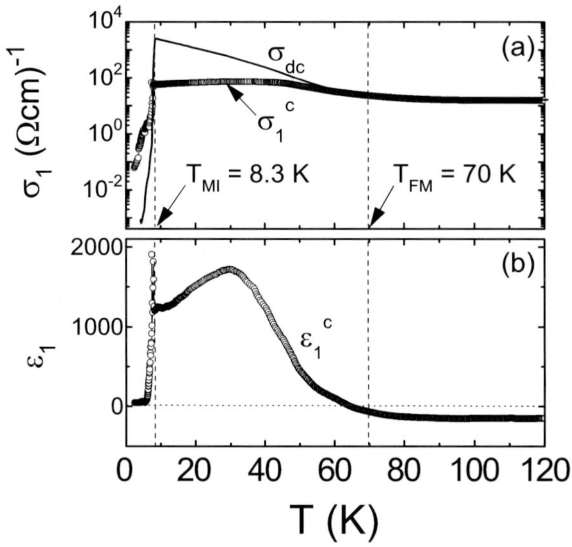

Fig. 7.10. Temperature dependence of the microwave conductivity σ_1 and the dielectric constant ε_1 at 44.5 GHz along the c-axis in λ-$(BETS)_2FeCl_4$. The dc conductivity σ_{dc} along the c-axis is plotted by a solid line. After [1095].

σ_1 becomes smaller than σ_{dc} around $T_{FM} \sim 70\,\mathrm{K}$ and, with further decreasing temperature, it decreases to a value almost as small as $0.01\sigma_{dc}$. On the other hand, the sign of ε_1 changes from negative to positive around T_{FM}. In particular, the large and positive ε_1 along the c-axis appears at lower temperatures in the PMM state and shows a sharp peak near T_{MI}, followed by a steep decrease in the AFMI ground state.

The broad peak around 30 K in ε_1 was speculated [1095] to be caused by the appearance of some relaxor ferroelectrics domains [1096, 1097] in the metallic background. It should be pointed out that the analysis, based on the electromagnetic formulation [1098, 271] standing for a homogeneous state, is in contradiction to the speculated heterogeneous state. Therefore the results in Fig. 7.10 are questionable. As will be described later, X-ray diffraction studies [1099] provided evidence for a domain structure above T_{MI}.

On the other hand, in ^{1}H-NMR [1100] spectra, an anomalous line shape in the metallic states has been found. Figure 7.11 shows the spin-echo intensity around the resonance field H_0, indicating a clear splitting of the resonance below 70 K ($\sim T_{FM}$). The Gaussian deconvolution analysis reproduces the resonance profiles with three independent components P_c, P_{s1} and P_{s2}. Asymmetric positions of two side peaks P_{s1} and P_{s2} suggest the existence of three types of proton sites experiencing different local fields below T_{FM}. Since the magnetic susceptibility in this temperature range shows a simple Curie-Weiss behavior without any observable anomalies (Fig. 7.8), the anomalous splitting was attributed not to the local fields produced by the interaction between the proton nucleus and d-spins on Fe^{3+} ions, but to the hyperfine

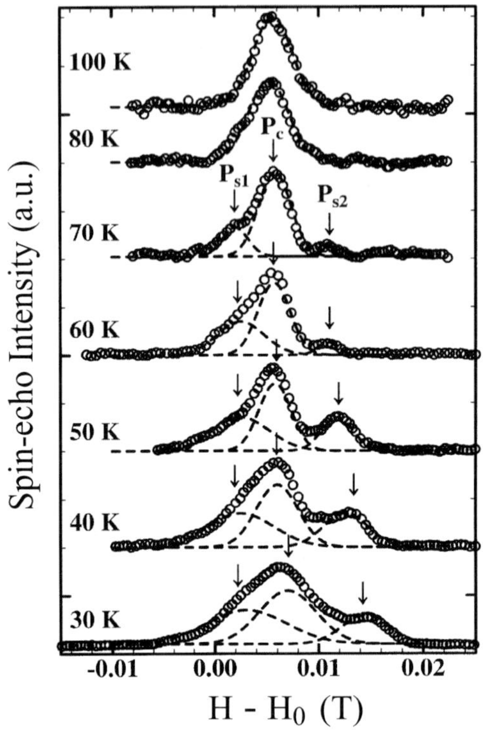

Fig. 7.11. Profiles of field-swept ^{1}H-NMR spectra (94.95 MHz) in λ-$(BETS)_2FeCl_4$. The dashed lines indicate the Gaussian deconvolution with three components, P_c, P_{s1}, and P_{s2}, the peak positions of which are shown by arrows. After [1100].

field between the proton nucleus and π-electrons. Here a possible change in the local charge density can be induced by a distribution of local fields. As a consequence, a charge order (or charge disproportionation) was discussed as a possible cause for the splitting below T_{FM}. To note, this speculation was supported by ^{13}C-NMR measurements detecting an anomalous line broadening [1101]. Taking into account the heterogeneous state speculated from the anomalous microwave conductivity, it may be reasonable to suppose that a uniform metallic state changes into some phase-separated states in which a domain pattern, consisting of metallic and dielectric regions, may appear. There, the metallic region seems to be responsible for the shift of the central peak, while the dielectric domains with modulated charge densities along the a-axis may be responsible for the side peaks.

Structural Anomalies

X-ray diffraction measurements of a peak profile of the (007) Bragg reflection revealed structural anomalies, both in the vicinity of T_{MI} and in the anomalous metallic region $T_{MI} < T < T_{FM}$ [1099, 1102]. Around T_{FM}, the reflection profile becomes broadened and deformed, clearly followed by a splitting at lower temperatures. Figure 7.12(a) shows the splitting seen in the intensity map obtained by a three-dimensional scan in the reciprocal space [1099]. The magnitude of the splitting increases with decreasing temperature, and the splitting direction and magnitude were determined to be $0.017\,\boldsymbol{a}^* - 0.010\,\boldsymbol{b}^* + 0.002\,\boldsymbol{c}^*$. This was assigned to be a change of 0.005 Å in the interplanar d-spacing along the c-axis. The splitting peaks giving different d^* values can be ascribed to a phase-separated, heterogeneous structure. The domain size was evaluated to be in a mesoscopic scale of about $0.3 - 0.4\,\mu$m.

These anomalies are consistent with the scenario for charge-ordered dielectric domains, as speculated from microwave conductivity [1094, 1095] and ^{1}H-NMR [1100]. Taking into account the relaxation time estimated to be in the order of 10^{-13} s or less, the shielding length of π-electrons in a unit cycle of the microwave is estimated to be in the order of $0.01\,\mu$m or less which is much smaller than the domain size. Therefore, the dielectric domain would absorb the microwave energy, and hence a positive dielectric constant should be observable, even in such a heterogeneous metallic state.

At T_{MI} a discontinuous structural change in the c^*-axis appears, while the splitting disappears so that the profile with a single peak becomes uniform again but remains broad. The temperature dependence of the lattice constant $1/c^*$, calculated from each splitting peak above T_{MI}, is shown in Fig.7.12(b). The solid circles are taken from earlier experiments [1102] by the same group with a lower resolution in which the high-temperature profiles were treated as a single broad peak. The high-angle data indicate a discontinuous change of

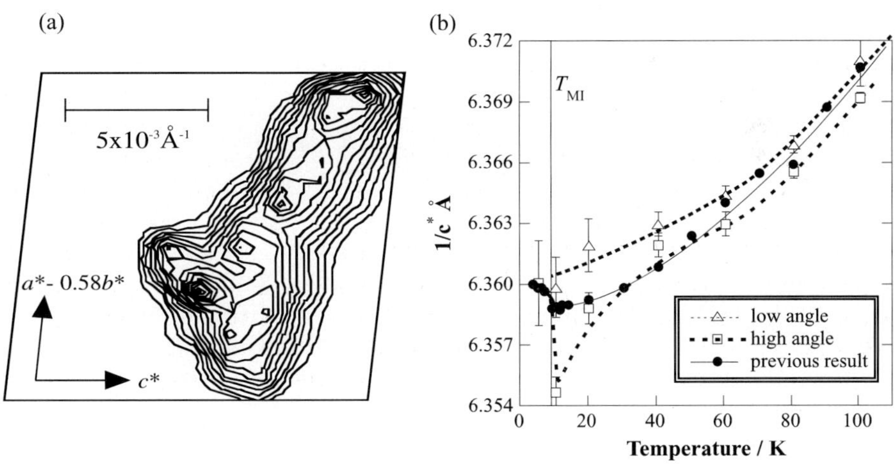

Fig. 7.12. (a) Sectional plane in the reciprocal space around the (007) Bragg reflection in λ-(BETS)$_2$FeCl$_4$ at 20 K. The data are obtained from a three-dimensional scan in the reciprocal space. (b) Temperature dependence of the lattice constant $1/c^*$, interplane (ab) spacing, derived from the high- and low-angle points in the splitting peaks in (a). The previous data with a lower resolution obtained by the same group [1102] are plotted by the solid circles for comparison. After [1099].

$1/c^*$ at T_{MI} more clearly, providing evidence for a first-order structural phase transition associated with the PMM-AFMI transition. On the other hand, the lattice constant, derived from the low-angle data, passes through T_{MI} rather smoothly within the experimental error, suggesting that the dielectric domain may have the same structure as that in the low-temperature AFMI state. Although the structural analysis is not conclusive, a uniform transformation from $P\bar{1}$ to the unique subgroup $P1$, the most primitive space group without the inversion symmetry of $P\bar{1}$, is most certainly expected to occur [1099].

7.2.5 Antiferromagnetic Insulating States

Nonlinear Transport and Switching Effect

The current-voltage (J-E) characteristics[7] along the c-axis have been measured in the well-ordered AFMI state [1103, 1104]. Figure 7.13(a) shows J-E curves at 4.2 K in magnetic fields up to 14 T applied along the interlayer b^*-direction normal to the ac-plane [1104]. A so-called negative-resistance (NR) effect is clearly seen at $B < 6\,\mathrm{T}$, followed by a monotonous nonlinear transport without NR at higher B and eventually by a conventional ohmic transport in the reentrant metallic states at $B = 12$ and $14\,\mathrm{T} > B_{MI} = 11\,\mathrm{T}$ (4.2 K). (Here B_{MI} is the critical field for the AFMI phase to reenter the PMM phase which will be described in Sect. 7.2.6.) The discontinuous drop of E in the low-J region is a switching phenomenon between low- and high-conducting states.

The empirical relation for the temperature and current-density dependent conductivity $\sigma(T, J)$ is expressed as

$$\sigma(T, J) = \sigma_A \exp\left(-\frac{\Delta}{k_B T}\right) + \sigma_B J^n, \tag{7.9}$$

where σ_A and σ_B are constants. This equation was successfully applied to similar phenomena in a variety of quasi-1D mixed- or segregated-stack molecular insulators [1105, 1106, 1107]. In (7.9), the total conductivity is given by an equivalent parallel circuit composed of two different conduction processes. The first term is a conventional J-independent ohmic conductivity caused by the thermal activation across an energy gap 2Δ characterizing the relevant insulating state. The second term represents a nonlinear conductivity with a power-law dependence of J. If $n > 1$, (7.9) is rewritten as

$$\sigma(T, J) = \sigma_0(T)\left[1 + \frac{1}{n-1}\left(\frac{J}{J_T}\right)^n\right], \tag{7.10}$$

[7] Nonlinear-transport measurements have proved an important tool in clarifying charge dynamics as in neutral-ionic transitions, density-wave and spin-Peierls states, etc., when localized π-electrons are so *soft* that even a small perturbation by electric fields easily drives electric charges into a decondensed state.

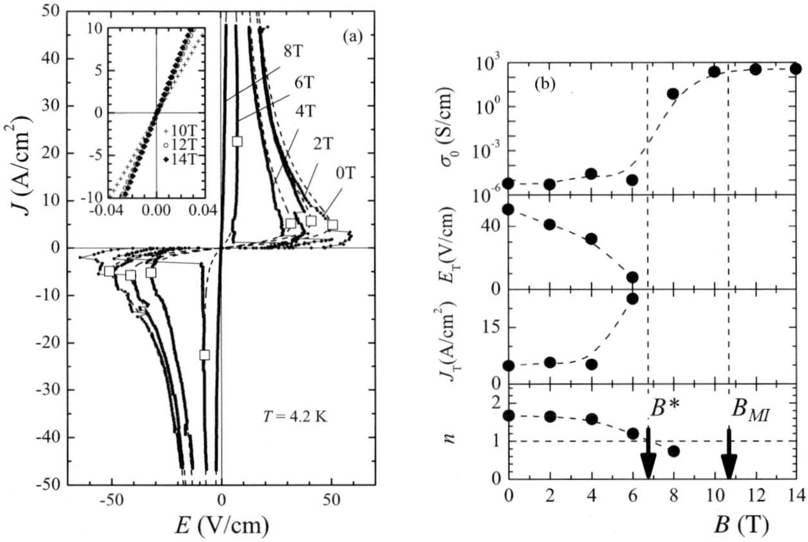

Fig. 7.13. (a) J-E curves at 4.2 K under magnetic fields applied along the interlayer b^*-direction of λ-$(BETS)_2FeCl_4$. Dashed lines and open squares represent the $J-E$ curves (7.9) and the threshold points (E_T, J_T), respectively. (b) Magnetic field dependence of the parameters σ_0, E_T, J_T, and n at 4.2 K. The dashed lines are for eye guide. After [1104].

where $\sigma_0(T) = \sigma_A \exp(-\Delta/k_BT)$ and $[J_T(T)]^n = \sigma_0(T)/[(n-1)\sigma_B]$, a threshold J_T is defined as $(\mathrm{d}E/\mathrm{d}J)_{J=J_T} = 0$, and $E_T = J_T^{1-n}/n\sigma_B$. This equation with $n > 1$ explains the NR quite well, and is appropriate in general for a positive feedback system [1106]. The systematic change in the J-E curves with magnetic fields is clearly seen in the field dependence of parameters such as σ_0, E_T, J_T and n as shown in Fig. 7.13(b). The thermally activated conductivity $\sigma_0(T)$ increases by seven orders of magnitude at $B < B_{MI}$, and enters the metallic state above B_{MI}. Just above $B^*(= 6-7\,\mathrm{T})$, a crossover from $n > 1$ to $n < 1$ occurs which is associated with diminishing E_T and divergent J_T. As mentioned above, it is only in the case of $n > 1$ that the NR effect can be observed with well-defined E_T and J_T. Since $n = (1 - \sigma_0 E_T/J_T)^{-1}$, the systematic changes in n, E_T, and J_T near B^* are consistent with each other. To summarize, the electronic transport phenomena appear differently in three regions: $n > 1$ with NR at $B < B^*$, $1 > n > 0$ without NR at $B^* < B < B_{MI}$, and $n = 0$ with ohmic conductivity at $B > B_{MI}$.

As mentioned above, the NR effect is necessarily induced once the switching instability is triggered. Quite similar phenomena were observed in the quasi-1D, half-filled Mott insulator K-TCNQ which undergoes a spin-Peierls (SP) transition to a spin-singlet ground state [1107]. The optical microscope observation for the NR state provided a photograph showing a periodic stripe pattern of the order of μm running perpendicular to the chain axis parallel to E. These stripes, which reversibly disappear below E_T, were identified to be a highly conducting pathway, where the localized π-electrons are forced back to higher conducting states above the SP transition. In analogy to these observations, it was suggested for the present salt that the switching and the subsequent NR effect may be, respectively, a local onset and its development of π charge carriers' self-decondensation process. Thus the magnetic and electric fields both make the AFMI ground state increasingly unstable to facilitate the inherently localized π-electrons to be mobile [1103, 1104].[8] These electronic instabilities, which are sensitively controlled by both electric and magnetic fields, are closely related to the dielectric properties.

Dielectric Properties and Charge-Ordering-Induced Polarization Model

We describe low-frequency ($10^1 - 10^4$ Hz) dielectric constants ($\tilde{\varepsilon}(\omega) \equiv \varepsilon_1 + i\varepsilon_2$) along the c-axis, which suggest a ferroelectric-like ordering in the AFMI states [1108]. These measurements, using a three-terminal method, can only be performed in the well-ordered AFMI states with a low conductance G less than 10^{-4} S. The dielectric constant ε_1^c, as large as 10^3 at 2.5 K, increases with increasing temperature, amounting to 10^4 at 6.5 K, the highest measurable temperature, above which G recovers abruptly, see Figs. 7.8 and 7.13(b). (The measurable regime in the B-T phase diagram is hatched in the inset of Fig. 7.15(a).) The bias field (E_{bias}) dependence of ε_1^c and σ_1^c is quite non-linear, where $\sigma_1^c(\equiv \omega\varepsilon_2/4\pi)$ is found to be exclusively determined by mobile π-electrons responsible for the steady, non-linear transport described in the preceding section. This means that the dielectric loss of electric polarization, i.e., the retardation effect is negligibly small.

Figure 7.14 shows the relative change of the electric polarization ΔP as

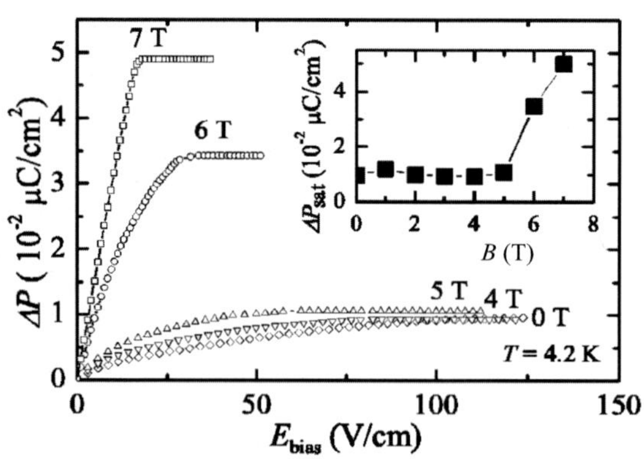

Fig. 7.14. Relative change of polarization ΔP in λ-$(BETS)_2FeCl_4$ as a function of E_{bias} at $T = 4.2$ K and $B = 0 - 7$ T. The inset shows the magnetic-field dependence of the saturation value ΔP_{sat}. After [1108].

[8] For the mixed-stack ionic salts, the logarithmic E_T is roughly proportional to Δ [1106]. The data of the present salt satisfy this relation [1103].

a function of E_{bias}, which was evaluated by integrating ε_1^c with E_{bias}. At $B \leq 5\,T$, the saturated polarization ΔP_{sat} is almost independent of B, while E_{bias} giving the saturation, decreases with B. At higher B, however, a sharp increase of ΔP occurs at low E_{bias}, and, consequently, ΔP_{sat} takes a sharp upturn as shown in the inset. To note, this crossover behavior in the dielectric response occurs around B^* just as seen in the nonlinear transport, see Fig. 7.13. Furthermore, the data in Fig. 7.15 disclose a colossal magnetodielectricity (CMD) effect. There ε_1^c, which is insensitive to low $B < B^*$, exhibits a huge increase by about two orders of magnitude at higher B up to 7.5 T, the highest measurable field. The magnetic field, where the diverging increase occurs, shifts systematically to lower values with increasing temperature. Together with the nonlinear transport, these results suggest that the

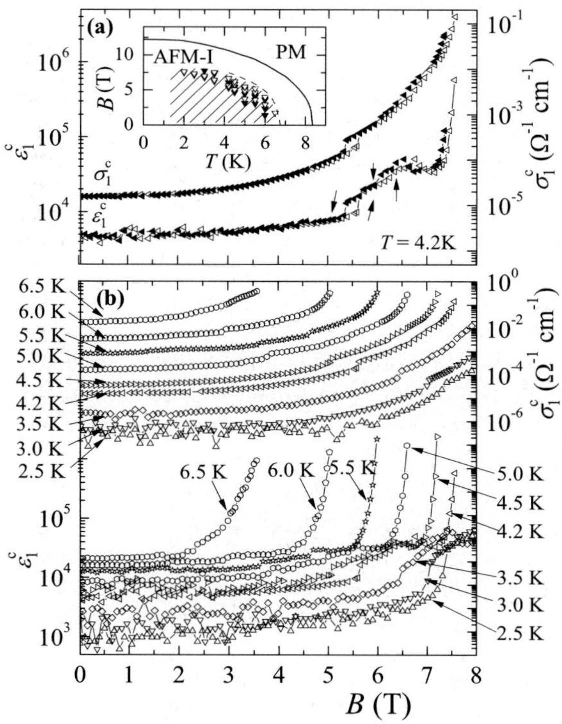

Fig. 7.15. Magnetic field dependence of ε_1^c and σ_1^c in λ-$(BETS)_2FeCl_4$; (a) at 4.2 K and (b) at 2.5 – 6.5 K. The inset shows the B-T phase diagram, in which the solid and dotted lines show a phase boundary B_{MI} and upper limiting fields for three-terminal capacitance measurements with $G < 10^{-4}$ S, respectively. The shaded region represents the measurable regime. After [1108].

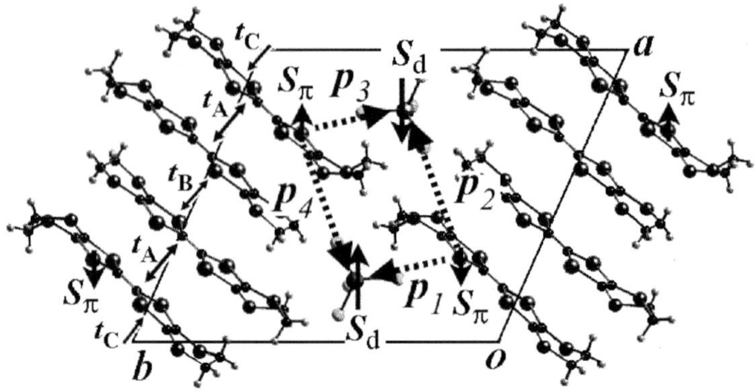

Fig. 7.16. AFM spin arrangement and possible electric dipoles proposed from a charge-ordering-induced polarization (COP) model for the AFMI ground state in λ-(BETS)$_2$FeCl$_4$. After [1108].

origin of CMD can be attributed to π charge carriers' *self-decondensation processes* providing a drastic dielectric response as well as a NR effect, which are triggered by both magnetic and electric fields.

On the basis of the described dielectric dynamics and the Mori-Katsuhara theory [1067], a charge-ordering-induced polarization model (COP) was proposed as illustrated in Fig. 7.16. Due to the exclusively large $J_{\pi d}$ through the coupling channel Se(7)–Cl(2) as described in Sect. 7.2.3, the charge as well as the spin of π-electrons is expected to be localized at Se(7) sites, forming a charge order of –1–0–0–1– sequence along the column. The AFM order between s_π and S_d, which is different from the dimer model (Fig. 7.9), is thus assumed to be realized by the gain in the superexchange energy which may overcompensate the cost due to the intersite repulsive Coulomb interaction between localized π-electrons. In this model, the charge ordering can be considered as the primary cause — not only magnetically, but also electrically — for the MI transition. Consequently, as shown in the figure, the charge ordered π-electrons (holes) are able to form electric dipole moments $\boldsymbol{p}_i$ $(i = 1-4)$ with a negative charge in the center of FeCl$_4$ anions. Provided that the inversion symmetry exists, $\sum_{i=1}^{4} \boldsymbol{p}_i = 0$, since $\boldsymbol{p}_1+\boldsymbol{p}_3 = \boldsymbol{p}_2+\boldsymbol{p}_4 = 0$. Some finite electric polarization is expected to emerge, however, since the low-temperature phase is supposed to lose the inversion symmetry as described in Sect. 7.2.4. Furthermore, the dielectric domains in the heterogeneous PMM state, in which a charge-ordered state may possibly be induced, can be considered to appear as a precursor of the AFMI ground state.

From the nonlinear transport and dielectric properties in the AFMI state, we may arrive at a comprehensive picture that the localized π-electrons tend to be *unlocked* or *melted* around B^* associated with the sharp upturn of the polarization, and eventually turn into band electrons in the PMM state at $B > B_{MI}$. Thus the present π-d system is considered to be a novel *ferroelectromagnet* in which both magnetic and dielectric order coexist and interact with each other [1109, 1110, 1111, 1112].

7.2.6 Magnetic Field-Induced Transitions

Reentrant Metallic Transition in λ*-*$(BETS)_2FeCl_4$

As described in the preceding sections, the AFMI state becomes unstable against reentering the PMM state with the 3d-spins forced to align along the magnetic field [1081, 1113]. Figure 7.17 shows the magnetic field ($\parallel b^*$) dependence of the in-plane resistance $R(\parallel c)$ [1077]. The critical field B_{MI}, defined as the onset of the resistance upturn, shifts to higher fields with decreasing temperature, approaching about 12 T at 1.8 K. These reentrant transitions are independent on the field orientation. As seen in the figure, the transition is associated by successive, step-like resistance jumps with hysteresis [1082]. This characteristic behavior is noticeable in the transit region of $B^* < B < B_{MI}$ ($B^* = 8 - 9$ T at 1.8 K), where the resistance decreases quite sharply.

In the reentrant metallic state above B_{MI}, a negative magnetoresistance is seen as demonstrated in Fig. 7.17(b). This phenomenon can be attributed to a reduction of π-electron scattering by d-spins that continue to be aligned by magnetic fields.

The reentrant transition also appears in the mixed $Fe_xGa_{1-x}Cl_4$ salts, in which $B_{MI}(0)$ can be scaled to T_{MI} with an empirical relation as $B_{MI}(0)$ (in Tesla) $= 1.5\,T_{MI}$ (in Kelvin) [1077, 1114, 1115, 1116]. Together with the fact

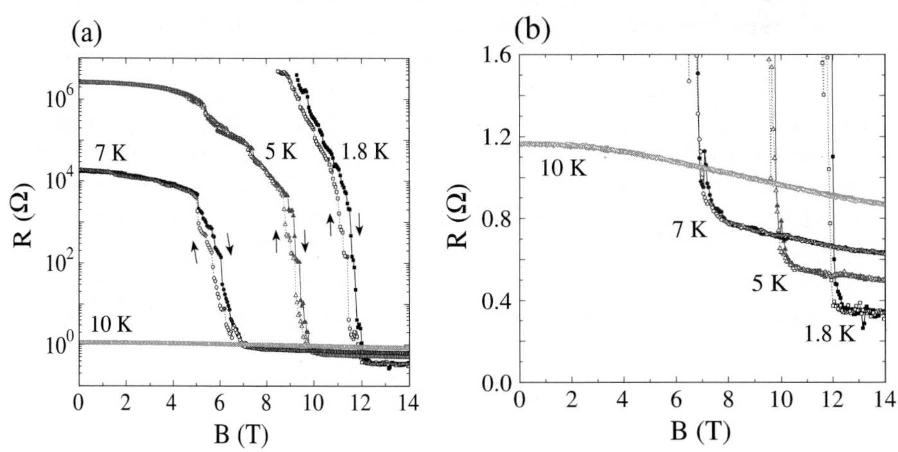

Fig. 7.17. Magnetic field ($\parallel b^*$) dependence of the in-plane resistance of λ-$(BETS)_2FeCl_4$. The onset of the resistance upturn is shown in (b) on an expanded scale. After [1077].

that the transition is isotropic with respect to the field orientation, this scaling suggests that the Pauli spin polarization of conducting π-electrons, which would lower the electronic energy by $\mu_B^2 N(E_F)B^2$, may play some important role in driving the transition [1077].[9] The mechanism for the transition, however, is not yet fully understood [1081, 677].

Jaccarino-Peter Effect in λ-*(BETS)*$_2$*Fe*$_x$*Ga*$_{1-x}$*Cl*$_4$ *and* κ-*(BETS)*$_2$*FeBr*$_4$

With further increasing B above B_{MI}, superconductivity is induced, this being termed field-induced superconductivity (FISC) [1114, 1115, 1116, 1117, 924], see also [676] for a recent review of the experimental results observed in the BETS systems and the theoretical model. This phenomenon has also been observed in metallic κ-$(BETS)_2FeBr_4$ [1118, 1119]. Figures 7.18 and 7.19 show

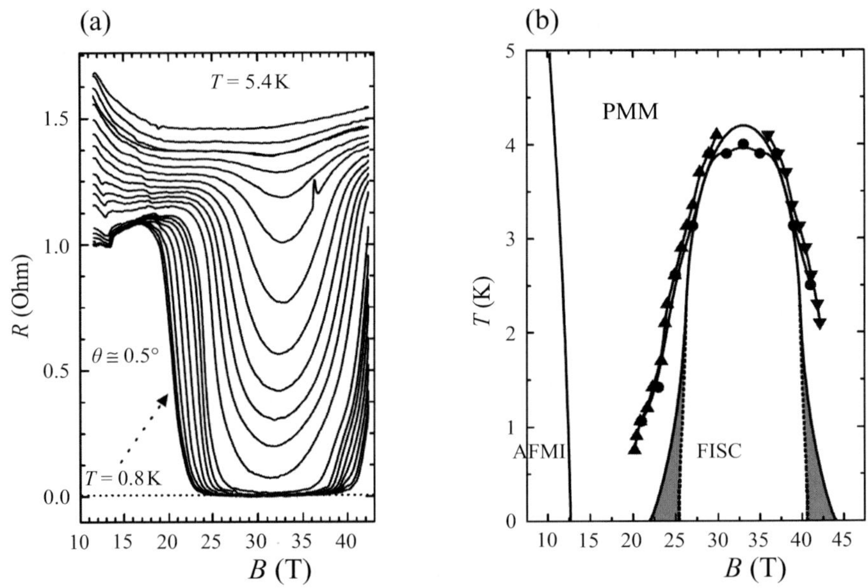

Fig. 7.18. (a) In-plane resistance as a function of magnetic field applied along the in-plane c-axis ($\pm 0.3°$) of λ-$(BETS)_2FeCl_4$ for temperature intervals of approximately 0.25 K between 5.4 and 0.8 K. (b) Magnetic field-temperature phase diagram showing the AFMI, PMM and FISC. Solid triangles indicate the midpoints of the resistive transitions from (a), while solid circles indicate the midpoint of the resistive transition as a function of T. The solid line is a theoretical fit to a second-order phase transition towards FISC. The dashed line indicates the phase boundary asserted from the assumption of an inhomogeneous Fulde-Ferrell-Larkin-Ovchinnikov state [735, 736] (shaded area). For details see also [1117]. After [1115].

[9] The energy gain is $(1/2)\chi_P B^2$, where the Pauli paramagnetic susceptibility $\chi_P = (1/2)(g\mu_B)^2 N(E_F) = 2\mu_B^2 N(E_F)$ $(g = 2)$.

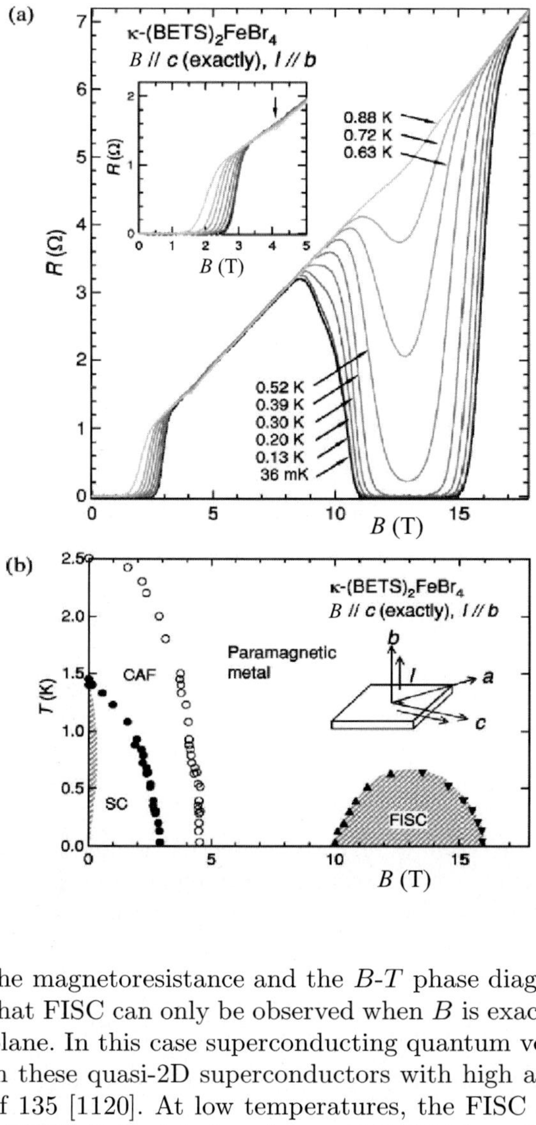

Fig. 7.19. (a) Magnetic field ($\parallel c$) dependence of the interlayer resistance at 36 mK $< T <$ 0.88 K. The inset shows the data in the low-field region. (b) Magnetic field-temperature phase diagram: closed circle, open circle, and closed triangle show the superconducting, canted AFM and the FISC transitions, respectively. Shaded areas show the calculated superconducting phases [1121]. After [1119].

the magnetoresistance and the B-T phase diagram. It has been pointed out that FISC can only be observed when B is exactly parallel to the conducting plane. In this case superconducting quantum vortices can be hardly induced in these quasi-2D superconductors with high anisotropy parameter (6.22) γ of 135 [1120]. At low temperatures, the FISC state in λ-(BETS)$_2$FeCl$_4$ (κ-(BETS)$_2$FeBr$_4$) is found at $\sim$16 T (10 T) $< B <\sim$ 45 T (16 T) centered at $B_0 = 33$ T (13 T) with $T_{c0} = 4$ K (0.7 K).

These findings are well explained by the so-called Jaccarino-Peter (JP) effect [1122]. This effect had been predicted in 1962, almost two decades before it was actually observed in resistance measurements on a Chevrel phase compound with localized $4f$ spins and narrow-d-band electrons [1123, 1124]. The essential point in the JP effect is that, for the case of the present π-d system, an external magnetic field $\boldsymbol{B}_e$ aligning the localized S_d spins can

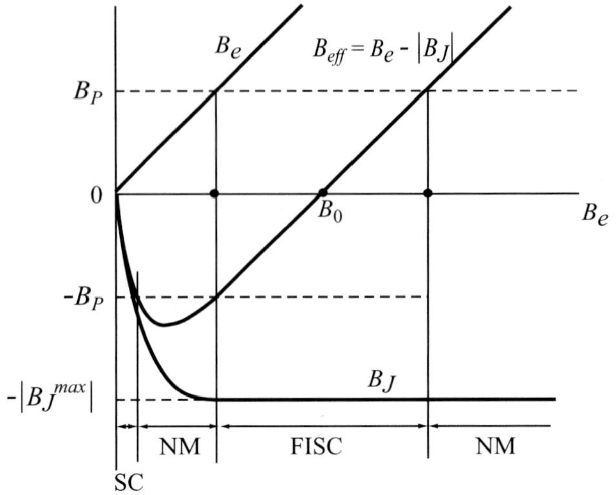

Fig. 7.20. Schematic diagram for the Jaccarino-Peter compensation effect. After [1093].

compensate the π-d interaction-induced internal field $\boldsymbol{B}_J = (J_{\pi d}/g\mu_{\mathrm{B}})\boldsymbol{S}_d$ (antiparallel to $\boldsymbol{B}_e$ since $J_{\pi d} < 0$) acting on π-electrons' spins.

Figure 7.20 illustrates the schematic diagram for the JP effect. For simplicity, consider a superconductor with localized moments and $B_P \ll B_{c2}$.[10] With increasing B_e, the magnetic moments of localized spins become aligned along the direction of $\boldsymbol{B}_e$ following a Brillouin function and eventually $\boldsymbol{B}_J$ (antiparallel to $\boldsymbol{B}_e$) arrives at the saturation maximum $|B_J^{max}| = (J_{\pi d}/g\mu_B)S_d = 1.86 J_{\pi d}/k_B$. If $|B_J| > B_P$, the effective magnetic field $B_{eff} = B_e - |B_J|$ can be larger than B_P, driving the superconductor into the normal state. After taking a negative maximum, B_{eff} starts to decrease and can drive it back to the superconducting state once again at $B_{eff} = -B_P$. With further increase of B_e, B_{eff} depends almost linearly on B_e, eventually becoming positive through the zero-crossing at the optimal external field $B_0 = |B_J^{max}|$ where an exact compensation of the internal field is realized. At $B_{eff} = B_P + |B_J|$, the field-induced superconducting state is driven into the reentrant normal state again.

[10] Apart from polarizing the magnetic moments, an external magnetic field can interact with conduction-electron spins both via the Zeeman interaction and the electromagnetic interaction of the vector potential. Both interactions with a broken time-reversal symmetry suppress superconductivity by breaking up the Cooper pairs of $(\boldsymbol{k}\uparrow, -\boldsymbol{k}\downarrow)$. Hence the former interaction provides the Chandrasekhar-Clogston limit $B_P = 1.84\,T_c$ (6.34) [845, 846], while the latter orbital interaction induces quantum vortices into a type-II superconductor, giving an upper critical field $B_{c_2} = \phi_0/2\pi\xi^2$ [847], see Sect. 6.4 and [587, 365].

Thus FISC can be expected around B_0, $|B_J^{max}| - B_P < B_e < |B_J^{max}| + B_P$. With $J_{\pi d} = 17.7$ K (7.9 K) evaluated for λ-$(BETS)_2FeCl_4$ (κ-$(BETS)_2FeBr_4$), $B_0 = 33$ T (15 T) which are close to the fields giving the optimal T_{c0} in Figs. 7.18 and 7.19 [1067]. It has been pointed out, however, that the low-temperature field regime ($2B_P$) occupied by FISC in λ-$(BETS)_2FeCl_4$ (κ-$(BETS)_2FeBr_4$) is about 29 T (6 T) in comparison with $2B_P(= 2 \times 1.84 T_{c0}) = 18 - 20$ T (5.5 T) expected from $T_{c0} = 4.2$ K (1.5 K). The JP compensation effect is more clearly established [677] by the direct evaluation for $B_J = 32$ T (13 T) from the observed splitting SdH frequencies [1125] due to the exchange field-induced modification of the spin splitting factor, see Sect. 5.2.5. To note, recent ^{77}Se-NMR measurements suggest $B_J = 23$ T [1101].

The above explanation holds for κ-$(BETS)_2FeBr_4$ with the superconducting ground state as well as Chevrel phase compounds. A question, however, arises as to why FISC is observed in λ-$(BETS)_2FeCl_4$ with the AFMI ground state. The reason may be attributed to the fact that the isostructural, non-magnetic $GaCl_4$ salt is a superconductor with $T_c = 4.8$ K being close to $T_c = 4$ K at B_J (Fig. 7.18). It is reasonable to suppose that the strength of an attractive electron-electron interaction mediated by phonons could be almost the same in both salts, where, except for the presence of localized spins, any differences in the electronic and phononic properties can hardly be expected.

8 Epilogue

The present monograph is intended to provide an introduction to some basic concepts of low-dimensional molecular conductors and to review some of their fascinating properties. Over the years, this field has been subject to tremendous changes triggered by outstanding discoveries and driven by the joint efforts from chemistry and physics to create new classes of conducting materials and to understand their exotic properties.

A first wave of excitement in this research area was generated in 1973 by the observation of metallic conductivity in the charge-transfer complex TTF-TCNQ, and speculations were expressed about traces of superconductivity at elevated temperatures. The latter possibility was particularly exciting in light of the proposal made by W.A. Little, in 1964, of an excitonic pairing mechanism giving rise to high-temperature superconductivity in some properly designed organic materials. Only a few years later in 1979, organic superconductivity — though of a different type — became reality in $(TMTSF)_2PF_6$. This discovery by D. Jérome and collaborators triggered the second wave of excitement which has spread far beyond the realm of this field — many more were to follow.

The main strategy in generating these molecular conductors has been to use molecular building blocks — the archetype TTF molecule and its derivatives being the most important ones — and their combinations to form charge-transfer complexes. A feature, tightly bound to the molecular nature of these substances, and which has played a central role in the field of molecular conductors, is that of the low dimensionality. Owing to the strongly directional-dependent chemical bonds, significant electronic interactions may develop only along certain directions. In addition, these interactions can be modified by chemical means which has stimulated organic chemists to introduce myriads of modifications to tune the systems in various directions.

In fact, molecular materials have provided unprecedented model systems for exploring the physics of low-dimensional systems, such as the Peierls- and spin-Peierls-instabilities and charge- or spin-density-wave ordered states. These studies have offered the exciting possibility for testing fundamental concepts of theoretical physics which were formerly only considered to be of academic interest.

As it has turned out, the low dimensionality was not the only fascinating property that these compounds show. The combination of low-dimensionality with other parameters, specific to molecular conductors, set the stage for Coulomb correlation effects to become relevant and, under certain circumstances, to dominate the properties of the π-electron system.

The combined effects of strong electron-electron and electron-phonon interactions — the latter are a consequence of the nature of the electronic states in these materials — together with the reduced dimensionality, give rise to the rich phenomenology encountered in these materials. A particularly interesting feature is the proximity to Mott insulating states, which, depending on the degree of frustrating interactions, may show long-range antiferromagnetic order or remain a disordered spin liquid down to low temperatures, correlated metallic phases, charge ordering and superconductivity. The variety of phenomena can be increased even further by the introduction of localized d-electron spins. Their interactions with the delocalized π-electrons can give rise to phenomena such as field-induced superconductivity or the hitherto unknown metal-superconductor-insulator transition.

As fine tuning the materials through chemical substitutions or physical pressure can be easily carried out, this has offered unique possibilities for studying the conditions which stabilize the various phases and for exploring in more detail the transitions between them. Recent achievements from both the experimental and theoretical sides indicate that indeed these molecular materials provide excellent model systems for exploring the fascinating physics expected near the Mott transition. These activities, now in motion, appear to manifest one of the latest waves of excitement generated by the class of molecular conductors.

The authors of this monograph are thoroughly convinced that molecular conductors — or, in a wider context, molecular materials — will play an important role in future research activities from the viewpoint of both fundamental research as well as applications, and that many more exciting discoveries will follow. It is our hope that this book will contribute towards stimulating young scientists to work in this promising field and to join the multi-facetted community as we have had the pleasure to do.

References

1. P. Day, Science **261**, 431 (1993)
2. W. Heitler, F. London, Z. Physik **44**, 455 (1927)
3. W. Heisenberg, Z. Physik **41**, 239 (1927)
4. L. Pauling, *The Nature of The Chemical Bond* (Cornell Univ. Press, Ithaca, New York, 1960)
5. O. Kahn, *Molecular Magnetism* (Wiley-VCH, Weinheim, 1993)
6. P. Day, A.E. Underhill, Phil. Trans. R. Soc. Lond. A **357**, 2851 (1999)
7. P.M. Lahti (ed.), *Magnetic Properties of Organic Materials* (Dekker, New York, 1999)
8. K. Itoh, M. Kinoshita (eds.), *Molecular Magnetism, New Magnetic Materials* (Kodansha, Gordon and Breach, Tokyo, Amsterdam, 2000)
9. J.S. Miller, M. Drillon (eds.), *Magnetism: Molecules to Materials*, vol. 1 - 3 (Wiley-VCH, Weinheim, 2001)
10. S.J. Blundell, F.L. Pratt, *Organic and Molecular Magnets*, J. Phys.: Condens. Matter **16**, R771 (2004)
11. K. Itoh, Chem. Phys. Lett. **1**, 235 (1967)
12. E. Wasserman, R.W. Murray, W.A. Yager, A.M. Trozzolo, G.J. Smolinsky, J. Am. Chem. Soc. **89**, 5076 (1967)
13. H.C. Longuet-Higgins, J. Chem. Phys. **18**, 265 (1950)
14. H.M. McConnell, J. Chem. Phys. **39**, 1910 (1963)
15. K. Mukai, H. Nishiguchi, Y. Deguchi, J. Phys. Soc. Jpn. **23**, 125 (1967)
16. M. Kinoshita, P. Turek, M. Tamura, K. Nozawa, D. Shiomi, Y. Nakazawa, M. Ishikawa, M. Takahashi, K. Awaga, T. Inabe, Y. Maruyama, Chem. Lett. **1991**, 1225 (1991)
17. M. Takahashi, P. Turek, Y. Nakazawa, M. Tamura, K. Nozawa, D. Shiomi, M. Ishikawa, M. Kinoshita, Phys. Rev. Lett. **67**, 746 (1991)
18. M. Tamura, Y. Nakazawa, D. Shiomi, K. Nozawa, Y. Hosokoshi, M. Ishikawa, M. Takahashi, M. Kinoshita, Chem. Phys. Lett. **186**, 401 (1991)
19. H. Akamatsu, H. Inokuchi, Y. Matsunaga, Nature **173**, 168 (1954)
20. R.S. Mulliken, J. Am. Chem. Soc. **74**, 811 (1952)
21. D.S. Acker, R.J. Harder, W.R. Hertler, W. Mahler, L.R. Melby, R.E. Benson, W.E. Mochel, J. Am. Chem. Soc. **82**, 6408 (1960)
22. D.L. Coffen, P.E. Garret, Tetrahedron. Lett. **10**, 2043 (1969)
23. D.L. Coffen, Tetrahedron. Lett. **30**, 2633 (1970)
24. D.L. Coffen, J.Q. Chambers, D.R. Williams, P.E. Garret, N.D. Canfield, J. Am. Chem. Soc. **93**, 2258 (1971)
25. F. Wudl, G.M. Smith, E.J. Hufnagel, J. Chem. Soc., Chem. Commun., p. 1453 (1970)

26. F. Wudl, D. Wobschal, E.J. Hufnagel, J. Am. Chem. Soc. **94**, 670 (1972)
27. R. Zahradnik, P. Carsky, S. Hünig, G. Kiesslich, D. Scheutzow, Int. J. Sulfur Chem. C **6**, 109 (1971)
28. J.M. Williams, J.R. Ferraro, R.J. Thorn, K.D. Carlson, U. Geiser, H.H. Wang, A.M. Kini, M.H. Whangbo, *Organic Superconductors (Including Fullerenes)* (Prentice Hall, Englewood Cliffs, N.J., 1992)
29. J. Yamada, T. Sugimoto (eds.), *TTF Chemistry: Fundamentals and Applications of Tetrathiafulvalene* (Kodansha, Springer, Tokyo, Berlin Heidelberg New York, 2002)
30. J. Ferraris, D.O. Cowan, J.V. Walatka, J.H. Perlstein, J. Am. Chem. Soc. **95**, 948 (1973)
31. L.B. Coleman, M.J. Cohen, D.J. Sandman, F.G. Yamagishi, A.F. Garito, A.J. Heeger, Solid State Commun. **12**, 1125 (1973)
32. R.E. Peierls, *Quantum Theory of Solids* (Oxford Univ., New York, London, 1955)
33. D. Jérome, A. Mazaud, M. Ribault, K. Bechgaard, J. Physique Lett. **41**, L95 (1980)
34. K. Bechgaard, C.S. Jacobsen, K. Mortensen, H.J. Pedersen, N. Thorup, Solid State Commun. **33**, 1119 (1979)
35. I.J. Lee, M.J. Naughton, G.M. Danner, P.M. Chaikin, Phys. Rev. Lett. **78**, 3555 (1997)
36. K. Bechgaard, K. Carneiro, M. Olsen, F.B. Rasmussen, C.S. Jacobsen, Phys. Rev. Lett. **46**, 852 (1981)
37. A. Aumüller, S. Hünig, Angew. Chem. Int. Ed. Engl. **23**, 447 (1984)
38. A. Aumüller, P. Erk, G. Klebe, S. Hünig, J.U. von Schütz, H.P. Werner, Angew. Chem. Int. Ed. Engl. **25**, 740 (1986)
39. S. Hünig, E. Herberth, *N,N'-Dicyanoquinone Diimines (DCNQIs): Versatile Acceptors for Organic Conductors*, Chem. Rev. **104**, 5535 (2004)
40. R. Kato, *Conductive Copper Salts of 2,5-Disubstituted N,N'-Dicyanobenzoquinonediimines (DCNQIs): Structural and Physical Properties*, Bull. Chem. Soc. Jpn. **73**, 515 (2000)
41. M. Mizuno, A.F. Garito, M.P. Cava, J. Chem. Soc. Chem. Commun. **1**, 18 (1978)
42. R. Kato, H. Kobayashi, A. Kobayashi, S. Moriyama, Y. Nishino, K. Kajita, W. Sasaki, Chem. Lett. **1986**, 507 (1986)
43. A. Kobayashi, R. Kato, H. Kobayashi, S. Moriyama, , Y. Nishino, K. Kajita, W. Sasaki, Chem. Lett. **1987**, 459 (1987)
44. H. Urayama, H. Yamochi, G. Saito, S. Sato, A. Kawamoto, J. Tanaka, T. Mori, Y. Murayama, H. Inokuchi, Chem. Lett. **1988**, 463 (1988)
45. H. Urayama, H. Yamochi, G. Saito, K. Nozawa, T. Sugano, M. Kinoshita, S. Sato, K. Oshima, A. Kawamoto, J. Tanaka, Chem. Lett. **1988**, 55 (1988)
46. A.M. Kini, U. Geiser, H.H. Wang, K.D. Carlson, J.M. Williams, W.K. Kwok, K.D. Vandervoort, J.E. Thompson, D.L. Supka, D. Jung, M.H. Whangbo, Inorg. Chem. **29**, 2555 (1990)
47. V.N. Laukhin, E.E. Kostynchenko, Yu. V. Sushko, I.F. Schegolev, E.B. Yagubskii, Sov. Phys. JETP Lett. **41**, 81 (1985)
48. K. Murata, M. Tokumoto, H. Anzai, H. Bando, G. Saito, K. Kajimura, T. Ishiguro, J. Phys. Soc. Jpn. **54**, 1236 (1985)
49. K. Kikuchi, M. Kikuchi, T. Namiki, G. Saito, K. Ikemoto, K. Murata, T. Ishiguro, K. Kobayashi, Chem. Lett. **1987**, 931 (1987)

50. G. Papavassiliou, G. Mousdis, J. Zambounis, A. Terzis, A. Hountas, B. Hilti, C. Mayer, J. Pfeiffer, Synth. Met. **27**, B379 (1988)
51. R. Kato, S. Aonuma, Y. Okano, H. Sawa, M. Tamura, M. Kinoshita, K. Oshima, A. Kobayashi, K. Bun, H. Kobayashi, Synth. Met. **61**, 199 (1993)
52. M.A. Beno, H.H. Wang, A.M. Kini, K.D. Carlson, U. Geiser, W.K. Kwok, J.E. Thompson, J.M. Williams, J. Ren, M.H. Whangbo, Inorg. Chem. **29**, 1599 (1990)
53. S. Kahlich, D. Schweitzer, I. Heinen, S.E. Lan, B. Nuber, H.J. Keller, K. Winzer, H.W. Helberg, Solid State Commun. **80**, 191 (1991)
54. L. Brossard, M. Ribault, M. Bousseau, L. Valade, P. Cassoux, C.R. Acad. Sci. Ser. 2 **302**, 205 (1986)
55. P. Cassoux, L. Valade, H. Kobayashi, A. Kobayashi, R.A. Clark, A.E. Underhill, *Molecular Metals and Superconductors Derived from Metal Complexes of 1,3-dithiol-2-thione-4,5-dithiolate (DMIT)*, Coord. Chem. Rev. **110**, 115 (1991)
56. H.W. Kroto, J.R. Heath, S.C. O'Brien, R.F. Curl, R.E. Smalley, Nature **318**, 162 (1985)
57. R.C. Haddon, A.F. Hebard, M.J. Rosseinsky, D.W. Murphy, S.J. Duclos, K.B. Lyons, B. Miller, J.M. Rosamilia, R.M. Fleming, A.R. Kortan, S.H. Glarum, A.V. Makhija, A.J. Muller, R.H. Elick, S.M. Zahurak, R. Tycko, G. Dabbagh, F.A. Thiel, Nature **350**, 320 (1991)
58. A.F. Hebard, M.J. Rosseinsky, R.C. Haddon, D.W. Murphy, S.H. Glarum, T.T.M. Palstra, A.P. Ramirez, A.R. Kortan, Nature **350**, 600 (1991)
59. M.J. Rosseinsky, A.P. Ramirez, S.H. Glarum, D.W. Murphy, R.C. Haddon, A.F. Hebard, T.T.M. Palstra, A.R. Kortan, S.M. Zahurak, A.V. Makhija, Phys. Rev. Lett. **66**, 2830 (1991)
60. K. Tanigaki, T.W. Ebbesen, S. Saito, J. Mizuki, J.S. Tsai, Y. Kubo, S. Kuroshima, Nature **352**, 222 (1991)
61. K. Tanigaki, K. Prassides, J. Mater. Chem. **5**, 1515 (1995)
62. O. Gunnarsson, *Alkali-Doped Fullerides–Narrow Band Solids with Unusual Properties* (World Scientific, Singapore, 2005)
63. E. Coronado, P. Day, *Magnetic Molecular Conductors*, Chem. Rev. **104**, 5419 (2004)
64. M. Kurmoo, A.W. Graham, P. Day, S.J. Coles, M.B. Hursthouse, J.L. Caufield, J. Singleton, F.L. Pratt, W. Hayes, L. Ducasse, P. Guionneau, J. Am. Chem. Soc. **117**, 12209 (1995)
65. A. Kobayashi, T. Udagawa, H. Tomita, T. Naito, H. Kobayashi, Chem. Lett. **1993**, 2179 (1993)
66. H. Kobayashi, A. Kobayashi, P. Cassoux, *BETS as a Source of Molecular Magnetic Superconductors (BETS = bis(ethylenedithio)tetraselenafulvalene)*, Chem. Soc. Rev. **29**, 325 (2000)
67. H. Kobayashi, H. Cui, A. Kobayashi, *Organic Metals and Superconductors Based on BETS (BETS = bis(ethylenedithio)tetraselenafulvalene*, Chem. Rev. **104**, 5265 (2004)
68. E. Coronado, J.R. Galan-Mascaros, C.J. Gomez-Garcia, V. Laukhin, Nature **408**, 447 (2000)
69. H. Fujiwara, K. Wada, T. Hiraoka, T. Hayashi, T. Sugimoto, H. Nakazumi, K. Yokogawa, M. Teramura, S. Yasuzuka, K. Murata, T. Mori, J. Am. Chem. Soc. **127**, 14166 (2005)

70. H. Tanaka, Y. Okano, H. Kobayashi, W. Suzuki, A. Kobayashi, Science **291**, 285 (2001)
71. A. Kobayashi, H. Tanaka, H. Kobayashi, J. Mater. Chem. **11**, 2078 (2001)
72. A. Kobayashi, E. Fujiwara, H. Kobayashi, *Single-component Molecular Metals with Extended-TTF Dithiolate Ligands*, Chem. Rev. **104**, 5243 (2004)
73. W.A. Little, Phys. Rev. **134**, A1416 (1964)
74. H. Shirakawa, E.J. Louis, A.G. MacDiarmid, C.K. Chiang, A.J. Heeger, J. Chem. Soc.- Chem. Commun. **16**, 578 (1977)
75. C.K. Chiang, C.R. Fincher, Y.W. Park, A.J. Heeger, H. Shirakawa, E.J. Louis, S.C. Gau, A.G. MacDiarmid, Phys. Rev. Lett. **39**, 1098 (1977)
76. C. Kittel, *Introduction to Solid State Physics*, 6th edn. (John Wiley and Sons, New York, 1986)
77. N.W. Ashcroft, N.D. Mermin, *Solid State Physics* (Saunders College Publishing, New York, 1976)
78. S.P. McGlynn, L.G. Vanquickenborne, M. Kinoshita, D.G. Carroll, *Introduction to Applied Quantum Chemistry* (Holt, Rinehart, and Winston, New York, 1972)
79. M.H. Whangbo, R. Hoffmann, J. Am. Chem. Soc. **100**, 6093 (1978)
80. P.M. Grant, Phys. Rev. B **26**, 6888 (1982)
81. E. Canadell, M.H. Whangbo, *Conceptual Aspects of Structure-Property Correlations and Electronic Instabilities, with Applications to Low-Dimensional Transition-Metal Oxides*, Chem. Rev. **91**, 965 (1991)
82. T. Mori, A. Kobayashi, Y. Sasaki, H. Kobayashi, G. Saito, H. Inokuchi, Bull. Chem. Soc. Jpn. **57**, 627 (1984)
83. P. Guionnau, C.J. Kepert, D. Chasseau, M.R. Turfer, P. Day, Synth. Met. **86**, 1973 (1997)
84. M. Watanabe, Y. Noda, Y. Nogami, H. Mori, J. Phys. Soc. Jpn. **73**, 116 (2004)
85. A.M. Kini, M.A. Beno, D. Son, H.H. Wang, K.D. Carlson, L.C. Porter, U. Welp, B.A. Vogt, J.M. Williams, D. Jung, M. Evain, M.H. Whangbo, D.L. Overmyer, J.E. Schirber, Solid State Commun. **69**, 503 (1989)
86. T. Ishiguro, K. Yamaji, G. Saito, *Organic Superconductors*, 2nd edn. (Springer, Berlin, Heidelberg, New York, 1998)
87. L.K. Montgomery, In *Organic Conductors - Fundamentals and Applications* (J.-P. Farges, Ed., M. Dekker, New York, 1994), chap. Chemical Synthesis and Crystal Growth Techniques, pp. 115 – 145
88. D. Pedron, G. Visentini, R. Bozio, J. Williams, J. Schlueter, Physica C **276**, 1 (1997)
89. J.E. Eldridge, C.C. Homes, J.M. Williams, A.M. Kini, H.H. Wang, Spectrochim. Acta A **51**, 947 (1995)
90. D. Jérome, *Organic Conductors: From Charge Density Wave TTF-TCNQ to Superconductivity*, Chem. Rev. **104**, 5565 (2004)
91. T.J. Kistenmacher, T.E. Phillips, D.O. Cowan, Acta Cryst. **B30**, 763 (1974)
92. D. Jérome, H.J. Schulz, *Organic Conductors and Superconductors*, Adv. Phys. **51**, 293 (2002)
93. C. Bourbonnais, D. Jérome, *One-dimensional conductors*, Physics World **September**, 41 (1998)
94. R. Moret, J.P. Pouget, In *Crystal Chemistry and Properties of Materials with Quasi-One-Dimensional Structures* (J. Rouxel, ed., D. Reidel, Dordrecht, The Netherlands, 1986), pp. 87 – 143

95. G. Saito, T. Enoki, K. Toriumi, H. Inokuchi, Solid State Commun. **42**, 557 (1982)
96. K. Oshima, T. Mori, H. Inokuchi, H. Urayama, H. Yamochi, G. Saito, Phys. Rev. B **38**, 938 (1988)
97. K. Bender, I. Hennig, D. Schweitzer, K. Diez, H. Endres, H.J. Keller, Mol. Cryst. Liq. Cryst. **108**, 359 (1984)
98. T. Mori, Bull. Chem. Soc. Jpn. **71**, 2509 (1998)
99. T. Mori, H. Mori, S. Tanaka, Bull. Chem. Soc. Jpn. **72**, 179 (1999)
100. T. Mori, Bull. Chem. Soc. Jpn. **72**, 2011 (1999)
101. R.P. Shibaeva, B. Yagubskii, *Molecular Conductors and Superconductors Based on Trihalides of BEDT-TTF and Some of its Analogues*, Chem. Rev. **104**, 5347 (2004)
102. H. Kobayashi, A. Kobayashi, Y. Sasaki, G. Saito, H. Inokuchi, Bull. Chem. Soc. Jpn. **59**, 301 (1986)
103. M. Lang, *Quasi-Twodimensional Organic Superconductors*, Superconductivity Review **2**, 1 (1996)
104. R.H. McKenzie, Science **278**, 820 (1997)
105. J.M. Williams, A.M. Kini, H.H. Wang, K.D. Carlson, U. Geiser, L.K. Montgomery, G.J. Pyrka, D.M. Watkins, J.M. Kommers, S.J. Boryschuk, A.V.S. Crouch, W.K. Kwok, J.E. Schirber, D.L. Overmyer, D. Jung, M.H. Whangbo, Inorg. Chem. **29**, 3272 (1990)
106. H.H. Wang, K.D. Carlson, U. Geiser, A.M. Kini, A.J. Schultz, J.M. Williams, L.K. Montgomery, W.K. Kwok, U. Welp, K.G. Vandervoort, S.J. Boryshuk, A.V.S. Crouch, J.M. Kommers, D.M. Watkins, J.E. Schirber, D.L. Overmyer, D. Jung, J.J. Novoa, M.H. Whangbo, Synth. Met. **41-43**, 1983 (1991)
107. Yu. V. Sushko, K. Andres, Phys. Rev. B **47**, 330 (1993)
108. M. Oshima, H. Mori, G. Saito, K. Oshima, Chem. Lett. **1989**, 1159 (1989)
109. H. Mori, S. Tanaka, M. Oshima, G. Saito, T. Mori, Y. Maruyama, H. Inokuchi, Bull. Chem. Soc. Jpn. **63**, 2183 (1990)
110. N.D. Kushch, L.I. Bravov, M.V. Kartsovnik, V.N. Lahkin, S.I. Pesotskii, R.P. Shibaeva, L.P. Rozenberhg, E.B. Yagubskii, A.V. Zvarikina, Synth. Met. **46**, 271 (1992)
111. H.H. Wang, K.D. Carlson, U. Geiser, W.K. Kwok, M.D. Vashon, J.E. Thompson, N.F. Larsen, G.D. McCabe, R.S. Hulscher, J.M. Williams, Physica C **166**, 57 (1990)
112. Y. Nakazawa, A. Kawamoto, K. Kanoda, Phys. Rev. B **52**, 12890 (1995)
113. J.S. Brooks, X. Chen, S.J. Klepper, S. Valfells, G.J. Athas, Y. Tanaka, T. Kinoshita, N. Kinoshita, M. Tokumoto, H. Anzai, C.C. Agosta, Phys. Rev. B **52**, 14457 (1995)
114. C.E. Campos, J.S. Brooks, P.J.M. van Bentum, J.A.A.J. Perenboom, S.J. Klepper, P.S. Sandhu, S. Valfells, Y. Tanaka, N.K. T. Kinoshita, M. Tokumoto, H. Anzai, Phys. Rev. B **52**, R7014 (1995)
115. J. Müller, *Thermodynamische Untersuchungen an Quasi-Zweidimensionalen Organischen Supraleitern.* Doctor thesis, TU Dresden, Shaker Verlag, Aachen (2002)
116. J.A. Schlueter, J.M. Williams, U. Geiser, J.D. Dudek, M.E. Kelly, S.A. Sirchio, K.D. Carlson, D. Naumann, T. Roy, C.F. Campana, Adv. Mater. **7**, 634 (1995)

117. J.A. Schlueter, U. Geiser, J.M. Williams, M.E.K. J. D. Dudek, J.P. Flynn, R.R. Wilson, H.I. Zakowicz, P.P. Sche, D. Naumann, T. Roy, P.G. Nixon, R.W. Winter, G.L. Gard, Synth. Met. **85**, 1453 (1997)
118. U. Geiser, J.A. Schlueter, H.H. Wang, A.M. Kini, J.M. Williams, P.P. Sche, H.I. Zakowicz, M.L. Vanzile, J.D. Dudek, P.G. Nixon, R.W. Winter, G.L. Gard, J. Ren, M.H. Whangbo, J. Am. Chem. Soc. **118**, 9996 (1996)
119. U. Geiser, J.A. Schlueter, *Conducting Organic Radical Cation Salts with Organic and Organometallic Anions*, Chem. Rev. **104**, 5203 (2004)
120. A. Kobayashi, R. Kato, T. Naito, H. Kobayashi, Synth. Met. **56**, 2078 (1993)
121. A. Kobayashi, T. Udagawa, H. Tomita, T. Naito, H. Kobayashi, Chem. Lett. **1993**, 2179 (1993)
122. L.K. Montgomery, T. Burgin, C. Husting, L. Tilley, J.C. Huffmann, K.D. Carlson, J.D. Dudek, G.A. Yaconi, U. Geiser, J.M. Williams, Mol. Cryst. Liq. Cryst. **211**, 283 (1992)
123. H. Kobayashi, H. Akutsu, E. Arai, H. Tanaka, A. Kobayashi, Phys. Rev. B **56**, R8526 (1997)
124. A. Sato, E. Ojima, H. Akutsu, Y. Nakazawa, H. Kobayashi, H. Tanaka, A. Kobayashi, P. Cassoux, Phys. Rev. B **61**, 111 (2000)
125. S. Uji, C. Terakura, T. Terashima, T. Yakabe, Y. Terai, M. Tokumoto, A. Kobayashi, F. Sakai, H. Tanaka, H. Kobayashi, Phys. Rev. B **65**, 113101 (2002)
126. N. Thorup, G. Rindorf, H. Soling, K. Bechgaard, Acta Cryst. B **37**, 1236 (1981)
127. G. Rindorf, H. Soling, N. Thorup, Acta Cryst. B **38**, 2805 (1981)
128. I. Hennig, K. Bender, D. Schweitzer, K. Diez, H. Endres, H.J. Keller, A. Gleiz, H.W. Helberg, Mol. Cryst. Liq. Cryst. **119**, 337 (1985)
129. J.M. Williams, H.H. Wang, M.A. Beno, T.J. Emge, L.M. Sowa, P.T. Copps, F. Behroozi, L.N. Hall, K.D. Carlson, G.W. Crabtree, Inorg. Chem. **23**, 3839 (1984)
130. H. Kobayashi, R. Kato, A. Kobayashi, Y. Nishino, K. Kajita, W. Sasaki, Chem. Lett. **1986**, 2017 (1986)
131. G. Saito, H. Urayama, H. Yamochi, K. Oshima, Synth. Met. **27**, A331 (1988)
132. J.A. Schlueter, U. Geiser, J.M. Williams, H.H. Wang, W.K. Kwok, J.A. Fendrich, K.D. Carlson, C.A. Achenbach, J.D. Dudek, D. Naumann, T. Roy, J.E. Schirber, W.R.J. Bayless, Chem. Soc., Chem. Commun. **1994**, 1599 (1994)
133. H. Mori, S. Tanaka, T. Mori, Phys. Rev. B **57**, 12023 (1998)
134. A. Kobayashi, Y. Okano, H. Kobayashi, *Molecular Design and Physical Properties of Single-Component Molecular Metals*, J. Phys. Soc. Jpn. **75**, 051002 (2006)
135. H. Tanaka, M. Tokumoto, S. Ishibashi, D. Graf, E.S. Choi, J.S. Brooks, S. Yasuzuka, Y. Okano, H. Kobayashi, A. Kobayashi, J. Am. Chem. Soc. **126**, 10518 (2004)
136. Y. Ueba, T. Mishima, H. Kusuhara, K. Tada, In *The Physics and Chemistry of Organic Superconductors* (G. Saito and K. Kagoshima, Eds., Springer, Berlin Heidelberg, 1990), *Proceedings in Physics*, vol. 51, chap. Crystal Growth and Properties of $(BEDT\text{-}TTF)_2Cu(NCS)_2$, pp. 284 – 289
137. M. Tokumoto, H. Anzai, K. Murata, K. Kajimura, T. Ishiguro, Japan. J. Appl. Phys. **26**, 1977 (1987)

138. J.G. Analytis, A. Ardavan, S.J. Blundell, R.L. Owen, E.F. Garman, C. Jeynes, B.J. Powell, Phys. Rev. Lett. **96**, 177002 (2006)
139. J.P. Pouget, S. Ravy, J. Phys. I (France) **6**, 1501 (1996)
140. S. Ravy, R. Moret, J. Pouget, Phys. Rev. B **38**, 4469 (1988)
141. S. Tomić, D. Jérome, P. Monod, K. Bechgaard, Journal de Physique **44 (NC-3)**, 1083 (1983)
142. H. Schwenk, K. Andres, F. Wudl, Phys. Rev. B **29**, 500 (1984)
143. T. Takahashi, D. Jérome, K. Bechgaard, J. Physique Lett. **43**, L565 (1982)
144. K. Mortensen, C.S. Jacobsen, A. Lindegaard-Andersen, K. Bechgaard, J. Physique **44**, C3 (1983)
145. H. Yang, J.C. Lasjaunias, P. Monceau, J. Phys.: Condens. Matter **12**, 7183 (2000)
146. D. Jérome, In *Organic Conductors - Fundamentals and Applications* (J.-P. Farges, Ed., M. Dekker, New York, 1994), chap. Organic Superconductors: From $(TMTSF)_2PF_6$ to Fullerenes, pp. 405 – 494
147. K. Saito, H. Akutsu, M. Sorai, Solid State Commun. **111**, 471 (1999)
148. H. Akutsu, K. Saito, M. Sorai, Phys. Rev. B **61**, 4346 (2000)
149. A. Sato, H. Akutsu, K. Saito, M. Sorai, Synth. Met. **120**, 1035 (2001)
150. M. Rahal, D. Chasseau, J. Gaultier, L. Ducasse, M. Kurmoo, P. Day, Acta Cryst. B **53**, 159 (1997)
151. C.A. Angell, Science **267**, 1924 (1995)
152. R.W. Cahn, In *Materials Science and Technology* (R.W. Cahn, Ed., Verlag Chemie, Weinheim, 1991), vol. 9, chap. Glasses and Amorphous Materials, p. 137
153. W. Kauzmann, *The Nature of the Glassy State and the Behavior of Liquids at Low Temperatures*, Chem. Rev. **43**, 219 (1948)
154. J. Müller, M. Lang, F. Steglich, J. Schlueter, J. Phys. IV (France) **114**, 341 (2004)
155. J. Müller, M. Lang, F. Steglich, J.A. Schlueter, A.M. Kini, T. Sasaki, Phys. Rev. B **65**, 144521 (2002)
156. M. Kund, H. Müller, W. Biberacher, K. Andres, Physica C **191**, 274 (1993)
157. M. Kund, K. Andres, H. Müller, G. Saito, Physica B **203**, 129 (1994)
158. U. Geiser, A.J. Schultz, H.H. Wang, D.M. Watkins, D.L. Stupka, J.M. Williams, J.E. Schirber, D.L. Overmyer, D. Jung, J.J. Novoa, M.H. Whangbo, Physica C **174**, 475 (1991)
159. U. Geiser, A.M. Kini, H.H. Wang, M.A. Beno, J.M. Williams, Acta Cryst. C **47**, 190 (1991)
160. Ch. Strack, C. Akinci, V. Pashchenko, B. Wolf, E. Uhrig, W. Assmus, M. Lang, J. Schreuer, L. Wiehl, J.A. Schlueter, J. Wosnitza, D. Schweitzer, J. Müller, J. Wykhoff, Phys. Rev. B **72**, 054511 (2005)
161. M.A. deBolt, A.J. Easteal, P.B. Macedo, C.T. Moynihan, J. Am. Chem. Soc. **59**, 16 (1976)
162. P. Nagel, V. Pasler, C. Meingast, A. Rykov, S. Tajima, Phys. Rev. Lett. **85**, 2376 (2000)
163. P. Wzietek, H. Mayaffre, D. Jérome, S. Brazovskii, J. Phys. IV (France) **6**, 2011 (1996)
164. T.F. Stalcup, J.S. Brooks, R.C. Haddon, Phys. Rev. B **60**, 9309 (1999)
165. M. Tanatar, T. Ishiguro, T. Kondo, G. Saito, Phys. Rev. B **59**, 3841 (1999)
166. X. Su, F. Zuo, J.A. Schlueter, M.E. Kelly, J.M. Williams, Solid State Commun. **107**, 731 (1998)

167. S. Senoussi, A. Tirbiyine, A. Ramzi, A. Haouam, F. Pesty, Phys. Rev. B **73**, 014525 (2006)
168. K. Miyagawa, A. Kawamoto, Y. Nakazawa, K. Kanoda, Phys. Rev. Lett. **75**, 1174 (1995)
169. R. Hoffmann, J. Chem. Phys. **39**, 1397 (1963)
170. M.H. Whangbo, R. Hoffmann, J. Am. Chem. Soc. **100**, 6093 (1978)
171. K.D. Carlson, U. Geiser, A.M. Kini, H.H. Wang, L.K. Montgomery, W.K. Kwok, M.A. Beno, J.M. Williams, C.S. Cariss, G.W. Crabtree, M.H. Whangbo, M. Evain, Inorg. Chem. **27**, 965 (1988)
172. D. Chasseau, J. Gaultier, M. Rahal, L. Ducasse, M. Kurmoo, P. Day, Synth. Met. **41-43**, 2039 (1991)
173. L. Ducasse, A. Fritsch, Solid State Commun. **91**, 201 (1994)
174. P. Auban-Senzier, A. Audouard, V.N. Laukhin, R. Rosseau, E. Canadell, L. Brossard, D. Jérome, N.D. Kushch, J. Phys. IV (France) **5**, 1301 (1995)
175. P.M. Grant, J. Phys. C3 **44**, 847 (1983)
176. H. Seo, C. Hotta, H. Fukuyama, *Toward Systematic Understanding of Diversity of Electronic Properties in Low-Dimensional Molecular Solids*, Chem. Rev. **104**, 5005 (2004)
177. M. Schiller, W. Schmidt, E. Balthes, D. Schweitzer, H.J. Koo, M.H. Whangbo, I. Heinen, T. Klausa, P. Kircher, W. Strunz, Europhys. Lett. **51**, 82 (2000)
178. T. Mori, *Organic Conductors with Unusual Band Fillings*, Chem. Rev. **104**, 4947 (2004)
179. Y. Shimizu, K. Miyagawa, K. Kanoda, M. Maesato, G. Saito, Phys. Rev. Lett. **91**, 107001 (2003)
180. H. Kino, H. Fukuyama, J. Phys. Soc. Jpn. **64**, 2726 (1995)
181. T. Komatsu, N. Matsukawa, T. Inoue, G. Saito, J. Phys. Soc. Jpn. **65**, 1340 (1996)
182. C. Campos, P. Sandhu, J. Brooks, T. Ziman., Phys. Rev. B **53**, 12725 (1996)
183. A. Fortunelli, A. Painelli, J. Chem. Phys. **106**, 8051 (1997)
184. A. Kawamoto, Y. Honma, K. Kumagai, Phys. Rev. B **70**, R060510 (2004)
185. Y. Kurosaki, Y. Shimizu, K. Miyagawa, K. Kanoda, G. Saito, Phys. Rev. Lett. **95**, 177001 (2005)
186. J. Kübler, M. Weger, C. Sommer, Solid State Commun. **76**, 507 (1987)
187. J. Kübler, C. Sommer, In *The Physics and Chemistry of Organic Superconductors* (G. Saito and S. Kagoshima, Eds., Springer, Heidelberg, 1990), p. 208
188. Y.N. Xu, W.Y. Ching, Y.C. Jean, Y. Lou, Phys. Rev. B **52**, 12946 (1995)
189. H. Kino, H. Fukuyama, J. Phys. Soc. Jpn. **65**, 2158 (1996)
190. L. Ducasse, A. Fritsch, F. Castet, Synth. Met. **85**, 1627 (1997)
191. F. Mila, Phys. Rev. B **52**, 4788 (1995)
192. H. Seo, H. Fukuyama, J. Phys. Soc. Jpn. **66**, 1249 (1997)
193. C. Hotta, H. Fukuyama, J. Phys. Soc. Jpn. **70**, 321 (2001)
194. C. Hotta, J. Phys. Soc. Jpn. **72**, 840 (2003)
195. T. Mori, A. Kobayashi, Y. Sasaki, H. Kobayashi, Chem. Lett. **1982**, 1923 (1982)
196. T. Mori, A. Kobayashi, Y. Sasaki, H. Kobayashi, G. Saito, H. Inokuchi, Chem. Lett. **1984**, 957 (1984)
197. H. Kobayashi, R. Kato, A. Kobayashi, Y. Nishio, K. Kajita, W. Sasaki, Chem. Lett. **1986**, 789 (1986)

198. H. Kobayashi, R. Kato, A. Kobayashi, Y. Nishio, K. Kajita, W. Sasaki, Chem. Lett. **1986**, 833 (1986)
199. D. Beckmann, S. Wanka, J. Wosnitza, D. Schweitzer, W. Strunz, Z. Phys. B **104**, 207 (1997)
200. T. Sasaki, H. Sato, N. Toyota, Synth. Met. **42**, 2211 (1991)
201. F.L. Pratt, T. Sasaki, N. Toyota, K. Nagamine, Phys. Rev. Lett. **74**, 3892 (1995)
202. P. Christ, W. Biberacher, M.V. Kartsovnik, E. Steep, E. Balthes, H. Weiss, H. Müller, JETP Lett. **71**, 303 (2000)
203. J.S. Qualls, L. Balicas, J.S. Brooks, N. Harrison, L.K. Montgomery, M. Tokumoto, Phys. Rev. B **62**, 10008 (2000)
204. A. Bjelis, D. Zanchi, G. Montambaux, J. Phys. IV (France) **9**, 203 (1999)
205. M.V. Kartsovnik, *High Magnetic Fields: A Tool for Studying Electronic Properties of Layered Organic Metals*, Chem. Rev. **104**, 5737 (2004)
206. K. Kajita, Y. Nishio, S. Moriyama, W. Sasaki, R. Kato, A. Kobayashi, H. Kobayashi, Solid State Commun. **64**, 1279 (1987)
207. M.H. Whangbo, J.J. Novoa, D. Jung, J.M. Williams, A.M. Kini, H.H. Wang, U. Geiser, M.A. Beno, K.D. Carlson, In *Organic Superconductivity* (V.Z. Kresin and W.A. Little, Eds., Plenum Press, New York, 1990), p. 243
208. J. Wosnitza, *Fermi Surfaces of Low-Dimensional Organic Metals and Superconductors* (Springer, Berlin Heidelberg New York, 1996)
209. J. Singleton, *Studies of Quasi-Twodimensional Organic Conductors Based on BEDT-TTF Using High Magnetic Fields*, Rep. Prog. Phys. **63**, 1111 (2000)
210. D. Jérome, H.J. Schulz, Adv. Phys. **31**, 299 (1982). This review article has been republished in: Adv. Phys. **51**, 293 (2002), refer to [92]
211. P. Wzietek, F. Creuzet, C. Bourbonnais, D. Jérome, K. Bechgaard, P. Batail, J. Phys. IV (France) **3**, 171 (1993)
212. T. Giarmachi, *Theoretical Framework for Quasi-One Dimensional Systems*, Chem. Rev. **104**, 5037 (2004)
213. A. Andrieux, D. Jérome, K. Bechgaard, J. Phys. (Paris) Lett. **42**, L87 (1981)
214. H. Wilhelm, D. Jaccard, R. Duprat, C. Bourbonnais, D. Jérome, J. Moser, C. Carcel, J.M. Fabre, Eur. Phys. J. B **21**, 175 (2001)
215. D. Jérome, Science **252**, 1509 (1991)
216. J.R. Cooper, B. Korin-Hamzić, In *Organic Conductors - Fundamentals and Applications* (J.-P. Farges, Ed., M. Dekker, New York, 1994), chap. Organic metals, pp. 359 – 404
217. K. Murata, Y. Honda, H. Anzai, M. Tokumoto, K. Takahashi, N. Kinoshita, T. Ishiguro, N. Toyota, T. Sasaki, Y. Muto, Synth. Met. **27**, A341 (1988)
218. H. Sato, T. Sasaki, N. Toyota, Physica C **185-189**, 2679 (1991)
219. J.E. Schirber, D.L. Overmyer, K.D. Carlson, J.M. Williams, A.M. Kini, H.H. Wang, H.A. Charlier, B.J. Love, D.M. Watkins, G.A. Yaconi, Phys. Rev. B **44**, 4666 (1991)
220. M. Oshima, H. Mori, G. Saito, K. Oshima, In *The Physics and Chemistry of Organic Superconductors* (G. Saito and S. Kagoshima, Eds., Springer-Verlag, Berlin, Heidelberg, 1990), p. 257
221. K. Kikuchi, Y. Honda, Y. Ishikawa, K. Saito, I. Ikemoto, K. Murata, H. Anzai, T. Ishiguro, Solid State Commun. **66**, 405 (1988)
222. K. Yamaji, Synth. Met. A **27**, 115 (1988)
223. S. Gärtner, E. Gogu, I. Heinen, H. Keller, T. Klutz, D. Schweitzer, Solid State Commun. **65**, 1531 (1998)

224. N. Toyota, T. Sasaki, H. Sato, Y. Watanabe, Physica C **178**, 339 (1991)
225. I. Parker, R. Friend, M. Kurmoo, P. Day, C. Lenoir, P. Batail, J. Phys.: Condens. Matter **1**, 4479 (1989)
226. N.L. Wang, B.P. Clayman, H. Mori, S. Tanaka, J. Phys.: Condens. Matter. **12**, 2867 (2000)
227. J. Merino, R.H. McKenzie, Phys. Rev. B **61**, 7996 (2000)
228. P. Limelette, P. Wzietek, S. Florens, A. Gorges, T.A. Costi, C. Pasquier, D. Jérome, C. Mézière, P. Batail, Phys. Rev. Lett. **91**, 016401 (2003)
229. A. Ugawa, G. Ojima, K. Yakushi, H. Kuroda, Phys. Rev. B **38**, R5122 (1988)
230. B. Thoma, Diplomarbeit (1996). Universität Stuttgart
231. L.K. Montgomery, R.M. Vestal, K.P. Starkey, B.W. Fravel, M.J. Samide, D.G. Peters, C.H. Mielke, J.D. Thompson, Synth. Met. **103**, 1878 (1999)
232. Ch. Strack, C. Akinci, V. Paschenko, B. Wolf, E. Uhrig, W. Assmus, J. Schreuer, L. Wiehl, J. Schlueter, J. Wosnitza, D. Schweitzer, M. Lang, Comptes Rendus Chimie, in press
233. K. Murata, M. Ishibashi, Y. Honda, N.A. Fortune, M. Tokumoto, N. Kinoshita, H. Anzai, Solid State Commun. **76**, 377 (1990)
234. Yu. V. Sushko, V.A. Bondarenko, R.A. Petrosov, N.D. Kushch, E.B. Yagubskii, J. Phys. IV (France) **1**, 1375 (1991)
235. M. Lang, J. Müller, F. Steglich, J.A. Schlueter, A.M. Kini, T. Sasaki, Synth. Met. **133 - 134**, 107 (2003)
236. T. Sasaki, N. Yoneyama, A. Matsuyama, N. Kobayashi, Phys. Rev. B **65**, R060505 (2002)
237. M. Weger, M. Tittelbach, E. Balthes, D. Schweitzer, H.J. Keller, J. Phys.: Condens. Matter **5**, 8569 (1993)
238. L. Bulaevskii, *Organic Layered Superconductors*, Adv. Phys. **37**, 443 (1988)
239. K. Kadowaki, S.B. Woods, Solid State Commun. **58**, 507 (1986)
240. K. Miyake, T. Matsuura, C.M. Varma, Solid State Commun. **71**, 1149 (1989)
241. J. Wosnitza, J. Hagel, O. Ignatchik, B. Bergk, V.M. Gvozdikov, J.A. Schlueter, R.W. Winter, G.L. Gard, J. Low Temp. Phys. **142**, 327 (2006)
242. J. Hagel, O. Ignatchik, J. Wosnitza, C. Pfleiderer, J.A. Schlueter, H. Davis, R. Winter, G. Gard, Physica C, in press
243. J. Hagel, J. Wosnitza, C. Pfleiderer, J.A. Schlueter, J. Mohtasham, G.L. Gard, Phys. Rev. B **68**, 104504 (2003)
244. M. Weger, M. Kaveh, H. Gutfreund, Solid State Commun. **37**, 421 (1981)
245. M. Weger, D. Schweitzer, Synth. Met. **70**, 889 (1995)
246. M. Weger, Acta Physica Polonica A **87**, 723 (1995)
247. L.I. Buravov, A.V. Zvarykina, N.D. Kushch, V.N. Laukhin, V.A. Merzhanov, A.G. Khomenko, E.B. Yagubskii, Sov. Phys. JETP **68**, 182 (1989)
248. R.H. McKenzie, P. Moses, Phys. Rev. Lett. **81**, 4492 (1998)
249. J. Singleton, P.A. Goddard, A. Ardavan, N. Harrison, S.J. Blundell, J.A. Schlueter, A.M. Kini, Phys. Rev. Lett. **88**, 037001 (2002)
250. J. Caulfield, W. Lubezynski, F.L. Pratt, J. Singleton, D.Y.K. Ko, W. Hayes, M. Karmoo, P. Day, J. Phys.: Condens. Matter **6**, 2911 (1994)
251. J. Wosnitza, G. Goll, D. Beckmann, S. Wanka, D. Schweitzer, W. Strunz, J. Physique I (France) **6**, 1597 (1996)
252. M.B. Maple, E.D. Bauer, V.S. Zapf, J. Wosnitza, In *The Physics of Superconductors* (K.H. Bennemann and J.B. Ketterson, Eds., Springer-Verlag, Berlin Heidelberg New York, 2004), vol. II, chap. Unconventional Superconductivity in Novel Materials, pp. 555 – 730

253. J. Wosnitza, S. Wanka, J. Hagel, R. Häussler, H. v. Löneysen, J.A. Schlueter, U. Geiser, P.G. Nixon, R.W. Winter, G.L. Gard, Phys. Rev. B **62**, R11973 (2000)
254. J. Wosnitza, S. Wanka, J. Hagel, E. Balthes, N. Harrison, J.A. Schlueter, A.M. Kini, U. Geiser, J. Mohtasham, R.W. Winter, G.L. Gard, Phys. Rev. B **61**, 7383 (2000)
255. J. Wosnitza, J. Hagel, J.S. Qualls, J.S. Brooks, E. Balthes, D. Schweitzer, J.A. Schlueter, U. Geiser, J. Mohtasham, R.W. Winter, G.L. Gard, Phys. Rev. B **65**, R180506 (2002)
256. J.W. Wilkins, In *Electrons at the Fermi Surfaces* (M. Springford, Ed., Cambridge Univ. Press, 1980), chap. Understanding Quasi-Particles: Observable Many-Body Effects in Metals and ^{3}He, p. 46
257. D. Pines, P. Nozières, *Theory of Quantum Liquids*, vol. I (Perseus Books Group, 1999)
258. L.D. Landau, Sov. Phys. JETP **3**, 920 (1957)
259. N. Grewe, F. Steglich, In *Handbook on the Physics and Chemistry of Rare Earths* (K.A Gschneidner Jr. and L. Eyring, Eds., Elsevier, Amsterdam, 1991), vol. 14, chap. Heavy Fermions, pp. 343 – 474
260. R.J. Cava, B. Batlogg, R.B. van Dover, D.W. Murphy, S. Sunshine, T. Siegrist, J.P. Remeika, E.A. Rietman, S. Zahurak, G.P. Espinosa, Phys. Rev. Lett. **58**, 1676 (1987)
261. H. v. Löhneysen, T. Pietrus, G. Portisch, H.G. Schlager, A. Schröder, M. Sieck, T. Trappmann, Phys. Rev. Lett. **72**, 3262 (1994)
262. H.J. Schulz, Phys. Rev. Lett. **64**, 2831 (1990)
263. J. Voit, *One-Dimensional Fermi Liquids*, Rep. Prog. Phys. **58**, 977 (1995)
264. M. Dressel, A. Schwartz, G. Grüner, L. Degiorgi, Phys. Rev. Lett. **77**, 398 (1996)
265. A. Schwartz, M. Dressel, G. Grüner, V. Vescoli, L. Degiorgi, T. Giarmachi, Phys. Rev. B **58**, 1261 (1998)
266. W. Henderson, V. Vescoli, P. Tran, L. Degiorgi, G. Grüner, Eur. Phys. J. B **11**, 365 (1999)
267. V. Vescoli, F. Zwick, W. Henderson, L. Degiorgi, M. Grioni, G. Grüner, L. Montgomery, Eur. Phys. J. B **13**, 503 (2000)
268. B. Dardel, D. Malterre, M. Grioni, P. Weibel, Y. Baer, J. Voit, D. Jérome, Europhys. Lett. **24**, 687 (1993)
269. F. Zwick, S. Brown, G. Margaritondo, C. Merlic, M. Onellion, J. Voit, M. Grioni, Phys. Rev. Lett. **79**, 3982 (1997)
270. F. Zwick, D. Jérome, G. Margaritondo, M. Onellion, J. Voit, M. Grioni, Phys. Rev. Lett. **81**, 2974 (1998)
271. M. Dressel, G. Grüner, *Electrodynamics of Solids* (Cambridge Univ. Press, Cambridge, 2002)
272. C.S. Jacobsen, In *Low-Dimensional Conductors and Superconductors* (D. Jérome and L.G. Caron, Eds., Plenum Press, New York, 1987), p. 253
273. A. Graja, In *Organic Conductors - Fundamentals and Applications* (J.-P. Farges, Ed., M. Dekker, New York, 1994), chap. Optical Properties, pp. 229 – 268
274. M. Dressel, Synth. Met. **85**, 1503 (1997)
275. L. Degiorgi, V. Vescoli, W. Henderson, G. Grüner, L.K. Montgomery, J. Phys. IV (France) **10**, Pr3 (2000)

276. M. Dressel, N. Drichko, *Optical Properties of Two-Dimensional Organic Conductors: Signatures of Charge Ordering and Correlation Effects*, Chem. Rev. **104**, 5689 (2004)
277. J. Singleton, C. Mielke, *Quasi-Two-Dimensional Organic Superconductors: A Review*, Contemp. Phys. **43**, 63 (2002)
278. M. Dressel, In *Studies of High Temperature Superconductors* (A. Narlikar, Ed., Nova Science Publisher, Huntington, 2000), vol. 34, chap. Optical Properties of Layered Organic Superconductors, pp. 1 –54
279. J.E. Eldridge, Y. Lin, H.H. Wang, J.M. Williams, A.M. Kini, Phys. Rev. B **57**, 597 (1998)
280. Y. Lin, J.E. Eldridge, J.A. Schlueter, H.H. Wang, A.M. Kini, Phys. Rev. B **64**, 024506 (2001)
281. J. Merino, R.H. McKenzie, Phys. Rev. B **62**, 2416 (2000)
282. C.S. Jacobsen, D.B. Danner, K. Bechgaard, Phys. Rev. Lett. **46**, 1142 (1981)
283. C.S. Jacobsen, K. Mortensen, M. Weger, K. Bechgaard, Solid State Commun. **38**, 423 (1981)
284. C.S. Jacobsen, D.B. Danner, K. Bechgaard, Phys. Rev. B **28**, 7019 (1983)
285. D. Pedron, R. Bozio, M. Meneghetti, C. Pecile, Phys. Rev. B **49**, 10893 (1994)
286. K. Kornelsen, J.E. Eldridge, C.C. Homes, H.H. Wang, J.M. Williams, Solid State Commun. **72**, 475 (1989)
287. J.E. Eldridge, K. Kornelson, H.H. Wang, J.M. Williams, A.V.D. Crouch, D.M. Watkins, Solid State Commun. **79**, 583 (1991)
288. K. Kornelsen, J.E. Eldridge, H.H. Wang, H.A. Charlier, J.M. Williams, Solid State Commun. **81**, 343 (1992)
289. H. Kuroda, K. Yakushi, H. Tajima, A. Ugawa, M. Tamura, Y. Okawa, A. Kobayashi, R. Kato, H. Kobayashi, Synth. Met. **27**, A491 (1988)
290. K. Kornelsen, J.E. Eldridge, H.H. Wang, J.M. Williams, Phys. Rev. B **44**, 5235 (1991)
291. N.O. Lipari, M.J. Rice, C.B. Duke, R. Bozio, A. Girlando, C. Pecile, Int. Quantum Chem. Symp. **11**, 583 (1977)
292. J.E. Eldridge, Y. Xie, H.H. Wang, J.M. Williams, A.M. Kini, J.A. Schlueter, Spectrochim. Acta A **52**, 45 (1996)
293. G. Visentini, M. Masino, C. Bellitto, A. Girlando, Phys. Rev. B **58**, 9460 (1998)
294. M.E. Kozlov, K.I. Pokhodnia, A.A. Yurchenko, Spectrochim. Acta A **43**, 323 (1987)
295. M.E. Kozlov, K.I. Pokhodnia, A.A. Yurchenko, Spectrochim. Acta A **45**, 437 (1989)
296. M. Meneghetti, R. Bozio, C. Pecile, J. Phys. **47**, 1377 (1986)
297. A. M.Kini, K.D. Carlson, H.H. Wang, J.A. Schlueter, J.D. Dudek, S.A. Sirchio, U. Geiser, K.R. Lykke, J.M. Williams, Physica C **264**, 81 (1996)
298. A. Nowack, M. Weger, D. Schweitzer, H. Keller, Solid State Commun. **60**, 199 (1986)
299. A. Nowack, U. Poppe, M. Weger, D. Schweitzer, H. Schwenk, Z. Phys. B **68**, 41 (1987)
300. S. Sugai, H. Mori, H. Yamochi, G. Saito, Phys. Rev. B **47**, 14374 (1993)
301. K.I. Pokhodnia, A. Graja, M. Weger, D. Schweitzer, Z. Phys. B **90**, 127 (1993)

302. M. Dressel, J.E. Eldridge, J.M. Williams, H.H. Wang, Physica C **203**, 247 (1992)
303. A. Girlando, M. Masino, G. Visentini, A. Brillante, R.G.D. Valle, E. Venuti, Synth. Met. **109**, 13 (2000)
304. L. Pintschovius, H. Rietschel, T. Sasaki, H. Mori, S. Tanaka, N. Toyota, M. Lang, F. Steglich, Europhys. Lett. **37**, 627 (1997)
305. D. Pedron, R. Bozio, J.A. Schlueter, M.E. Kelly, A.M. Kini, J.M. Williams, Synth. Met. **103**, 2220 (1999)
306. E. Faulques, V.G. Ivanov, C. Mézière, P. Batail, Phys. Rev. B **62**, R9291 (2000)
307. R. Zeyher, G. Zwicknagl, Z. Phys. B **78**, 175 (1990)
308. J.E. Eldridge, Y. Xie, Y. Lin, H.H. Wang, J.M. Williams, J.A. Schlueter, Synth. Met. **86**, 2067 (1997)
309. J. Moldenhauer, C. Horn, I.K. Pokhodnia, D. Schweitzer, I. Heinen, J.H. Keller, Synth. Met. **60**, 31 (1993)
310. H.H. Wang, R. Ferraro, J.M. Williams, U. Geiser, J.A. Schlueter, J. Chem. Soc. Chem. Commun. **1994**, 1893 (1994)
311. K. Yamamoto, K. Yakushi, K. Miyagawa, K. Kanoda, A. Kawamoto, Phys. Rev. B **65**, 085110 (2002)
312. V. Vescoli, L. Degiorgi, W. Henderson, G. Grüner, K.P. Starkey, L.K. Montgomery, Science **218**, 1181 (1998)
313. S.S.P. Parkin, J.C. Scott, J.B. Torrance, , E.H. Engler, J. Phys. Colloq. **44**, C3 (1983)
314. F. Creuzet, Mol. Crys. Liq. Cryst. **119**, 289 (1985)
315. J.P. Pouget, R. Comes, K. Bechgaard, J.M. Fabre, L. Giral, Physica B **108**, 1197 (1981)
316. M. Dumm, A. Loidl, B.W. Fravel, K.P. Starkey, L.K. Montgomery, M. Dressel, Phys. Rev. B **61**, 511 (2000)
317. J.C. Bonner, M.E. Fisher, Phys. Rev. B **135**, A640 (1964)
318. L.N. Bulaevskii, Sov. Phys. Solid State, Fiz. Tverd. Tela (Leningrad) **11**, 1132 and 921 (1969)
319. E. Pytte, Phys. Rev. B **10**, 4637 (1974)
320. M. Lang, J. Müller, F. Steglich, A. Brühl, B. Wolf, M. Dressel, J. Phys. IV (France) **114**, 111 (2004)
321. K. Mortensen, Y. Tomkiewicz, T.D. Schultz, E.M. Engler, Phys. Rev. Lett. **46**, 1234 (1981)
322. J.B. Torrance, H.J. Pedersen, , K. Bechgaard, Phys. Rev. Lett. **49**, 881 (1982)
323. L.P. Le, G.M. Luke, B.J. Sternlieb, W.D. Wu, Y.J. Uemura, J.H. Brewer, T.M. Riseman, R.V. Upasani, L.Y. Chiang, P.M. Chaikin, Europhys. Lett. **15**, 547 (1991)
324. L.P. Le, A. Keren, G.M. Luke, B.J. Sternlieb, W.D. Wu, Y.J. Uemura, J.H. Brewer, T.M. Riseman, R.V. Upasani, L.Y. Chiang, W. Kang, P.M. Chaikin, T. Csiba, G. Grüner, Phys. Rev. B **48**, 7284 (1993)
325. J.M. Delrieu, M. Roger, Z. Toffano, E.W. Mbougue, R.S. James, K. Bechgaard, Physica B **143**, 412 (1986)
326. T. Takahashi, Y. Maniwa, H. Kawamura, G. Saito, Physica B **143**, 417 (1986)
327. J.P. Pouget, S. Ravy, Synth. Met. **85**, 1523 (1997)
328. W.A. Seitz, D.J. Klein, Phys. Rev. B **9**, 2159 (1974)
329. V.J. Emery, J. Phys. (Paris) **44**, C3 (1983)

330. U. Welp, S. Fleshler, W.K. Kwok, G.W. Crabtree, K.D. Carlson, H.H. Wang, U. Geiser, J.M. Williams, V.M. Hitsman, Phys. Rev. Lett. **69**, 840 (1992)
331. A. Kawamoto, K. Miyagawa, Y. Nakazawa, K. Kanoda, Phys. Rev. Lett. **74**, 3455 (1995)
332. A. Kawamoto, K. Miyagawa, Y. Nakazawa, K. Kanoda, Phys. Rev. B **52**, 15522 (1995)
333. K. Kanoda, Hyperfine Int. **104**, 235 (1997)
334. K. Miyagawa, A. Kawamoto, K. Uchida, K. Kanoda, Physica B **284-288**, 1589 (2000)
335. K. Miyagawa, A. Kawamoto, K. Kanoda, Phys. Rev. Lett. **89**, 017003 (2002)
336. H. Mayaffre, P. Wzietek, C. Lenoir, D. Jérome, P. Batail, Europhys. Lett. **28**, 205 (1994)
337. K. Miyagawa, K. Kanoda, A. Kawamoto, Chem. Rev. **104**, 5635 (2004)
338. K. Kanoda, *NMR Studies on Two-Dimensional Molecular Conductors and Superconductors: Mott transition in κ-(BEDT-TTF)*$_2$ *X*, Physica C **282 - 287**, 299 (1997)
339. T. Timusk, B.W. Statt, *The Pseudogap in High-Temperature Superconductors: An Experimental Survey*, Rept. Prog. Phys. **62**, 61 (1999)
340. V. Kataev, G. Winkel, D. Khomskii, D. Wohlleben, W. Crump, K. Tebbe, J. Hahn, Solid State Commun. **83**, 435 (1992)
341. K. Frikach, M. Poirier, M. Castonguay, K. Truong, Phys. Rev. B **61**, R6491 (2000)
342. T. Shimizu, N. Yoshimoto, M. Nakamura, M. Yoshizawa, Physica B **281 & 282**, 896 (2000)
343. D. Fournier, M. Poirier, M. Castonguay, K. Truong, Phys. Rev. Lett. **90**, 127002 (2003)
344. G. Kotliar, S. Murthy, M.J. Rozenberg, Phys. Rev. Lett. **89**, 046401 (2002)
345. S.R. Hassan, A. Georges, H.R. Krishnamurthy, Phys. Rev. Lett. **94**, 036402 (2005)
346. Y. Lin, J.E. Eldridge, H.H. Wang, A.M. Kini, M.E. Kelly, J.M. Williams, J.A. Schlueter, Phys. Rev. B **58**, R599 (1998)
347. J. Merino, R.H. McKenzie, Phys. Rev. B **62**, 16442 (2000)
348. J. Müller, M. Lang, F. Steglich, J.A. Schlueter, A.M. Kini, U. Geiser, J. Mohtasham, R.W. Winter, G.L. Gard, T. Sasaki, N. Toyota, Phys. Rev. B **61**, 11739 (2000)
349. J. Singleton, C.H. Mielke, W. Hayes, J.A. Schlueter, J. Phys.: Condens. Matter **15**, L203 (2003)
350. N. Harrison, J. Caulfield, J. Singleton, P.H.P. Reinders, F. Herlach, W. Hayes, M. Kurmoo, P. Day, J. Phys.: Condens. Matter **8**, 5415 (1996)
351. Y. Tsunoda, M. Mori, N. Kunitomi, Y. Teraoka, J. Kanamori, Solid State Commun. **14**, 287 (1974)
352. C.Y. Young, J.B. Sokoloff, J. Phys. F **4**, 1304 (1974)
353. E. Fawcett, *Spin-Density-Wave Antiferromagnetism in Chromium*, Rev. Mod. Phys. **60**, 209 (1988)
354. E. Fawcett, H.L. Alberts, V.Y. Galkin, D.R. Noakes, J.V. Yakhmi, *Spin-Density-Wave Antiferromagnetism in Chromium Alloys*, Rev. Mod. Phys. **66**, 25 (1994)
355. A.W. Overhauser, Phys. Rev. Lett. **4**, 462 (1960)
356. A.W. Overhauser, Phys. Rev. **128**, 1437 (1962)

357. A.W. Overhauser, Phys. Rev. Lett. **27**, 343 (1978)
358. P. Monceau, *Electronic Properties of Inorganic Quasi-One-dimensional Materials* (Reidel, Dordrecht, 1985)
359. S. Kagoshima, H. Nagasawa, T. Sambongi, *One-Dimensional Conductors* (Springer-Verlag, Berlin New York, 1988)
360. L.P. Gor'kov, G. Grüner (eds.), *Density Waves in Solids* (North-Holland, Amsterdam, 1989)
361. G. Grüner, *Density Waves in Solids* (Addison Wesley, Massachusetts, 1994)
362. G. Grüner, *The Dynamics of Charge-Density Waves*, Rev. Mod. Phys. **60**, 1129 (1988)
363. G. Grüner, *The Dynamics of Spin-Density Waves*, Rev. Mod. Phys. **66**, 1 (1994)
364. J. Bardeen, L.N. Cooper, J.R. Schrieffer, Phys. Rev. **108**, 1175 (1957)
365. M. Tinkham, *Introduction to Superconductivity*, 2nd edn. (McGraw-Hill, New York, 1996)
366. W.D. Grobman, B.D. Silverman, Solid State Commun. **19**, 319 (1976)
367. M.J. Cohen, L.B. Coleman, A.F. Garito, A.J. Heeger, Phys. Rev. B **10**, 1298 (1974)
368. S. Etemad, Phys. Rev. B **13**, 2254 (1976)
369. T. Ishiguro, S. Kagoshima, H. Anzai, J. Phys. Soc. Jpn. **41**, 351 (1976)
370. M.J. Cohen, A.J. Heeger, Phys. Rev. B **16**, 688 (1977)
371. T. Ishiguro, H. Sumi, S. Kagoshima, K. Kajimura, H. Anzai, J. Phys. Soc. Jpn. **48**, 456 (1980)
372. A.A. Bright, A.F. Garito, A.J. Heeger, Phys. Rev. B **10**, 1328 (1974)
373. F. Denoyer, R. Comes, A.F. Garito, A.J. Heeger, Phys. Rev. Lett. **35**, 445 (1975)
374. S. Kagoshima, H. Anzai, K. Kajimura, T. Ishiguro, J. Phys. Soc. Jpn. **39**, 1143 (1975)
375. S. Kagoshima, T. Ishiguro, H. Anzai, J. Phys. Soc. Jpn. **41**, 2061 (1976)
376. R. Còmes, S.M. Shapiro, G. Shirane, A.F. Garito, A.J. Heeger, Phys. Rev. Lett. **35**, 1518 (1975)
377. R. Còmes, G. Shirane, S.M. Shapiro, A.F. Garito, A.J. Heeger, Phys. Rev. B **14**, 2376 (1976)
378. G. Shirane, S.M. Shapiro, R. Comes, A.F. Garito, A.J. Heeger, Phys. Rev. B **14**, 2325 (1976)
379. J.P. Pouget, S.M. Shapiro, G. Shirane, A.F. Garito, A.J. Heeger, Phys. Rev. B **19**, 1792 (1979)
380. Z. Wang, J. Girard, C. Pasquier, D. Jérome, K. Bechgaard, Phys. Rev. B **67**, R121401 (2003)
381. P. Bak, V.J. Emery, Phys. Rev. Lett. **36**, 978 (1976)
382. T.D. Schultz, S. Etemad, Phys. Rev. B **13**, 4928 (1976)
383. T.D. Schultz, Solid State Commun. **22**, 289 (1977)
384. S. Megtert, R. Còmes, C. Vettier, R. Pynn, A.F. Garito, Solid State Commun. **31**, 977 (1979)
385. A. Andrieux, H.J. Schulz, D. Jérome, K. Bechgaard, Phys. Rev. Lett. **43**, 227 (1979)
386. P.A. Lee, T.M. Rice, P.W. Anderson, Solid State Commun. **14**, 703 (1974)
387. H. Fukuyama, P.A. Lee, Phys. Rev. B **17**, 535 (1978)
388. T. Ohmi, H. Yamamoto, Prog. Theor. Phys. **58**, 743 (1977)

389. H. Fröhlich, Proc. Roy. Soc. London, Ser. A **223**, 296 (1954)
390. J.M. Delrieu, M. Roger, Z. Toffano, A. Moradpour, K. Bechgaard, J. Phys. (France) **47**, 839 (1986)
391. T. Takahashi, Y. Maniwa, H. Kawamura, G. Saito, J. Phys. Soc. Jpn. **55**, 1364 (1986)
392. K. Yamaji, J. Phys. Soc. Jpn. **51**, 2787 (1982)
393. K. Yamaji, J. Phys. Soc. Jpn. **52**, 1361 (1983)
394. W. Kang, S. Tomić, J.R. Cooper, D. Jérome, Phys. Rev. B **41**, 4862 (1990)
395. S. Tomić, J.R. Cooper, W. Kang, D. Jérome, K. Maki, J. Phys. I (France) **1**, 1603 (1991)
396. S. Donovan, Y. Kim, L. Degiorgi, M. Dressel, G. Grüner, W. Wonneberger, Phys. Rev. B **49**, 3363 (1994)
397. K. Maki, A. Virosztek, Phys. Rev. B **39**, 9640 (1989)
398. Y. Tomio, Y. Suzumura, J. Phys. Soc. Jpn. **71**, 2742 (2002)
399. S. Kagoshima, Y. Saso, M. Maesato, R. Kondo, T. Hasegawa, Solid State Commun. **110**, 479 (1999)
400. T. Takahashi, T. Harada, Y. Kobayashi, K. Kanoda, K. Suzuki, K. Murata, G. Saito, Synth. Met. **41-43**, 3985 (1991)
401. K. Nomura, Y. Hosokawa, N. Matsunaga, M. Nagasawa, T. Sambongi, H. Anzai, Synth. Met. **70**, 1295 (1995)
402. J.C. Lasjaunias, K. Bilijakovic, P. Monceau, Phys. Rev. B **53**, 7699 (1996)
403. F. Nad', P. Monceau, K. Bechgaard, Solid State Commun. **655**, 95 (1996)
404. N. Matsunaga, H. Takashige, K. Keitoku, M. Nagasawa, K. Nomura, T. Sambongi, Physica B **194-196**, 1265 (1994)
405. J.P. Ulmet, P. Auban, A. Khmou, S. Askenazy, A. Moradpour, J. Phys. Lett. **46**, 535 (1985)
406. S. Uji, J.S. Brooks, M. Chaparala, S. Takasaki, J. Yamada, H. Anzai, Phys. Rev. B **55**, 12446 (1997)
407. I.S. Jacobs, J.W. Bray, H.R. Hart Jr., L.V. Interrante, J.S. Kasper, G.D. Watkins, D.E. Prober, J.C. Bonner, Phys. Rev. B **14**, 3036 (1976)
408. D.J. Klein, W.A. Seitz, Phys. Rev. B **10**, 3217 (1974)
409. M. Hase, I. Terasaki, K. Uchinokura, Phys. Rev. Lett. **70**, 3651 (1993)
410. J.P. Boucher, L.P. Regnault, J. Phys. I (France) **6**, 1939 (1996)
411. M.C. Cross, D.S. Fisher, Phys. Rev. B **19**, 402 (1979)
412. D.S. Chow, P. Wzietek, D. Fogliatti, B. Alavi, D.J. Tantillo, C.A. Merlic, S.E. Brown, Phys. Rev. Lett. **81**, 3984 (1998)
413. D.S. Chow, F. Zamborsky, B. Alavi, D.J. Tantillo, A. Baur, C.A. Merlic, S.E. Brown, Phys. Rev. Lett. **85**, 1698 (2000)
414. F.D.M. Haldane, Phys. Rev. B **25**, 4925 (1982)
415. S. Huizinga, J. Kommandeur, G. Sawatzky, K. Kopinga, W.J.M. de Jonge, In *Quasi One-Dimensional Conductors* (S. Barisic, S. Bjelis, J. R. Cooper, and B. Leontic, Eds., Springer-Verlag, Berlin, 1979), vol. II, p. 45
416. B. van Bodegom, B.C. Larson, H.A. Mook, Phys. Rev. B **24**, 1520 (1981)
417. N. Kobayashi, M. Ogata, K. Yonemitsu, J. Phys. Soc. Jpn. **67**, 1098 (1998)
418. Y. Tomio, Y. Suzumura, J. Phys. Soc. Jpn. **69**, 796 (2000)
419. Y. Suzumura, J. Phys. Soc. Jpn. **66**, 3244 (1997)
420. Y. Suzumura, N. Tanemura, J. Phys. Soc. Jpn. **64**, 2298 (1995)
421. Y. Tomio, Y. Kurihara, Y. Suzumura, J. Phys. Soc. Jpn. **71**, 1174 (2002)
422. T. Miyazaki, K. Terakura, Phys. Rev. B **54**, 10452 (1996)

423. K. Hiraki, K. Kanoda, Phys. Rev. B **54**, 17276 (1996)
424. K. Hiraki, K. Kanoda, Phys. Rev. Lett. **80**, 4737 (1998)
425. Y. Nogami, Y. Yamamoto, S. Hayashi, K. Oshima, K. Hiraki, K. Kanoda, Synth. Met. **103**, 2252 (1999)
426. L. Ducasse, M. Abderrabba, J. Hoarau, M. Pesquer, B. Gallois, J. Gaultier, J. Phys. C **19**, 3805 (1986)
427. S. Fujiyama, T. Nakamura, J. Phys. Soc. Jpn. **75**, 014705 (2006)
428. F. Zamborzsky, W. Yu, W. Raas, S. Brown, B. Alavi, C.A. Merlic, A. Bauer, Phys. Rev. B **66**, 081103 (2002)
429. F. Nad, P. Monceau, T. Nakamura, K. Furukawa, J. Phys.: Condens. Matter **17**, L399 (2005)
430. F. Nad, P. Monceau, C. Carcel, J. Fabre, Phys. Rev. B **62**, 1753 (2000)
431. P. Monceau, F. Nad, S. Brasovsky, Phys. Rev. Lett. **86**, 4080 (2001)
432. K. Furukawa, K. Hara, T. Nakamura, J. Phys. Soc. Jpn. **74**, 3288 (2005)
433. T. Nakamura, K. Furukawa, T. Hara, J. Phys. Soc. Jpn. **75**, 013707 (2006)
434. K. Miyagawa, A. Kawamoto, K. Kanoda, Phys. Rev. B **62**, R7679 (2000)
435. R. Chiba, H. Yamamoto, K. Hiraki, T. Takahashi, T. Nakamura, J. Phys. Chem. Solids **62**, 389 (2001)
436. H. Tajima, S. Kyoda, H. Mori, S. Tanaka, Synth. Met. **120**, 757 (2001)
437. H. Seo, J. Phys. Soc. Jpn. **69**, 805 (2000)
438. R. Wojciechowski, K. Yamamoto, K. Yakushi, M. Inokuchi, A. Kawamoto, Phys. Rev. B **67**, 224105 (2003)
439. Y. Takano, K. Hiraki, H.M. Yamamoto, T. Nakamura, T. Takahashi, J. Phys. Chem. Solid **62**, 393 (2001)
440. M. Weger, L. Burlachkov, M. Kaveh, Europhys. Lett. **19**, 505 (1992)
441. M. Weger, M. Tittelbach, E. Balthes, D. Schweitzer, H.J. Keller, J. Phys.: Condens. Matter **5**, 8569 (1993)
442. K. Bender, K. Dietz, H. Enders, H. Helberg, I. Henning, H.J. Keller, H.W. Schaffer, D. Schweitzer, Mol. Cryst. Liq. Cryst. **107**, 45 (1984)
443. N. Harrison, A. House, I. Deckers, J. Caulfield, J. Singleton, F. Herlach, W. Hayes, P. Day, Phys. Rev. B **52**, 5584 (1995)
444. W. Kang, G. Montambaux, J.R. Cooper, D. Jerome, P. Batail, C. Lenoir, Phys. Rev. Lett. **62**, 2559 (1989)
445. M. Heinecke, K. Winzer, D. Schweitzer, Z. Phys. B **93**, 45 (1993)
446. E. Balthes, D. Schweitzer, I. Heinen, H.J. Keller, W. Strunz, W. Biberacher, A.G.M. Jansen, E. Steep, Z. Phys. B **99**, 163 (1996)
447. T. Sasaki, H. Sato, N. Toyota, Solid State Commun. **76**, 507 (1990)
448. C.H. Mielke, N. Harrison, D.G. Rickel, A.H. Lacerda, R. Vestal, L.K. Montgomery, Phys. Rev. B **56**, R4309 (1997)
449. H. Weiss, M.V. Kartsovnik, W. Biberacher, E. Steep, E. Balthes, A.G.M. Janssen, K. Andres, N.D. Kushch, Phys. Rev. B **59**, 12370 (1999)
450. M.V. Kartsovnik, W. Biberacher, K. Andres, N.D. Kushch, Pis'ma Zh. Eksp. Teor. **62**, 890 (1995)
451. M.V. Kartsovnik, W. Biberacher, K. Andres, N.D. Kushch, JETP Lett. **62**, 905 (1995)
452. C. Mielke, J. Singleton, M.S. Nam, N. Harrison, C.C. Agosta, B. Fravel, L.K. Montgomery, J. Phys.: Condens. Matter **13**, 8325 (2001)
453. J. Merino, R.H. McKenzie, Phys. Rev. B **62**, 2416 (2000)
454. G.R. Stewart, J. O'Rourke, G.W. Crabtree, K.D. Carlson, H.H. Wang, J.M. Williams, F. Gross, K. Andres, Phys. Rev. B **33**, 2046 (1985)

455. J. Wosnitza, X. Liu, D. Schweitzer, H.J. Keller, Phys. Rev. B **50**, 12747 (1994)
456. J. Müller, M. Lang, R. Helfrich, F. Steglich, T. Sasaki, Phys. Rev. B **65**, R140509 (2002)
457. J. Wosnitza, S. Wanka, J. Hagel, M. Reibelt, D. Schweitzer, J.A. Schlueter, Synth. Met. **133 - 134**, 201 (2003)
458. H. Elsinger, J. Wosnitza, S. Wanka, J. Hagel, D. Schweitzer, W. Strunz, Phys. Rev. Lett. **84**, 6098 (2000)
459. S. Wanka, J. Hagel, D. Beckmann, J. Wosnitza, J.A. Schlueter, J.M. Williams, P.G. Nixon, R.W. Winter, G.L. Gard, Phys. Rev. B **57**, 3084 (1998)
460. Y. Ishizaki, H. Uozaki, H. Tsuchiya, Y. Abe, E. Negishi, H. Matsui, S. Endo, N. Toyota, Synth. Met. **133-134**, 219 (2003)
461. D.R. Harshman, L.F. Schneemeyer, J.V. Waszczak, G. Aeppli, R.J. Cava, B. Batlogg, L.W. Rupp, E.J. Ansaldo, D.L. Williams, Phys. Rev. B **39**, R851 (1989)
462. T. Biggs, A.K. Klehe, J. Singleton, D. Bakker, J. Symington, P. Goddard, A. Ardavan, W. Hayes, J.A. Schlueter, T. Sasaki, M. Kurmoo, J. Phys.: Condens. Matter **14**, L495 (2002)
463. H. Weiss, M.V. Kartsovnik, W. Biberacher, E. Steep, A.G.M. Janssen, N.D. Kushch, JETP Lett. **66**, 202 (1997)
464. A.K. Klehe, R.D. McDonald, A.F. Goncharov, V.V. Struzhkin, Ho-Kwang Mao, R.J. Hemley, T. Sasaki, W. Hayes, J. Singleton, J. Phys.: Condens. Matter **12**, L247 (2000)
465. T. Adachi, E. Ojima, K. Kato, H. Kobayashi, J. Am. Chem. Soc. **122**, 3238 (2000)
466. D. Jaccard, H. Wilhelm, D. Jérome, J. Moser, C. Carcel, J.M. Fabre, J. Phys.: Condens. Matter **13**, L1 (2001)
467. J. Moser, M. Gabay, P. Auban-Senzier, D. Jérome, K. Bechgaard, J.M. Fabre, Eur. Phys. J. B **1**, 39 (1998)
468. M. Lang, M. de Souza, A. Brühl, C. Strack, B. Wolf, J.A. Schlueter, J. Müller, D. Schweitzer, Physica C, in press
469. J.E. Schirber, E.L. Venturini, A.M. Kini, H.H. Wang, J.R. Witworth, J.M. Williams, Physica C **152**, 157 (1988)
470. J.E. Schirber, D.L. Overmyer, J.M. Williams, A.M. Kini, H.H. Wang, Physica C **170**, 231 (1990)
471. T. Sasaki, N. Yoneyama, A. Suzuki, N. Kobayashi, Y. Ikemoto, H. Kimura, J. Phys. Soc. Jpn. **74**, 2351 (2005)
472. A. Kawamoto, H. Taniguchi, K. Kanoda, J. Am. Chem. Soc. **120**, 10984 (1998)
473. H. Taniguchi, K. Kanoda, Synth. Met. **103**, 1967 (1999)
474. H. Taniguchi, A. Kawamoto, K. Kanoda, Physica B **284-288**, 519 (2000)
475. Y. Nakazawa, H. Taniguchi, A. Kawamoto, K. Kanoda, Phys. Rev. B **61**, R16295 (2000)
476. Yu. V. Sushko, V.A. Bondarenko, R.A. Petrosov, N.D. Kushch, E.B. Yagubskii, J. Phys. I (France) **1**, 1015 (1991)
477. L.I. Buravov, N.D. Kushch, V.A. Merzhanov, M.V. Osherov, A.G. Khomenko, E.B. Yagubskii, J. Phys. I (France) **2**, 1257 (1992)
478. S.M. De Soto, C.P. Slichter, A.M. Kini, H.H. Wang, U. Geiser, J.M. Williams, Phys. Rev. B **52**, 10364 (1995)

479. T. Sasaki, I. Ito, N. Yoneyama, N. Kobayashi, N. Hanasaki, H. Tajima, T. Ito, Y. Isawa, Phys. Rev. B **69**, 064508 (2004)
480. T. Sasaki, N. Yoneyama, N. Kobayashi, Y. Ikemoto, H. Kimura, Phys. Rev. Lett. **92**, 227001 (2004)
481. K.M. Lang, V. Madhavan, J.E. Hoffman, E.W. Hudson, H. Eisaki, S. Uchida, J.C. Davis, Nature (London) **415**, 412 (2002)
482. J.M. Tranquada, B.J. Sternlieb, J.D. Axe, Y. Nakamura, S. Uchida, Nature (London) **375**, 561 (1995)
483. M. Fäth, S. Freisem, A.A. Menovsky, Y. Tomioka, J. Aarts, J.A. Mydosh, Science **285**, 1540 (1999)
484. S. Lefebvre, P. Wzietek, S. Brown, C. Bourbonnais, D. Jérome, C. Mézière, M. Fourmigué, P. Batail, Phys. Rev. Lett. **85**, 5420 (2000)
485. Yu. V. Sushko, V.A. Bondarenko, R.A. Petrosov, N.D. Kushch, E.B. Yagubskii, Physica C **185 - 189**, 2683 (1991)
486. Yu. V. Sushko, H. Ito, T. Ishiguro, S. Horiuchi, G. Saito, Solid State Commun. **87**, 997 (1993)
487. H. Ito, M. Kubota, Yu. V. Sushko, N. Kojima, G. Saito, T. Ishiguro, Synth. Met. **70**, 925 (1995)
488. H. Ito, T. Ishiguro, M. Kubota, G. Saito, J. Phys. Soc. Jpn. **65**, 2987 (1996)
489. F. Kagawa, T. Itou, K. Miyagawa, K. Kanoda, Phys. Rev. B **69**, 064511 (2004)
490. F. Kagawa, K. Miyagawa, K. Kanoda, Nature **436**, 534 (2005)
491. K. Kanoda, *Metal-Insulator Transition in* κ*-(ET)*$_2$*X and (DCNQI)*$_2$*M: Two Contrasting Manifestation of Electron Correlation*, J. Phys. Soc. Jpn. **75**, 051007 (2006)
492. A.A. Abrikosov, *Fundamentals of Theory of Metals* (Elsevier Science Publisher B.V., Amsterdam, 1988)
493. J. Singleton, *Band Theory and Electronic Properties of Solids* (Oxford Univ. Press, Oxford, 2001)
494. L.W. Shubnikov, W.J. de Haas, Proc. Netherlands Roy. Acad. Sci. **33**, 130 and 163 (1930)
495. W.J. de Haas, P.M. van Alphen, Proc. Netherlands Roy. Acad. Sci. **33**, 680 and 1106 (1930)
496. I.M. Lifshitz, M.Y. Azbel, M.I. Kaganov, *Electron Theory of Metals* (Consultant Bureau, New York, 1973)
497. D. Shoenberg, *Magnetic Quantum Oscillations in Metals* (Cambridge Univ. Press, Cambridge, 1984)
498. A.B. Pippard, *Magnetoresistance in Metals* (Cambridge Univ. Press, Cambridge, 1989)
499. J.F. Cochran, R.R. Haering (eds.), *Electrons in Metals*, vol. Solid State Physics 1 (Gordon and Breach, New York, London, Paris, 1968)
500. M. Springford (ed.), *Electrons at the Fermi Surface* (Cambridge Univ. Press, Cambridge, 1980)
501. J.R. Anderson, D.R. Stone, In *Methods of Experimental Physics* (R. V. Coleman, eds., Academic Press, New York, 1974), vol. II, chap. *Experimental Methods for the de Haas-van Alphen Effect*, p. 33
502. K. von Klitzing, G. Dorda, M. Pepper, Phys. Rev. Lett. **45**, 449 (1980)
503. T. Ando, A.B. Fowler, F. Stern, *Electronic Properties of Two-Dimensional Systems*, Rev. Mod. Phys. **54**, 437 (1982)

504. R.E. Prange, S.M. Girvin (eds.), *The Quantum Hall Effect* (Springer, Berlin, Heidelberg, New York, 1990)
505. T. Chakraborty, P. Pietiläinen, *The Quantum Hall Effects* (Springer, Berlin, Heidelberg, New York, 1995)
506. D. Yoshioka, *The Quantum Hall Effect* (Springer, Berlin, Heidelberg, New York, 1998)
507. Z.F. Ezawa, *Quantum Hall Effects: Field Theoretic Approach and Related Topics* (World Scientific, Singapore, New Jersey, London, Hong Kong, 2000)
508. M.V. Kartsovnik, V.G. Peschansky, *Galvanomagnetic Phenomena in Layered Organic Conductors*, Low Temp. Phys. **31**, 185 (2005)
509. R.R. Gerhardts, D. Weiss, K. von Klitzing, Phys. Rev. Lett. **62**, 1173 (1989)
510. R.W. Winkler, J.P. Kotthaus, K. Ploog, Phys. Rev. Lett. **62**, 1177 (1989)
511. C.W. Beenakker, Phys. Rev. Lett. **62**, 2020 (1989)
512. A.G. Lebed, JETP Lett. **43**, 174 (1986)
513. G.S. Boebinger, G. Montambaux, M.L. Kaplan, R.C. Haddon, S.V. Chichester, L.Y. Chiang, Phys. Rev. Lett. **64**, 591 (1990)
514. M.J. Naughton, O.H. Chung, M. Chaparala, X. Bu, P. Coppens, Phys. Rev. Lett. **67**, 3712 (1991)
515. T. Osada, A. Kawasumi, S. Kagoshima, N. Miura, G. Saito, Phys. Rev. Lett. **66**, 1525 (1991)
516. W. Kang, S.T. Hannah, P.M. Chaikin, Phys. Rev. Lett. **69**, 2827 (1992)
517. T. Osada, S. Kagoshima, N. Miura, Phys. Rev. B **46**, 1812 (1992)
518. L.I. Buravov, Technol. Phys. **39**, 743 (1994)
519. G.M. Danner, W. Kang, P.M. Chaikin, Phys. Rev. Lett. **72**, 3714 (1994)
520. T. Osada, S. Kagoshima, N. Miura, Phys. Rev. Lett. **77**, 5261 (1996)
521. H. Yoshino, K. Saito, K. Kikuchi, H. Nishikawa, K. Kobayashi, I. Ikemoto, J. Phys. Soc. Jpn. **64**, 2307 (1995)
522. E.I. Chashechkina, P.M. Chaikin, Phys. Rev. Lett. **80**, 2181 (1998)
523. I.J. Lee, M.J. Naughton, Phys. Rev. B **57**, 7423 (1998)
524. E.I. Chashechkina, P.M. Chaikin, Phys. Rev. B **65**, 012405 (2002)
525. W. Wu, I.J. Lee, P.M. Chaikin, Phys. Rev. Lett. **91**, 056601 (2003)
526. H. Kang, Y.J. Jo, S. Uji, W. Kang, Phys. Rev. B **68**, 132508 (2003)
527. H. Kang, Y.J. Jo, W. Kang, Phys. Rev. B **69**, 033103 (2004)
528. M.V. Kartsovnik, V.N. Laukhin, J. Phys. I (France) **6**, 1753 (1996)
529. P.A. Goddard, S.J. Blundell, J. Singleton, R.D. McDonald, A. Ardavan, A. Narduzzo, J.A. Schlueter, A.M. Kini, T. Sasaki, Phys. Rev. B **69**, 174509 (2004)
530. N. Biškup, J.S. Brooks, R. Kato, K. Oshima, Phys. Rev. B **62**, 21 (2000)
531. Y. Shimojo, M.A. Tanatar, T. Ishiguro, R. Koto, J. Phys. Soc. Jpn. **71**, 393 (2002)
532. A.G. Lebed, M.J. Naughton, Phys. Rev. Lett. **91**, 187003 (2003)
533. A.G. Lebed, N.N. Bagmet, M.J. Naughton, Phys. Rev. Lett. **93**, 157006 (2004)
534. A.G. Lebed, H.I. Ha, M.J. Naughton, Phys. Rev. B **71**, 132504 (2005)
535. K. Kobayashi, M. Saito, E. Ohmichi, T. Osada, Phys. Rev. Lett. **96**, 126601 (2006)
536. S.P. Strong, D.G. Clarke, P.W. Anderson, Phys. Rev. Lett. **73**, 1007 (1994)
537. G.M. Danner, P.M. Chaikin, Phys. Rev. Lett. **75**, 4690 (1995)

538. D.G. Clarke, S.P. Strong, E.I. Chashechkina, P.M. Chaikin, Science **279**, 2071 (1998)
539. P. Moses, R.H. McKenzie, Phys. Rev. B **60**, 7998 (1999)
540. M. Kuraguchi, E. Ohmichi, T. Osada, Y. Shiraki, Synth. Met. **133-134**, 113 (2003)
541. T. Osada, E. Ohmichi, *Magnetotransport and Magnetic-Field-Induced Density Waves in Low-Dimensional Layered Conductors*, J. Phys. Soc. Jpn. **75**, 051006 (2006)
542. M.V. Kartsovnik, P.A. Kononovich, V.N. Lauhkin, I.F. Schegolev, JETP Lett. **48**, 541 (1988)
543. K. Kajita, Y. Nishio, T. Takahashi, W. Sasaki, R. Kato, H. Kobayashi, A. Kobayashi, Y. Iye, Solid State Commun. **70**, 1189 (1989)
544. K. Yamaji, J. Phys. Soc. Jpn. **58**, 1520 (1989)
545. R. Yagi, Y. Iye, T. Osada, S. Kagoshima, J. Phys. Soc. Jpn. **59**, 3069 (1990)
546. Y. Kurihara, J. Phys. Soc. Jpn. **61**, 975 (1992)
547. V.G. Peschansky, J.A.R. Lopez, T.G. Yao, J. Phys. I (France) **1**, 1469 (1991)
548. M.V. Kartsovnik, V.N. Laukhin, S.I. Pesotskii, I.F. Schegolev, V.M. Yakovenko, J. Phys. I (France) **2**, 89 (1991)
549. N. Hanasaki, S. Kagoshima, T. Hasegawa, T. Osada, N. Miura, Phys. Rev. B **57**, 1336 (1998)
550. N.E. Hussey, M. Abdel-Jawad, A. Carrington, A.P. Mackenzie, L. Balicas, Nature **425**, 814 (2003)
551. E. Ohmichi, H. Adachi, Y. Mori, Y. Maeno, T. Ishiguro, T. Oguchi, Phys. Rev. B **59**, 7263 (1999)
552. E. Ohmichi, Y. Maeno, T. Ishiguro, J. Phys. Soc. Jpn. **68**, 24 (1999)
553. Y. Yoshida, A. Mukai, R. Settai, K. Miyake, Y. Inada, Y. Ōnuki, K. Betsuyaku, H. Harima, T.D. Matsuda, Y. Aoki, H. Sato, J. Phys. Soc. Jpn. **68**, 3041 (1999)
554. C. Bergemann, A.P. Mackenzie, S.R. Julian, D. Forsythe, E. Ohmichi, *Quasi-Two-Dimensional Fermi Liquid Properties of the Unconventional Superconductor* Sr_2RuO_4, Adv. Phys. **52**, 639 (2003)
555. Y. Iye, M. Baxendale, V.Z. Mordkovich, J. Phys. Soc. Jpn. **63**, 1643 (1994)
556. M. Baxendale, V.Z. Mordkovich, S. Yoshimura, Solid State Commun. **107**, 165 (1998)
557. R. Yagi, Y. Iye, Y. Hashimoto, T. Odagiri, H. Noguchi, H. Sasaki, T. Ikoma, J. Phys. Soc. Jpn. **60**, 3784 (1990)
558. L.P. Gor'kov, A.G. Lebed, Phys. Rev. Lett. **71**, 3874 (1993)
559. M.Y. Azbel, E.A. Kaner, Sov. Phys. JETP **5**, 730 (1957)
560. K.F. Kip, D.N. Langenberg, T.W. Moore, Phys. Rev. B **124**, 359 (1961)
561. L.P. Gor'kov, A.G. Lebed, J. Phys. (Paris) Lett. **45**, L433 (1984)
562. S. Hill, Phys. Rev. B **62**, 8699 (2000)
563. M. Mola, S. Hill, P. Goy, M. Gros, Rev. Sci. Instrum. **71**, 186 (2000)
564. S. Hill, Phys. Rev. B **55**, 4931 (1997)
565. S.J. Blundell, J. Singleton, Phys. Rev. B **53**, 5609 (1996)
566. S.J. Blundell, A. Ardavan, J. Singleton, Phys. Rev. B **55**, R6129 (1997)
567. A. Ardavan, S.J. Blundell, J. Singleton, Phys. Rev. B **60**, 15500 (1999)
568. R.H. McKenzie, P. Moses, Phys. Rev. B **60**, R11241 (1999)
569. S. Hill, N. Harrison, M. Mola, J. Wosnitza, Phys. Rev. Lett. **86**, 3451 (2001)
570. C. Palassis, M. Mola, J. Tritz, S. Hill, S. Uji, K. Kawano, M. Tamura, T. Naito, H. Kobayashi, Synth. Met. **120**, 999 (2001)

571. A.E. Kovalev, S. Hill, K. Kawano, M. Tamura, T. Naito, H. Kobayashi, Phys. Rev. Lett. **91**, 216402 (2003)
572. M. Tamura, H. Tajima, K. Yakushi, H. Kuroda, A. Kobayashi, R. Kato, H. Kobayashi, J. Phys. Soc. Jpn. **60**, 3861 (1991)
573. N. Harrison, C.H. Mielke, D.G. Rickel, J. Wosnitza, J.S. Qualls, J.S. Brooks, E. Balthes, D. Schweitzer, I. Heinen, W. Strunz, Phys. Rev. B **58**, 10248 (1998)
574. A.E. Kovalev, S. Hill, S. Takahashi, T.N. Dhakal, S. Takasaki, J. Yamada, H. Anzai, J.S. Brooks, J. Appl. Phys. **93**, 8665 (2003)
575. S. Takahashi, S. Hill, S. Takasaki, J. Yamada, H. Anzai, Phys. Rev. B **72**, 024540 (2005)
576. J.M. Schrama, J. Singleton, R.S. Edwards, A. Ardavan, E. Rzepniewski, R. Harris, P. Goy, M. Gross, J. Schlueter, M. Kurmoo, P. Days, J. Phys.: Condens. Matter **13**, 2235 (2001)
577. S.V. Demishev, A.V. Semeno, N.E. Sluchanko, N.A. Samarin, I.B. Voskoboinikov, V.V. Glushkov, J. Singleton, S.J. Blundell, S.O. Hill, W. Hayes, M.K. Kartsovnik, A.E. Kovalev, M. Kurmoo, P. Day, N.D. Kushch, Phys. Rev. B **53**, 12794 (1996)
578. H. Ohta, Y. Yamamoto, K. Akioka, M. Motokawa, T. Sasaki, T. Fukase, Synth. Met. **86**, 2011 (1997)
579. A. Ardavan, J.M. Schrama, S.J. Blundell, J. Singleton, W. Hayes, M. Kurmoo, P. Day, P. Goy, Phys. Rev. Lett. **81**, 713 (1998)
580. A.E. Kovalev, S. Hill, J.S. Qualls, Phys. Rev. B **66**, 134513 (2002)
581. A. Polisskii, J. Singleton, P. Goy, W. Hayes, M. Kurmoo, P. Day, J. Phys.: Condens. Matter. **8**, L195 (1996)
582. A. Ardavan, J. Singleton, W. Hayes, A. Polisski, P. Goy, M. Kurmoo, P. Day, Synth. Met. **85**, 1501 (1997)
583. K. Akioka, H. Ohta, Y. Yamamoto, M. Motokawa, K. Kanoda, Synth. Met. **86**, 2051 (1997)
584. Y. Oshima, H. Ohta, K. Koyama, M. Motokawa, H.M. Yamamoto, R. Kato, M. Tamura, Y. Nishio, K. Kajita, J. Phys. Soc. Jpn. **72**, 143 (2003)
585. S. Hill, J.S. Brooks, Z.Q. Mao, Y. Maeno, Phys. Rev. Lett. **84**, 3374 (2000)
586. L.M. Roth, Phys. Rev. **145**, 434 (1966)
587. P.G. de Gennes, *Superconductivity of Metals and Alloys*, 4th edn. (Addison-Wesley, 1992)
588. L. Onsager, Phil. Mag. **43**, 1006 (1952)
589. L. Landau, Z. Phys. **64**, 629 (1930)
590. I.M. Lifshitz, A.M. Kosevich, Sov. Phys. JETP **2**, 636 (1956)
591. R.B. Dingle, Proc. Roy. Soc. A **211**, 517 (1952)
592. E.H. Sondheimer, A.H. Wilson, Proc. Roy. Soc. A **210**, 173 (1951)
593. H. Matsui, M. Yamaguchi, S. Endo, T. Inuzuka, H. Uozaki, N. Toyota, J. Phys. Soc. Jpn. **67**, 2586 (1998)
594. H. Matsui, Y. Yoshida, A. Mukai, R. Settai, Y. Ōnuki, H. Takei, N. Kimura, H. Aoki, N. Toyota, J. Phys. Soc. Jpn. **69**, 3769 (2000)
595. M. Kataoka, T. Goto, J. Phys. Soc. Jpn. **62**, 4352 (1993)
596. M. Kataoka, T. Goto, Physica B **219-220**, 92 (1996)
597. V.A. Bondarenko, S. Uji, T. Terashima, C. Terakura, S. Tanaka, S. Maki, J. Yamada, S. Nakatsuji, Synth. Met. **120**, 1039 (2001)
598. E.N. Adams, T.D. Holstein, J. Phys. Chem. Solid **10**, 254 (1959)

599. M.J. Naughton, J.P. Ulmet, A. Narjis, S. Askenazy, A.P. Hope, Rev. Sci. Instrum. **68**, 4061 (1997)
600. E. Ohmichi, E. Komatsu, T. Osada, Rev. Sci. Instrum. **75**, 2094 (2004)
601. N. Toyota, T. Sasaki, K. Murata, Y. Honda, M. Tokumoto, H. Bando, N. Kinoshita, H. Anzai, T. Ishiguro, Y. Muto, J. Phys. Soc. Jpn. **57**, 2616 (1988)
602. H. Müller, C.P. Heidemann, A. Lerf, W. Biberacher, R. Sieburger, K. Andres, In *The Physics and Chemistry of Organic Superconductors* (G. Saito and S. Kagoshima, Eds., Springer, Berlin, Heidelberg, 1990), p. 195
603. C.P. Heidemann, H. Müller, W. Biberacher, K. Neumeier, Ch. Probst, K. Andres, A.G.M. Jansen, W. Joss, Synth. Met. **41-43**, 2029 (1991)
604. J. Wosnitza, G.W. Crabtree, H.H. Wang, U. Geiser, J.M. Williams, K.D. Carlson, Phys. Rev. B **45**, 3018 (1992)
605. N. Toyota, Y. Watanabe, T. Sasaki, Synth. Met. **55-57**, 2536 (1993)
606. Y. Watanabe, H. Sato, T. Sasaki, N. Toyota, J. Phys. Soc. Jpn. **60**, 3608 (1991)
607. J.M. Luttinger, Phys. Rev. **119**, 1153 (1960)
608. J.M. Luttinger, Phys. Rev. **121**, 1251 (1961)
609. N. Toyota, E.W. Fenton, T. Sasaki, M. Tachiki, Solid State Commun. **72**, 859 (1989)
610. J. Caulfield, W. Lubezynski, W. Lee, J. Singleton, F.L. Pratt, W. Hayes, M. Karmoo, P. Day, Synth. Met. **70**, 815 (1995)
611. S. Uji, M. Chaparala, S. Hill, P.S. Sandhu, J. Qualls, L. Seger, J.S. Brooks, Synth. Met. **85**, 1573 (1997)
612. K. Murata, M. Tokumoto, H. Anzai, Y. Honda, N. Kinoshita, T. Ishiguro, N. Toyota, T. Sasaki, Y. Muto, Synth. Met **27**, A263 (1988)
613. J. Singleton, F.L. Pratt, M. Doporto, W. Hayes, T.J.B.M. Janssen, J.A.A.J. Perenboom, M. Kurmoo, P. Day, Phys. Rev. Lett. **68**, 2500 (1992)
614. W. Kohn, Phys. Rev. **123**, 1242 (1961)
615. K. Kanki, K. Yamada, J. Phys. Soc. Jpn. **66**, 1103 (1997)
616. T. Moss, M. Balkanski (eds.), In *Optical Properties of Semiconductors (Handbook on Semiconductors)* (Elsevier, Amsterdam, 1994), vol. 2, chap. 6 and 7 and 11
617. M.I. Kaganov, A.A. Slutskin, *Coherent Magnetic Breakdown*, Phys. Rep. **98**, 189 (1983)
618. T. Sasaki, H. Sato, N. Toyota, Physica C **185-189**, 2687 (1991)
619. F.L. Pratt, M. Doporto, W. Hayes, J. Singleton, T. Jansen, M. Kurmoo, P. Day, Synth. Met. **41-43**, 2195 (1991)
620. T. Sasaki, N. Toyota, Synth. Met. **55-57**, 2303 (1993)
621. A.B. Pippard, Proc. Roy. Soc. **A270**, 1 (1962)
622. A.B. Pippard, Phil. Trans. Roy. Soc. **A256**, 317 (1964)
623. L.M. Falicov, H. Stachoviak, Phys. Rev. **147**, 505 (1966)
624. W.G. Chambers, Phys. Rev. **165**, 799 (1968)
625. H. Shiba, H. Fukuyama, J. Phys. Soc. Jpn. **26**, 910 (1969)
626. Y. Aharonov, D. Bohm, Phys. Rev. **115**, 485 (1959)
627. R.W. Stark, C.B. Friedberg, Phys. Rev. Lett. **26**, 556 (1971)
628. R.W. Stark, C.B. Friedberg, J. Low Temp. Phys. **14**, 111 (1974)
629. R.W. Stark, R. Reifenberger, J. Low Temp. Phys. **26**, 763, 819 (1977)
630. J. Caulfield, J. Singleton, F.L. Pratt, M. Doporto, W. Lyubczynski, W. Hayes, M. Kurmoo, P. Day, P.T.J. Hendriks, J.A.A.J. Perenboom, Synth. Met. **61**, 63 (1993)

631. N. Harrison, J. Caulfield, J. Singleton, P.H.P. Reinders, F. Herlach, W. Hayes, M. Kurmoo, P. Day, J. Phys.: Condens. Matter **8**, 5415 (1996)
632. M.V. Kartsovnik, G.Y. Logvenov, T. Ishiguro, W. Biberacher, H. Anzai, N. Kushch, Phys. Rev. Lett. **77**, 2530 (1996)
633. N. Harrison, J. Singleton, Phys. Rev. Lett. **82**, 4148 (1999)
634. R.A. Webb, S. Washburn, C.P. Umbach, R.B. Laibowitz, Phys. Rev. Lett. **54**, 2696 (1985)
635. F.A. Meyer, E. Steep, W. Biberacher, P. Christ, A. Lerf, A.G.M. Jansen, W. Joss, P. Wyder, K. Andres, Europhys. Lett. **32**, 681 (1995)
636. J.W. Eddy, R.W. Stark, Phys. Rev. Lett. **48**, 275 (1982)
637. A.P. Mackenzie, S.R. Julian, A.J. Diver, G.J. McMullan, M.P. Ray, G.G. Lonzarich, Y. Maeno, S. Nishizaki, T. Fujita, Phys. Rev. Lett. **76**, 3786 (1996)
638. R.A. Deutschmann, W. Wegscheider, M. Rother, M. Bichler, G. Abstreiter, C. Albrecht, J.H. Smet, Phys. Rev. Lett. **86**, 1857 (2001)
639. R.A. Shepherd, M. Elliott, W.G. Herrenden-Harker, M. Zervos, P.R. Morris, M. Beck, M. Ilegems, Phys. Rev. B **60**, R11277 (1999)
640. K. Machida, K. Kishigi, Y. Hori, Phys. Rev. B **51**, 8946 (1995)
641. N. Harrison, R. Bogaerts, P.H.P. Reinders, J. Singleton, S.J. Blundell, F. Herlach, Phys. Rev. B **54**, 9977 (1996)
642. M. Nakano, J. Phys. Soc. Jpn. **66**, 19 (1997)
643. K. Kishigi, J. Phys. Soc. Jpn. **66**, 910 (1997)
644. P.S. Sandhu, J.H. Kim, J.S. Brooks, Phys. Rev. B **56**, 11566 (1997)
645. N. Harrison, E. Rzepniewsky, J. Singleton, P.J. Gee, M.M. Honold, P. Day, M. Kurmoo, J. Phys.: Condens. Matter **11**, 7227 (1999)
646. N. Harrison, Phys. Rev. Lett. **83**, 1395 (1999)
647. K. Kishigi, Y. Hasegawa, Phys. Rev. B **65**, 205405 (2002)
648. V.M. Gvozdikov, Y.V. Pershin, E. Steep, A.G.M. Jansen, P. Wyder, Phys. Rev. B **65**, 165102 (2002)
649. V.M. Gvozdikov, A.G.M. Jansen, D.A. Pesin, I.D. Vagner, P. Wyder, Phys. Rev. B **68**, 155107 (2003)
650. V.M. Gvozdikov, A.G.M. Jansen, D.A. Pesin, I.D. Vagner, P. Wyder, Phys. Rev. B **70**, 245114 (2004)
651. K. Kishigi, Y. Hasegawa, Synth. Met. **153**, 381 (2005)
652. D. Schoenberg, J. Low Temp. Phys. **56**, 417 (1984)
653. D. Beckmann, S. Wanka, J. Wosnitza, J.A. Schlueter, J.M. Williams, P.G. Nixon, R.W. Winter, G.L. Gard, J. Ren, M.H. Whangbo, Eur. Phys. J. B **1**, 295 (1998)
654. I.D. Vagner, T. Maniv, E. Ehrenfreund, Phys. Rev. Lett. **51**, 1700 (1983)
655. K. Jauregui, V.I. Marchenko, I.D. Vagner, Phys. Rev. B **41**, 12922 (1990)
656. E. Balthes, C. Proust, A. Audouard, L. Brossard, D. Schweitzer, I. Heinen, H.J. Keller, W. Strunz, A.G.M. Jansen, E. Steep, Synth. Met. **94**, 3 (1998)
657. E. Balthes, M. Schiller, W. Schmidt, D. Schweitzer, A.G.M. Jansen, P. Wyder, Synth. Met. **133-134**, 87 (2003)
658. D. Schweitzer, E. Balthes, S. Kahlich, I. Heinen, H.J. Keller, W. Strunz, W. Biberacher, A.G.M. Jansen, E. Steep, Synth. Met. **69-71**, 857 (1995)
659. E. Balthes, D. Schweitzer, I. Heinen, H.J. Keller, W. Biberacher, A.G.M. Jansen, E. Steep, Synth. Met. **70**, 841 (1995)
660. J.E. Graebner, M. Robbins, Phys. Rev. Lett. **36**, 422 (1976)

661. T.J.B.M. Janssen, C. Haworth, S.M. Hayden, P. Meeson, M. Springford, Phys. Rev. B **57**, 11698 (1998)
662. R. Corcoran, N. Harrison, S.M. Hayden, P. Messon, M. Springford, P.J. van der Wel, Phys. Rev. Lett. **72**, 701 (1994)
663. N. Harrison, S.M. Hayden, P. Messon, M. Springford, P.J. van der Wel, A.A. Menovsky, Phys. Rev. B **50**, 4208 (1994)
664. T. Terashima, C. Haworth, H. Takeya, S. Uji, H. Aoki, Phys. Rev. B **56**, 5120 (1997)
665. M. Inada, H. Yamagami, Y. Haga, K. Sakurai, Y. Tokiwa, T. Honma, E. Yamamoto, Y. Ōnuki, T. Yanagisawa, J. Phys. Soc. Jpn. **68**, 3643 (1999)
666. T. Sasaki, T. Fukuda, N. Yoneyama, N. Kobayashi, Phys. Rev. B **67**, 144521 (2003)
667. N.J. Clayton, H. Ito, S.M. Hayden, P.J. Meeson, M. Springford, G. Saito, Phys. Rev. B **65**, 064515 (2002)
668. T. Maniv, V. Zhuravlev, I. Vagner, P. Wyder, *Vortex States and Quantum Magnetic Oscillations in Conventional Type-II Superconductors*, Rev. Mod. Phys. **73**, 867 (2001)
669. P.J. van der Wel, J. Caulfield, R. Corcoran, P. Day, S.M. Hayden, W. Hayes, M. Kurmoo, P. Meeson, J. Singleton, M. Springford, Physica C **235-240**, 2453 (1994)
670. T. Sasaki, W. Biberacher, K. Neumaier, W. Hehn, K. Andres, T. Fukase, Phys. Rev. B **57**, 10889 (1998)
671. H. Ito, S.M. Hayden, P.J. Meeson, M. Springford, G. Saito, J. Supercond. **12**, 525 (1999)
672. T. Sasaki, T. Fukuda, T. Nishizaki, T. Fujita, N. Yoneyama, N. Kobayashi, W. Biberacher, Phys. Rev. B **66**, 224513 (2002)
673. A. Wasserman, M. Springford, Physica B **194**, 1801 (1994)
674. R.J. Higgins, D.H. Lowndes, In *Electrons at the Fermi Surfaces* (M. Springford, Ed., Cambridge Univ. Press, 1980), chap. Waveshape Analysis in the de Haas-van Alphen Effect, pp. 393 – 478
675. Y. Ōnuki, A. Hasegawa, In *Handbook of the Physics and Chemistry of Rare Earths* (K. A. Gschneider Jr. and L. Eyring .Eds., Elsevier Science B. V., Amsterdam, 1995), vol. 20, chap. Fermi Surface of Intermetallic Compounds, p. 135
676. S. Uji, J.S. Brooks, *Magnetic-Field-Induced Superconductivity in Organic Conductors*, J. Phys. Soc. Jpn. **75**, 051014 (2006)
677. O. Cépas, R.H. McKenzie, J. Merino, Phys. Rev. B **65**, R100502 (2002)
678. J.F. Kwak, J.E. Schirber, R.L. Greene, E.M. Engler, Phys. Rev. Lett. **46**, 1296 (1981)
679. H. Bando, K. Oshima, M. Suzuki, H. Kobayashi, G. Saito, J. Phys. Soc. Jpn. **51**, 2711 (1982)
680. K. Kajimura, H. Tokumoto, M. Tokumoto, K. Murata, T. Ukachi, H. Anzai, T. Ishiguro, G. Saito, J. Physique **44**, C3. 1059 (1983)
681. L.J. Azevedo, J.M. Williams, S.J. Compton, Phys. Rev. B **28**, 6600 (1983)
682. T. Takahashi, D. Jérome, K. Bechgaard, J. Physique **45**, 945 (1984)
683. M.J. Naughton, J.S. Brooks, L.Y. Chiang, R.V. Chamberlin, P.M. Chaikin, Phys. Rev. Lett. **55**, 969 (1985)
684. F. Pesty, P. Garoche, K. Bechgaard, Phys. Rev. Lett. **55**, 2495 (1985)
685. G. Montambaux, M. Héritier, P. Lederer, Phys. Rev. Lett. **55**, 2078 (1985)

686. K. Yamaji, J. Phys. Soc. Jpn. **54**, 1034 (1985)
687. P.M. Chaikin, Phys. Rev. B **31**, 4770 (1985)
688. K. Maki, Phys. Rev. B **33**, 4826 (1986)
689. K. Yamaji, J. Phys. Soc. Jpn. **56**, 1841 (1987)
690. A. Virosztek, L. Chen, K. Maki, Phys. Rev. B **34**, 3371 (1986)
691. L. Chen, K. Maki, Phys. Rev. B **35**, 8462 (1987)
692. G. Montambaux, D. Poilblanc, Phys. Rev. B **37**, 1913 (1988)
693. A.G. Lebed, P. Bak, Phys. Rev. B **40**, 11433 (1989)
694. T. Osada, S. Kagoshima, N. Miura, Phys. Rev. Lett. **69**, 1117 (1992)
695. P.M. Chaikin, J. Phys. I (France) **6**, 1875 (1996)
696. M. Ribault, D. Jérome, J. Tuchendler, C. Weyl, K. Bechgaard, J. Phys. Lett. **44**, 953 (1983)
697. P.M. Chaikin, M.Y. Choi, J.F. Kwak, J.S. Brooks, K.P. Martin, M.J. Naughton, E.M. Engler, R.L. Greene, Phys. Rev. Lett. **51**, 2333 (1983)
698. J.R. Cooper, W. Kang, P. Auban, G. Montambaux, D. Jérome, K. Bechgaard, Phys. Rev. Lett. **63**, 1984 (1989)
699. S.T. Hannahs, J.S. Brooks, W. Kang, L.Y. Chiang, P.M. Chaikin, Phys. Rev. Lett. **63**, 1988 (1989)
700. S.K. McKernan, S.T. Hannahs, U.M. Scheven, G.M. Danner, P.M. Chaikin, Phys. Rev. Lett. **75**, 1630 (1995)
701. S. Uji, S. Yasuzuka, T. Konoike, K. Enomoto, J. Yamada, E.S. Choi, D. Graf, J.S. Brooks, Phys. Rev. Lett. **94**, 077206 (2005)
702. L. Gor'kov, A.G. Lebed, Phys. Rev. B **51**, R3285 (1995)
703. K. Kishigi, J. Phys. Soc. Jpn. **67**, 3825 (1998)
704. Y. Hasegawa, K. Kishigi, M. Miyazaki, J. Phys. Soc. Jpn. **67**, 964 (1998)
705. K. Sengupta, N. Dupuis, Phys. Rev. B **65**, 035108 (2001)
706. S. Haddad, S. Charfi-Kaddour, C. Nickel, M. Héritier, R. Bennaceur, Phys. Rev. Lett. **89**, 087001 (2002)
707. D. Radic, A. Bjelis, D. Zanchi, Phys. Rev. B **69**, 014411 (2004)
708. T. Sasaki, N. Toyota, M. Tokumoto, N. Kinoshita, H. Anzai, Solid State Commun. **75**, 93 and 97 (1990)
709. R. Tsuchiya, T. Nakamura, T. Takahashi, T. Sasaki, N. Toyota, Synth. Met. **70**, 965 (1995)
710. T. Miyagawa, A. Kawamoto, K. Kanoda, Phys. Rev. B **56**, R8487 (1997)
711. T. Osada, R. Yagi, A. Kawasumi, S. Kagoshima, N. Miura, M. Oshima, G. Saito, Phys. Rev. B **41**, R5428 (1990)
712. M.V. Kartsovnik, A.E. Kovalev, N.D. Kushch, J. Phys. I (France) **3**, 1187 (1993)
713. Y. Iye, R. Yagi, N. Hanasaki, S. Kagoshima, H. Mori, H. Fujimoto, G. Saito, J. Phys. Soc. Jpn. **63**, 674 (1994)
714. T. Sasaki, N. Toyota, Phys. Rev. B **49**, 10120 (1994)
715. P. Foury-Leylekian, S. Ravy, J.P. Pouget, H. Müller, Synth. Met. **137**, 1271 (2003)
716. K. Maki, B. Dóra, M.V. Kartsovnik, A. Virosztek, B. Korin-Hamzić, M. Basletić, Phys. Rev. Lett. **90**, 256402 (2003)
717. B. Dóra, K. Maki, A. Ványolos, A. Virosztek, Phys. Rev. B **68**, R241102 (2003)
718. B. Dóra, K. Maki, B. Korin-Hamzić, M. Basletić, A. Virosztek, M.V. Kartsovnik, H. Müller, Europhys. Lett. **60**, 737 (2002)

719. E.S. Choi, J.S. Brooks, J.S. Qualls, Phys. Rev. B **65**, 205119 (2002)
720. N. Harrison, L. Balicas, J.S. Brooks, M. Tokumoto, Phys. Rev. B **62**, 14212 (2000)
721. P. Christ, W. Biberacher, D. Andres, M.V. Kartsovnik, E. Balthes, H. Weiss, H. Müller, Synth. Met. **120**, 1019 (2001)
722. A. Kovalev, H. Müller, M. Kartsovnik, JETP **86**, 578 (1998)
723. W. Dietrich, P. Fulde, Z. Phys. **265**, 239 (1973)
724. T. Tiedje, J.F. Carolan, A.J. Berlinsky, Can. J. Phys. **53**, 1593 (1975)
725. N. Harrison, A. House, M.V. Kartsovnik, A.V. Polisski, J. Singleton, F. Herlach, W. Hayes, N. Kushch, Phys. Rev. Lett. **77**, 1576 (1996)
726. D.P. Datars, J.E. Sipe, Phys. Rev. B **51**, 4312 (1995)
727. M.V. Kartsovnik, W. Biberacher, E. Steep, P. Christ, K. Andres, A.G.M. Jansen, H. Müller, Synth. Met. **86**, 1933 (1997)
728. N. Biškup, J.A.A.J. Perenboom, J.S. Brooks, J.S. Qualls, Solid State Commun. **107**, 503 (1998)
729. T. Sasaki, A.G. Lebed, T. Fukase, N. Toyota, Phys. Rev. B **54**, 12969 (1996)
730. A.A. House, S.J. Blundell, M.M. Honold, J. Singleton, J.A.A.J. Perenboom, W. Hayes, M. Kurmoo, P. Day, J. Phys.: Condens. Matter **8**, 8829 (1996)
731. T. Sasaki, N. Toyota, Phys. Rev. B **48**, 11457 (1993)
732. N. Harrison, N. Biškup, J.S. Brooks, L. Balicas, M. Tokumoto, Phys. Rev. B **63**, 195102 (2001)
733. D. Zanchi, A. Bjeliš, G. Montambaux, Phys. Rev. B **53**, 1240 (1996)
734. D. Andres, M.V. Kartsovnik, P.D. Grigoriev, W. Biberacher, H. Müller, Phys. Rev. B **68**, R2011018 (2003)
735. P. Fulde, R.A. Ferrell, Phys. Rev. **135**, A550 (1964)
736. A.I. Larkin, Y.N. Ovchinnikov, Sov. Phys. JETP **20**, 762 (1965)
737. N. Harrison, C.H. Mielke, J. Singleton, J.S. Brooks, M. Tokumoto, J. Phys.: Condens. Matter **13**, L389 (2001)
738. H. Mielke, N. Harrison, A. Ardavan, P. Goddard, J. Singleton, A. Narduzzo, L.K. Montgomery, L. Balicas, J.S. Brooks, M. Tokumoto, Synth. Met. **133**, 99 (2003)
739. S. Hill, P.S. Sandhu, J.S. Qualls, J.S. Brooks, M. Tokumoto, N. Kinoshita, T. Kinoshita, T. Tanaka, Phys. Rev. B **55**, R4891 (1997)
740. N. Harrison, M.M. Honold, M.V. Kartsovnik, J. Singleton, S.T. Hannahs, D.G. Rickel, N. Kushch, Phys. Rev. B **55**, 16005 (1997)
741. M.M. Honold, N. Harrison, J. Singleton, H. Yaguchi, C.H. Mielke, D. Rickel, I. Deckers, P.H.P. Reinders, F. Herlach, M. Kurmoo, P. Day, J. Phys.: Condens. Matter **9**, L533 (1997)
742. M.M. Honold, N. Harrison, J. Singleton, M.S. Nam, S.J. Blundell, C.H. Mielke, M.V. Kartsovnik, N.D. Kushch, Phys. Rev. B **59**, R10417 (1999)
743. N. Harrison, J. Singleton, A. Bangura, A. Ardavan, P.A. Goddard, R.D. McDonald, L.K. Montgomery, Phys. Rev. B **69**, 165103 (2004)
744. D. Andres, M.V. Kartsovnik, W. Biberacher, K. Neumaier, E. Schuberth, H. Müller, Phys. Rev. B **72**, 174513 (2005)
745. N. Tajima, M. Tamura, Y. Nishio, K. Kajita, Y. Iye, J. Phys. Soc. Jpn. **69**, 543 (2000)
746. J.S. Brooks, R.G. Clark, R.P. Starrett, R. Newbury, R.H. McKenzie, A.V. Skougarevsky, M. Tokumoto, N. Kinoshita, T. Kinoshita, Y. Tanaka, H. Anzai, S. Takasaki, J. Yamada, M.V. Kartsovnik, A.I. Schegolev, G.J. Athas, P. Sandhu, Physica B **216**, 380 (1996)

747. T. Konoike, K. Iwashita, H. Yoshino, K. Murata, T. Sasaki, G.C. Papavassiliou, Phys. Rev. B **66**, 245308 (2002)
748. G.C. Papavassiliou, D.J. Lagouvardos, J.S. Zambounis, A. Terzis, C.P. Raptopoulou, K. Murata, N. Shirakawa, L. Ducasse, P. Delhaes, Mol. Cryst. Liq. Cryst. **285**, 83 (1996)
749. G.C. Papavassiliou, K. Murata, J.P. Ulmet, A. Terzis, G.A. Mousdis, H. Yoshino, A. Oda, D. Vignolles, C.P. Raptopoulou, Synth. Met. **103**, 1921 (1999)
750. J.S. Zambounis, J. Pfeiffer, G.C. Papavassiliou, D.J. Lagouvardos, A. Terzis, C.P. Raptopoulou, P. Delhaes, L. Ducasse, N.A. Fortune, K. Murata, Solid State Commun. **95**, 211 (1995)
751. G.C. Papavassiliou, G.A. Mousdis, A. Terzis, C.P. Raptopoulou, K. Murata, T. Konoike, H. Yoshino, Synth. Met. **120**, 743 (2001)
752. J.S. Brooks, D. Graf, E.S. Choi, L. Balicas, K. Storr, C.H. Mielke, G.C. Papavassiliou, Phys. Rev. B **67**, 153104 (2003)
753. D. Graf, E.S. Choi, J.S. Brooks, N. Harrison, K. Murata, T. Konoike, G.A. Mousdis, G.C. Papavassiliou, Phys. Rev. B **71**, 045117 (2005)
754. M.S. Nam, A. Ardavan, J.A. Symington, J. Singleton, N. Harrison, C.H. Mielke, J.A. Schlueter, R.W. Winter, G.L. Gard, Phys. Rev. Lett. **87**, 117001 (2001)
755. J. Wosnitza, S. Wanka, J. Hagel, H. v. Löhneysen, J.S. Qualls, J.S. Brooks, E. Balthes, J.A. Schlueter, U. Geiser, J. Mohtasham, R.W. Winter, G.L. Gard, Phys. Rev. Lett. **86**, 508 (2001)
756. T. Champel, V.P. Mineev, Phys. Rev. B **66**, 195111 (2002)
757. K. Murata, M. Tokumoto, H. Anzai, H. Bando, G. Saito, K. Kajimura, T. Ishiguro, J. Phys. Soc. Jpn. **54**, 2084 (1985)
758. H. Mori, *Materials Viewpoint of Organic Superconductors*, J. Phys. Soc. Jpn. **75**, 051003 (2006)
759. K. Bechgaard, Mol. Cryst. Liq. Cryst. **79**, 1 (1982)
760. E.B. Yagubskii, I.F. Schegolev, V.N. Laukhin, P.A. Kononovich, M.V. Kartsovnik, A.V. Zvarykina, L.I. Bubarov, Sov. Phys. JETP Lett. **39**, 12 (1984)
761. R. Lyubovskaya, R. Luyubovskii, R. Shibaeva, M. Aldoshina, L. Gol'denberg, L. Rozenberg, M. Khidekel', Y. Shul'pyakov, Sov. Phys. JETP Lett. **42**, 468 (1985)
762. T. Komatsu, T. Nakamura, N. Matsukawa, H. Yamochi, G. Saito, H. Ito, T. Ishiguro, M. Kusunoki, K. Sakaguchi, Solid State Commun. **80**, 843 (1991)
763. H. Taniguchi, M. Miyashita, K. Uchiyama, K. Satoh, N. Môri, H. Okamoto, K. Miyagawa, K. Kanoda, M. Hedo, Y. Uwatoko, J. Phys. Soc. Jpn. **72**, 468 (2003)
764. H. Kobayashi, T. Udagawa, H. Tomita, K. Bun, T. Naito, A. Kobayashi, Chem. Lett. **1993**, 1559 (1993)
765. K. Kikuchi, K. Murata, Y. Honda, T. Namiki, G. Saito, H. Anzai, K. Kobayashi, T. Ishiguro, I. Ikemoto, J. Phys. Soc. Jpn. **56**, 4241 (1987)
766. A. Kobayashi, H. Kobayashi, A. Miyamoto, R. Kato, R. Clark, A. Underhill, Chem. Lett. **1991**, 2163 (1991)
767. L.D. Landau, E.M. Lifshitz, *Statistical Physics* (Pergamon Press, Elmsford, 1969)
768. V.L. Ginzburg, L.D. Landau, Zh. Eksp. Teor. Fiz. **20**, 1064 (1950)

769. L.P. Gor'kov, Zh. Eksperim. i. Teor. Fiz. **36**, 1918 (1959)
770. A.A. Abrikosov, L.P. Gor'kov, I.E. Dzyaloshinski, *Methods of Quantum Field Theory in Statistical Physics* (Dover, New York, 1963)
771. J.R. Schrieffer, *Theory of Superconductivity* (Benjamin/Cummings, Reading, MA, 1964)
772. A.L. Fetter, J.D. Walecka, *Quantum Many-Particle Theory* (McGraw Hill, New York, 1971)
773. J.F. Annett, *Superconductivity, Superfluids and Condensates in Condensed Matter Physics.* Oxford Master Series (Oxford University Press, Oxford, 2004)
774. B. Mühlschlegel, Z. Phys. **155**, 313 (1959)
775. W.L. McMillan, Phys. Rev. **167**, 331 (1968)
776. W.K. Kwok, U. Welp, K.D. Carlson, G.W. Crabtree, K.G. Vandervoort, H.H. Wang, A.M. Kini, J.M. Williams, D.L. Stupka, L.K. Montgomery, J.E. Thompson, Phys. Rev. B **42**, 8686 (1990)
777. H. Ito, M. Watanabe, Y. Nogami, T. Ishiguro, T. Komatsu, G. Saito, N. Hosoito, J. Phys. Soc. Jpn. **60**, 3230 (1992)
778. T.T.M. Palstra, B. Batlogg, R.B.V. Dover, L.F. Schneemeyer, J.V. Waszczak, Phys. Rev. Lett. **61**, 1662 (1988)
779. H.R. Ott, In *The Physics of Superconductors* (K.H. Bennemann and J.B. Ketterson, Eds., Springer-Verlag, Berlin Heidelberg New York, 2004), vol. I, chap. High-T_c Superconductivity, pp. 385 – 494
780. V.G. Kogan, Phys. Rev. B **24**, 1572 (1981)
781. W.E. Lawrence, S. Doniach, in *Proceedings to the Twelfth International Conference on Low-Temperature Physics, Kyoto 1970*, ed. by E. Kanada (Keigaku, Tokyo, 1971), p. 361
782. R. Kleiner, F. Steinmeyer, G. Kunkel, P. Müller, Phys. Rev. Lett. **68**, 2394 (1992)
783. P. Müller, *Advances in Solid State Physics*, vol. 34 (Vieweg-Verlag, Braunschweig, 1994)
784. G. Bergmann, Z. Phys. **225**, 430 (1969)
785. P.A. Lee, S.R. Shenoy, Phys. Rev. Lett. **28**, 1025 (1972)
786. W.J. Skocpol, M. Tinkham, *Fluctuations near Superconducting Phase Transitions*, Rep. Prog. Phys. **38**, 1049 (1975)
787. G. Blatter, M.V. Feigel'man, V.B. Geshkenbein, A.I. Larkin, V.M. Vinokur, *Vortices in High-Temperature Superconductors*, Rev. Mod. Phys. **66**, 1125 (1994)
788. U. Welp, S. Fleshler, W.K. Kwok, R.A. Klemm, V.M. Vinokur, J. Downey, B. Veal, G.W. Crabtree, Phys. Rev. Lett. **67**, 3180 (1991)
789. M. Lang, F. Steglich, N. Toyota, T. Sasaki, Phys. Rev. B **49**, 15227 (1994)
790. G. Blatter, V.B. Geshkenbein, In *The Physics of Superconductors* (K.H. Bennemann and J.B. Ketterson, Eds., Springer-Verlag, Berlin Heidelberg New York, 2003), vol. I, chap. Vortex Matter, pp. 725 – 736
791. M. Lang, J. Müller, In *The Physics of Superconductors* (K.H. Bennemann and J.B. Ketterson, Eds., Springer-Verlag, Berlin Heidelberg New York, 2004), vol. II, chap. Organic Superconductors, pp. 453 – 554
792. R. Ikeda, T. Ohmi, T. Tsuneto, J. Phys. Soc. Jpn. **59**, 1397 (1990)
793. S. Ullah, A.T. Dorsey, Phys. Rev. B **44**, 262 (1991)

794. A.I. Larkin, A.A. Varlamov, In *The Physics of Superconductors* (K.H. Bennemann and J.B. Ketterson, Eds., Springer-Verlag, Berlin Heidelberg New York, 2004), vol. I, chap. Fluctuation Phenomena in Superconductors, pp. 95 – 232
795. S. Friemel, C. Pasquier, Y. Loirat, D. Jérome, Physica C **259**, 181 (1996)
796. V. Vulcanescu, B. Janossy, P. Batail, L. Fruchter, Phys. Rev. B **53**, 2590 (1996)
797. T.P. Orlando, E.J.M. Jr., S. Foner, M.R. Beasly, Phys. Rev. B **19**, 4545 (1979)
798. Y. Shimojo, T. Ishiguro, H. Yamochi, G. Saito, J. Phys. Soc. Jpn. **71**, 1716 (2002)
799. T. Takahashi, T. Tokiwa, K. Kanoda, H. Urayama, H. Yamochi, G. Saito, Synth. Met. **27**, A319 (1988)
800. D. Schweitzer, K. Polychroniadis, T. Klutz, H.J. Keller, I. Hennig, I. Hennen, U. Haeberlen, E. Gogu, S. Gärtner, Synth. Met. **27**, A465 (1988)
801. J.E. Graebner, R.C. Haddon, S.V. Chichester, S.H. Glarum, Phys. Rev. B **41**, 4808 (1990)
802. P.A. Mansky, P.M. Chaikin, R.C. Haddon, Phys. Rev. Lett. **70**, 1323 (1993)
803. D.E. Farrell, C.J. Allen, R.C. Haddon, S.V. Chichester, Phys. Rev. B **42**, 8694 (1990)
804. S. Kawamata, K. Okuda, T. Sasaki, N. Toyota, Solid State Commun. **89**, 955 (1994)
805. J.C. Martinez, S.H. Brongersma, A. Koshelev, B. Ivlev, P.H. Kes, R.P. Griessen, D.G. de Groot, Z. Tarnavski, A.A. Menovsky, Phys. Rev. Lett. **69**, 2276 (1992)
806. Y. Matsuda, M.B. Gaifullin, K. Kumagai, M. Kosugi, K. Hirata, Phys. Rev. Lett. **78**, 1972 (1997)
807. K. Murata, S. Kagoshima, S. Yasuzuka, H. Yoshino, R. Kondo, *High-Pressure Research in Organic Conductors*, J. Phys. Soc. Jpn. **75**, 051011 (2006)
808. M.K. Wu, J.R. Ashburn, C.J. Torng, P.H. Hor, R.L. Meng, L. Gao, Z.J. Huang, Y. Wang, C.W. Chu, Phys. Rev. Lett. **58**, 908 (1987)
809. C.W. Chu, P.H. Hor, R.L. Meng, L. Gao, Z.J. Huang, Science **235**, 567 (1987)
810. A.K. Klehe, T. Biggs, C.A. Kuntscher, A.M. Kini, J.A. Schlueter, J.Phys.: Condens. Matter **16**, 6109 (2004)
811. R.L. Greene, E.M. Engler, Phys. Rev. Lett. **45**, 1587 (1980)
812. G. Gladstone, M.A. Jensen, J.R. Schrieffer, In *Superconductivity* (R.D. Parks, Ed., M. Dekker, New York, 1969), vol. 2, p. 655
813. S. Kagoshima, R. Kondo, *Control of Electronic Properties of Molecular Conductors by Uniaxial Strain*, Chem. Rev. **104**, 5593 (2004)
814. H. Kusuhara, Y. Sakata, Y. Ueba, K. Tada, M. Kaji, T. Ishiguro, Solid State Commun. **74**, 251 (1990)
815. S. Kagoshima, M. Maesato, R. Kondo, T. Shibata, H. Hirai, Synth. Met. **120**, 683 (2001)
816. M. Maesato, Y. Kaga, R. Kondo, H. Hirai, S. Kagoshima, Synth. Met. **120**, 941 (2001)
817. E.S. Choi, J.S. Brooks, S.Y. Han, L.B.J.S. Qualls, Phil. Mag. B **81**, 399 (2001)
818. M. Kund, H. Veith, H. Müller, K. Andres, G. Saito, Physica C **221**, 119 (1994)

819. M. Kund, J. Lehrke, W. Biberacher, A. Lerf, K. Andres, Synth. Met. **70**, 949 (1995)
820. B. Andraka, J.S. Kim, G.R. Stewart, K.D. Carlson, H.H. Wang, J.M. Williams, Phys. Rev. B **40**, 11345 (1989)
821. S. Sadewasser, C. Looney, J.S. Schilling, J.A. Schlueter, J.M. Williams, P.G. Nixon, R.W. Winter, G.L. Gard, Solid State Commun. **104**, 571 (1997)
822. J. Müller, M. Lang, J.A. Schlueter, U. Geiser, D. Schweitzer, Synth. Met. **120**, 855 (2001)
823. A.K. Klehe, J.Phys.: Condens. Matter **17**, S937 (2005)
824. A. Painelli, A. Girlando, A. Fortunelli, Phys. Rev. B **64**, 054509 (2001)
825. A. Filhol, In *Organic Conductors - Fundamentals and Applications* (J.-P. Farges, Ed., M. Dekker, New York, 1994), chap. Organic Conductors: The Crystallographic Approach, pp. 147 – 228
826. X. Su, F. Zuo, J.A. Schlueter, M.E. Kelly, J.M. Williams, Phys. Rev. B **57**, R14056 (1998)
827. X. Su, F. Zuo, J.A. Schlueter, A.M. Kini, J.M. Williams, Phys. Rev. B **58**, R2944 (1998)
828. A. Aburto, L. Fruchter, C. Pasquier, Physica C **303**, 185 (1998)
829. M. Pinterić, S. Tomić, M. Prester, D. Drobac, K. Maki, Phys. Rev. B **66**, 174521 (2002)
830. A. Kawamoto, K. Miyagawa, K. Kanoda, Phys. Rev. B **55**, 14140 (1998)
831. H. Ito, T. Ishiguro, T. Kondo, G. Saito, Phys. Rev. B **61**, 3243 (2000)
832. N. Yoneyama, T. Sasaki, T. Nishizaki, N. Kobayashi, J. Phys. Soc. Jpn. **73**, 184 (2004)
833. N. Yoneyama, T. Sasaki, N. Kobayashi, Y. Ikemoto, H. Kimura, Phys. Rev. B **72**, 214519 (2005)
834. P.W. Anderson, J. Phys. Chem. Solids **11**, 26 (1959)
835. M.B. Maple, Appl. Phys. **9**, 179 (1976)
836. A.A. Abrikosov, L.P. Gor'kov, Zh. Eksp. Teor. Fiz. **39**, 1782 (1960)
837. A.I. Larkin, Sov. Phys. JETP Lett **2**, 130 (1965)
838. B.J. Powell, R.H. McKenzie, Phys. Rev. B **69**, 024519 (2004)
839. H. Taniguchi, A. Kawamoto, K. Kanoda, Phys. Rev. B **59**, 8424 (1998)
840. H. Taniguchi, A. Kawamoto, K. Kanoda, Phys. Rev. B **67**, 014510 (2003)
841. N. Yoneyama, A. Higashihara, T. Sasaki, T. Nojima, N. Kobayashi, J. Phys. Soc. Jpn. **73**, 1290 (2004)
842. J.P. Pouget, S. Kagoshima, T. Tamegai, Y. Nogami, K. Kubo, T. Nakajima, K. Bechgaard, J. Phys. Soc. Jpn. **59**, 2036 (1990)
843. N. Joo, P. Auban-Sezier, C. Pasquir, P. Monod, D. Jérome, K. Bechgaard, Eur. Phys. J. B **40**, 43 (2004)
844. Y. Hasegawa, H. Fukuyama, J. Phys. Soc. Jpn. **55**, 3717 (1986)
845. A.M. Clogston, Phys. Rev. Lett. **9**, 266 (1962)
846. B.S. Chandrasekhar, Appl. Phys. Lett. **1**, 7 (1962)
847. N.R. Werthamer, E. Helfand, P.C. Hohenberg, Phys. Rev. **147**, 295 (1966)
848. K. Maki, Phys. Rev. **148**, 362 (1966)
849. D. Mailly, M. Ribault, K. Bechgaard, J. Physique **44**, C3 (1983)
850. H. Schwenk, K. Andres, F. Wudl, E. Aharon-Shalom, Solid State Commun. **45**, 767 (1983)
851. P. Garoche, R. Brusetti, D. Jérome, K. Bechgaard, J. Physique Lett. **43**, L147 (1982)

852. K. Murata, M. Tokumoto, H. Anzai, K. Kajimura, T. Ishiguro, Japan. J. Appl. Phys. **26**(Suppl. 26-3), 1367 (1987)
853. R.L. Greene, P. Haen, S.Z. Huang, E.M. Engler, M.Y. Choi, P.M. Chaikin, Mol. Cryst. Liq. Cryst. **79**, 539 (1982)
854. P.M. Chaikin, M.Y. Choi, R.L. Greene, J. Magn. Magn. Mater. **31**, 1268 (1983)
855. J.I. Oh, M.J. Naughton, Phys. Rev. Lett. **92**, 067001 (2004)
856. Y. Shimojo, A.E. Kovalev, S. Kamiya, E. Ohmichi, T. Ishiguro, H. Yamochi, G. Saito, A. Ayari, P. Monceau, Physica B **294-295**, 427 (2001)
857. A. Kovalev, T. Ishiguro, T. Kondo, G. Saito, Phys. Rev. B **62**, 103 (2000)
858. F.L. Pratt, S.J. Blundell, Phys. Rev. Lett. **94**, 097006 (2005)
859. S. Wanka, D. Beckmann, J. Wosnitza, E. Balthes, D. Schweitzer, W. Strunz, H.J. Keller, Phys. Rev. B **53**, 9301 (1996)
860. J. Hagel, S. Wanka, D. Beckmann, J. Wosnitza, D. Schweitzer, W. Strunz, M. Thumfart, Physica C **291**, 213 (1997)
861. F. Zuo, P. Zhang, X. Su, J.S. Brooks, J.A. Schlueter, J. Mohtasham, R.W. Winter, G.L. Gard, J. Low Temp. Phys. **117**, 1711 (1999)
862. H. Mayaffre, P. Wzietek, D. Jérome, C. Lenoir, P. Batail, Phys. Rev. Lett. **75**, 4122 (1995)
863. T. Kawasaki, M.A. Tanatar, T. Ishiguro, H. Tanaka, A. Kobayashi, H. Kobayashi, Synth. Met. **120**, 771 (2001)
864. T. Suzuki, E. Negishi, H. Uozaki, H. Matsui, N. Toyota, Physica C **440**, 17 (2006)
865. F. Zuo, J.S. Brooks, R.H. McKenzie, J.A. Schlueter, J.M. Williams, Phys. Rev. B **61**, 750 (2000)
866. J. Singleton, J.A. Symington, M.S. Nam, A. Ardavan, M. Kurmoo, P. Day, J. Phys.: Condens. Matter **12**, L641 (2000)
867. M. Tanatar, T. Ishiguro, H. Tanaka, A. Kobayashi, H. Kobayashi, J. Supercond. **12**, 511 (1999)
868. M. Dressel, O. Klein, G. Grüner, K.D. Carlson, H.H. Wang, J.M. Williams, Phys. Rev. B **50**, 13603 (1994)
869. R. Prozorov, R.W. Giannetta, J.A. Schlueter, A.M. Kini, J. Mohtasham, R.W. Winter, G.L. Gard, Phys. Rev. B **65**, 052506 (2001)
870. P.A. Mansky, P.M. Chaikin, R.C. Haddon, Phys. Rev. B **50**, 15929 (1994)
871. M. Pinterić, S. Tomić, M. Prester, D. Drobac, O. Milat, K. Maki, D. Schweitzer, I. Heinen, W. Strunz, Phys. Rev. B **61**, 7033 (2000)
872. D.R. Harshman, R.N. Kleiman, R.C. Haddon, S.V. Chichester-Hicks, M.L. Kaplan, L.W.R. Jr., T. Pfiz, D.L. Williams, D.B. Mitzi, Phys. Rev. Lett. **64**, 1293 (1990)
873. M. Lang, N. Toyota, T. Sasaki, H. Sato, Phys. Rev. Lett. **69**, 1443 (1992)
874. M. Lang, N. Toyota, T. Sasaki, H. Sato, Phys. Rev. B **46**, R5822 (1992)
875. D. Achkir, M. Poirier, C. Bourbonnais, G. Quirion, C. Lenoir, P. Batail, D. Jérome, Phys. Rev. B **47**, 11595 (1993)
876. F.L. Pratt, S.J. Blundell, I.M. Marshall, T. Lancaster, S.L. Lee, A. Drew, U. Divakar, H. Matsui, N. Toyota, Polyhedron **22**, 2307 (2003)
877. E. Helfand, N.R. Werthamer, Phys. Rev. **147**, 288 (1966)
878. J.A. Symington, J. Singleton, M.S. Nam, A. Ardavan, M. Kurmoo, P. Day, Physica B **294 - 295**, 418 (2001)
879. T. Ishiguro, J. Phys. IV (France) **10**, 186 (2000)

880. A.A. Abrikosov, Sov. Phys. JETP **5**, 1174 (1957)
881. R.P. Huebener, *Magnetic Flux Structures in Superconductors*, 2nd edn. (Springer, Berlin Heidelberg New York, 2001)
882. E. Brandt, *The Flux-Line Lattice in Superconductors*, Rep. Prog. Phys. **58**, 1465 (1995)
883. R.P. Huebener, N. Schopohl, G.E. Volovik (eds.), *Vortices in Unconventional Superconductors and Superfluids* (Springer, Berlin Heidelberg, 2002)
884. J.R. Clem, M.W. Coffey, Phys. Rev. B **42**, 6209 (1990)
885. J.R. Clem, Phys. Rev. B **43**, 7837 (1991)
886. D.S. Fisher, M.P.A. Fisher, D. Huse, Phys. Rev. B **43**, 130 (1991)
887. L.I. Glazman, A.E. Koshelev, Phys. Rev. B **43**, 2835 (1991)
888. L.I. Glazman, A.E. Koshelev, Physica C **173**, 180 (1991)
889. S.L. Lee, F.L. Pratt, S.J. Blundell, C.M. Aegerter, P.A. Pattenden, K.H. Chow, E.M. Forgan, T. Sasaki, W. Hayes, H. Keller, Phys. Rev. Lett. **79**, 1563 (1997)
890. F. Pratt, S. Lee, C. Aegerter, C. Ager, S. Lloyd, S. Blundell, F. Ogrin, E. Forgan, H. Keller, W. Hayes, T. Sasaki, N. Toyota, S. Endo, Synth. Met. **120**, 1015 (2001)
891. S.J. Blundell, *Muon-Spin Rotation Studies of Electronic Properties of Molecular Conductors and Superconductors*, Chem. Rev. **104**, 5717 (2004)
892. T. Nishizaki, T. Sasaki, T. Fukase, N. Kobayashi, Phys. Rev. B **54**, R3760 (1996)
893. L.Y. Vinnikov, T.L. Barkov, M.V. Kartsovnik, N.D. Kushch, Phys. Rev. B **61**, 14358 (2000)
894. F.L. Barkov, L.Y. Vinnikov, M.V. Kartsovnik, N.D. Kushch, Physica C **385**, 568 (2003)
895. M. Mola, J.T. King, C.P. McRaven, S. Hill, J.S. Qualls, J.S. Brooks, Phys. Rev. B **62**, 5965 (2000)
896. S. Schilling, R. Jin, J.D. Guo, H.R. Ott, Phys. Rev. Lett. **71**, 1899 (1993)
897. M. Inada, T. Sasaki, T. Nishizaki, N. Kobayashi, S. Yamada, T. Fukase, J. Low Temp. Phys. **117**, 1423 (1999)
898. M. Mola, S. Hill, J.S. Brooks, J.S. Qualls, Phys. Rev. Lett. **86**, 2130 (2001)
899. T. Shibauchi, M. Sato, S. Ooi, T. Tamegai, Phys. Rev. B **57**, R5622 (1998)
900. C. Pasquier, S. Friemel, Synth. Met. **103**, 1845 (1999)
901. S. Friemel, C. Pasquier, Physica C **265**, 121 (1996)
902. C. Pasquier, N. Joo, P. Batail, D. Schweitzer, D. Jérome, J. Phys. IV (France) **114**, 239 (2004)
903. H. Mayaffre, P. Wzietek, D. Jérome, S. Brazovskii, Phys. Rev. Lett. **76**, 4951 (1995)
904. A. Van-Quynh, C. Berthier, H. Mayaffre, P. Ségransan, P. Batail, Phys. Rev. B **59**, 12064 (1999)
905. S. Friemel, C. Pasquier, D. Jérome, Physica C **292**, 273 (1997)
906. S. Endo, T. Nojima, N. Toyota, Physica C **378-381**, 508 (2002)
907. E. Zeldov, D. Majer, M. Konczykowski, V.B. Geshkenbein, V.M. Vinokur, H. Shtrikman, Nature **375**, 373 (1995)
908. S. Schilling, R.A. Fisher, N.E. Phillips, U. Welp, D. Dasgupta, W.K. Kwok, G.W. Crabtree, Nature **382**, 791 (1996)
909. U. Welp, J.A. Fendrich, W.K. Kwok, G.W. Crabtree, B.W. Veal, Phys. Rev. Lett. **76**, 4809 (1996)

910. S. Schilling, R.A. Fisher, N.E. Phillips, U. Welp, W.K. Kwok, G.W. Crabtree, Phys. Rev. Lett. **78**, 4833 (1997)
911. L. Fruchter, A. Aburto, C. Pham-Phu, Phys. Rev. B **56**, R2936 (1997)
912. N. Yoneyama, A. Higashihara, T. Sasaki, N. Kobayashi, S. Yamada, Synth. Met. **133 - 134**, 223 (2003)
913. S.M. De Soto, C.P. Slichter, H.H. Wang, U. Geiser, J.M. Williams, Phys. Rev. Lett. **70**, 2956 (1993)
914. T. Takahashi, T. Tokiwa, K. Kanoda, H. Urayama, H. Yamochi, G. Saito, Physica C **153-155**, 487 (1988)
915. K. Kanoda, K. Miyagawa, A. Kawamoto, Physica C **282 - 287**, 2063 (1997)
916. T. Sasaki, T. Fukuda, T. Nishizaki, T. Fujita, N. Yoneyama, N. Kobayashi, Physica C **388-389**, 609 (2003)
917. T. Sasaki, T. Fukuda, N. Yoneyama, N. Kobayashi, Physica B **346-347**, 354 (2004)
918. M. Tachiki, S. Takahashi, Solid State Commun. **70**, 291 (1989)
919. W.K. Kwok, U. Welp, V.M. Vinokur, S. Fleshler, J. Downey, G.W. Crabtree, Phys. Rev. Lett. **67**, 390 (1991)
920. D. Feinberg, C. Villard, Phys. Rev. Lett. **65**, 919 (1990)
921. L.N. Bulaevskii, M. Ledvij, V.G. Kogan, Phys. Rev. B **46**, 366 (1992)
922. F. Steinmeyer, R. Kleiner, P. Müller, H. Müller, K. Winzer, Europhys. Lett. **25**, 459 (1994)
923. R.W. Giannetta, N.H. Tea, F.A.B. Chaves, S. Rao, M.B. Salamon, A.M. Kini, H.H. Wang, U. Geiser, J. Schlueter, M.W. Trawick, J.C. Garland, Physica C **321**, 207 (1999)
924. S. Uji, T. Terashima, Y. Terai, S. Yasuzuka, M. Tokumoto, H. Tanaka, A. Kobayashi, H. Kobayashi, Phys. Rev. B **71**, 104525 (2005)
925. H. Shimahara, Phys. Rev. B **50**, 1994 (1994)
926. H. Shimahara, J. Phys. Soc. Jpn. **66**, 541 (1997)
927. H. Shimahara, J. Phys. Soc. Jpn. **67**, 1872 (1998)
928. M. Tachiki, S. Takahashi, P. Gegenwart, M. Weiden, M. Lang, C. Geibel, F. Steglich, R. Modler, C. Paulsen, Y. Ōnuki, Z. Phys. B **100**, 369 (1996)
929. P. Gegenwart, M. Deppe, M. Köppen, F. Kromer, M. Lang, R. Modler, M. Weiden, C. Geibel, F. Steglich, T. Fukase, N. Toyota, Ann. Physik **5**, 307 (1996)
930. C. Martin, C.C. Agosta, S.W. Tozer, H.A. Radovan, T. Kinoshota, M. Tokumoto, J. Low Temp. Phys. **138**, 1025 (2005)
931. P.W. Anderson, *Basic Notions of Condensed Matter Physics* (Benjamin-Cummings, San Francisco, 1984)
932. G.E. Volovik, L.P. Gor'kov, *Superconducting Classes in Heavy-Fermion System*, Sov. Phys. JETP **61**, 843 (1985)
933. J.F. Annett, *Symmetry of the Order Parameter for High-Temperature Superconductivity*, Advances in Physics **39**, 83 (1991)
934. M. Sigrist, K. Ueda, *Phenomenological Theory of Unconventional Superconductivity*, Rev. Mod. Phys. **63**, 239 (1991)
935. C.C. Tsuei, J.R. Kirtley, In *The Physics of Superconductors* (K. H. Bennemann and J. B. Ketterson, Eds., Springer-Verlag, Berlin, Heidelberg, 2004), vol. I, chap. Pairing Symmetry in Cuprate Superconductors: Phase-Sensitive Tests, pp. 647 – 724
936. B. Obst, In *Anisotropic Effects in Superconductors* (H.W. Weber, Ed., Plenum Press, New York, 1977), p. 139

937. R. Balian, N.R. Werthamer, Phys. Rev. **131**, 1553 (1963)
938. L.C. Hebel, C.P. Slichter, Phys. Rev. **107**, 901 (1957)
939. R.A. Hein, J.W. Gibson, Phys. Rev. **131**, 1105 (1963)
940. J.M. Schrama, E. Rzepniewski, R.S. Edwards, J. Singleton, A. Ardavan, M. Kurmoo, P. Day, Phys. Rev. Lett. **83**, 3041 (1999)
941. T. Arai, K. Ichimura, K. Nomura, S. Takasaki, J. Yamada, S. Nakatsuji, H. Anzai, Phys. Rev. B **63**, 104518 (2001)
942. K. Izawa, H. Yamaguchi, T. Sasaki, Y. Matsuda, Phys. Rev. Lett. **88**, 027002 (2002)
943. J.M. Schrama, J. Singleton, Phys. Rev. Lett. **86**, 3453 (2001)
944. J. Schmalian, Phys. Rev. Lett. **81**, 4232 (1998)
945. T. Shibauchi, Y. Matsuda, M.B. Gaifullin, T. Tamegai, Phys. Rev. Lett. **86**, 3452 (2001)
946. K. Ichimura, K. Nomura, *d-wave Pair Symmetry in the Superconductivity of* κ*-(BEDT-TTF)*$_2X$, J. Phys. Soc. Jpn. **75**, 051012 (2006)
947. H. Bando, S. Kashiwaya, H. Tokumoto, H. Anzai, N. Kinoshita, K. Kajimura, J. Vac. Sci. Technol. A **8**, 479 (1990)
948. H. Won., K. Maki, Physica B **312-313**, 44 (2002)
949. D.J. Scalapino, J. E. Loh, J.E. Hirsch, Phys. Rev. B **35**, 6694 (1987)
950. J. Marino, R.H. McKenzie, Phys. Rev. Lett. **87**, 237002 (2001)
951. K. Yosida, Phys. Rev. **110**, 769 (1958)
952. A.J. Leggett, *A Theoretical Description of the New Phases of Liquid* ^3He, Rev. Mod. Phys. **47**, 331 (1975)
953. K. Kanoda, K. Miyagawa, A. Kawamoto, Y. Nakazawa, Phys. Rev. B **54**, 76 (1996)
954. S. Kamiya, Y. Shimojo, M.A. Tanatar, T. Ishiguro, H. Yamochi, G. Saito, Phys. Rev. B **65**, 134510 (2002)
955. S. Belin, K. Behnia, A. Deluzet, Phys. Rev. Lett. **81**, 4728 (1998)
956. M.J. Graf, S. Yip, J.A. Sauls, D. Rainer, Phys. Rev. B **53**, 15147 (1996)
957. M.R. Norman, P.J. Hirschfeld, Phys. Rev. B **53**, 5706 (1996)
958. P.A. Lee, Phys. Rev. Lett. **71**, 1887 (1993)
959. K. Kanoda, K. Akiba, K. Suzuki, T. Takahashi, G. Saito, Phys. Rev. Lett. **65**, 1271 (1990)
960. D.R. Harshman, A.T. Fiory, R.C. Haddon, M.L. Kaplan, T. Pfiz, E. Koster, I. Shinkoda, D.L. Williams, Phys. Rev. B **49**, 12990 (1994)
961. L.P. Le, G.M. Luke, B.J. Sternlieb, W.D. Wu, Y.J. Uemura, J.H. Brewer, T.M. Riseman, C.E. Stronach, G. Saito, H. Yamochi, H.H. Wang, A.M. Kini, K.D. Carlson, J.M. Williams, Phys. Rev. Lett. **68**, 1923 (1992)
962. K. Holczer, D. Quinlivan, O. Klein, G. Grüner, F. Wudl, Solid State Commun. **76**, 499 (1990)
963. O. Klein, K. Holczer, G. Grüner, J. Chang, F. Wudl, Phys. Rev. Lett. **66**, 655 (1991)
964. M. Dressel, L. Degiorgi, O. Klein, G. Grüner, J. Phys. Chem. Solids **54**, 1411 (1993)
965. A. Carrington, I.J. Bonalde, R. Prozorov, R.W. Giannetta, A.M. Kini, J.A. Schlueter, H.H. Wang, U. Geiser, J.M. Williams, Phys. Rev. Lett. **83**, 4172 (1999)
966. I. Kosztin, Q.J. Chen, B. Janko, K. Levin, Phys. Rev. B **58**, R5936 (1998)
967. Q.J. Chen, I. Kosztin, B. Janko, K. Levin, Phys. Rev. Lett. **81**, 4708 (1998)

968. Q. Chen, I. Kosztin, K. Levin, Phys. Rev. Lett. **85**, 2801 (2000)
969. A.F. Hebard, A.T. Fiory, M.P. Siegal, J.M. Phillips, R.C. Haddon, Phys. Rev. B **44**, 9753 (1991)
970. Y. Hasegawa, H. Fukuyama, J. Phys. Soc. Jpn. **56**, 2619 (1987)
971. Z. Hao, J.R. Clem, Phys. Rev. Lett. **67**, 2371 (1991)
972. Y. Nakazawa, K. Kanoda, Phys. Rev. B **55**, R8670 (1997)
973. N. Momono, M. Ido, Physica C **264**, 311 (1996)
974. H. Padamsee, J.E. Neighbor, C.A. Shiffman, J. Low Temp. Phys. **12**, 387 (1973)
975. J. Wosnitza, *Recent Aspects of Organic Metals and Superconductors*, Curr. Opin. Solid State Mater. Sci. **5**, 131 (2001)
976. F. Marsiglio, J.P. Carbotte, Phys. Rev. B **33**, 6141 (1986)
977. P.B. Allen, R.C. Dynes, Phys. Rev. B **12**, 905 (1975)
978. J. Sólyom, *The Fermi Gas Model of One-Dimensional Conductors*, Adv. Phys. **28**, 201 (1979)
979. M.Y. Choi, P.M. Chaikin, S.Z. Huang, P. Haen, E.M. Engler, R.L. Greene, Phys. Rev. B **25**, 6208 (1982)
980. S. Bouffard, M. Ribault, R. Brusetti, D. Jérome, K. Bechgaard, J. Phys. C **15**, 2951 (1982)
981. C. Coulon, P. Delhaes, J. Amiell, J.P. Manceau, J.M. Fabre, L. Giral, J. Phys. (Paris) Colloq. **43**, 1721 (1982)
982. S. Tomić, D. Jérome, D. Mailly, M. Ribault, K. Bechgaard, J. Phys. (Paris) Colloq. **44 (C-3)**, 1075 (1983)
983. A.A. Abrikosov, J. Low Temp. Phys. **53**, 359 (1983)
984. L.P. Gor'kov, D. Jérome, J. Phys. (Paris) Lett. **46**, L643 (1985)
985. M. Dressel, Physica C **317-318**, 89 (1999)
986. M. Takigawa, H. Yasuoka, G. Saito, J. Phys. Soc. Jpn. **56**, 873 (1987)
987. S. Belin, K. Behnia, Phys. Rev. Lett. **79**, 2125 (1997)
988. I.J. Lee, M.J. Naughton, P.M. Chaikin, Phys. Rev. Lett. **88**, 207002 (2002)
989. B.S. Chandrasekhar, Appl. Phys. Lett. **1**, 7 (1962)
990. R.A. Klemm, A. Luther, M.R. Beasley, Phys. Rev. B **12**, 877 (1975)
991. X. Huang, K. Maki, Phys. Rev. B **39**, 6459 (1989)
992. A.G. Lebed, JETP Lett. **44**, 114 (1986)
993. N. Dupuis, G. Montambaux, C.A.R.S. de Melo, Phys. Rev. Lett. **70**, 2613 (1993)
994. I.J. Lee, P.M. Chaikin, M.J. Naughton, Phys. Rev. B **65**, 180502 (2002)
995. P. Fulde, K. Maki, Phys. Rev. **139**, A788 (1965)
996. I.J. Lee, S.E. Brown, W.G. Clark, M.J. Strouse, M.J. Naughton, W. Kang, P.M. Chaikin, Phys. Rev. Lett. **88**, 017004 (2002)
997. I.J. Lee, D.S. Chow, W.G. Clark, J. Strouse, M.J. Naughton, P.M. Chaikin, S.E. Brown, Phys. Rev. B **68**, 92510 (2003)
998. I.J. Lee, S.E. Brown, M.J. Naughton, *Unconventional Superconductivity in a Quasi-One-Dimensional System (TMTSF)$_2$X*, J. Phys. Soc. Jpn. **75**, 051011 (2006)
999. W. Little, In *Organic Conductors - Fundamentals and Applications* (J.-P. Farges, Ed., M. Dekker, New York, 1994), chap. Organic Conductors: An Overview, pp. 1 – 24
1000. K. Yamaji, Solid State Commun. **61**, 413 (1987)
1001. A. Girlando, M. Masino, A. Brillante, R.G.D. Valle, E. Venuti, Phys. Rev. B **66**, R100507 (2002)

1002. A. Girlando, M. Masino, G. Visentini, R.G.D. Valle, A. Brillante, E. Venuti, Phys. Rev. B **62**, 14476 (2000)
1003. G. Varelogiannis, Phys. Rev. Lett. **88**, 117005 (2002)
1004. E. Demiralp, W.A. Goddard, Phys. Rev. B **56**, 11907 (1997)
1005. H. Kondo, T. Moriya, J. Phys. Soc. Jpn. **67**, 3695 (1998)
1006. H. Kondo, T. Moriya, J. Phys.: Condens. Matter **11**, L363 (1999)
1007. H. Kondo, T. Moriya, J. Phys. Soc. Jpn. **68**, 3170 (1999)
1008. H. Kondo, T. Moriya, Physica B **281 - 282**, 940 (2000)
1009. H. Kino, H. Kontani, J. Phys. Soc. Jpn. **67**, 3691 (1998)
1010. M. Vojta, E. Dagotto, Phys. Rev. B **59**, R713 (1999)
1011. K. Kuroki, H. Aoki, Phys. Rev. B **60**, 3060 (1999)
1012. T. Jujo, K. Yamada, J. Phys. Soc. Jpn. **68**, 1331 (1999)
1013. H. Kino, H. Kontani, J. Phys. Soc. Jpn. **68**, 1481 (1999)
1014. T. Jujo, K. Yamada, J. Phys. Soc. Jpn. **68**, 2198 (1999)
1015. B.J. Powell, R.H. McKenzie, Phys. Rev. Lett **94**, 047004 (2005)
1016. R.H. McKenzie, *A Strongly Correlated Electron Model for the Layered Organic Superconductors* κ*-(BEDT-TTF)*$_2X$, Comments Cond. Mat. Phys. **18**, 309 (1998)
1017. J.Y. Gan, Y. Chen, Z.B. Su, F.C. Zhang, Phys. Rev. Lett **94**, 067005 (2005)
1018. R. Louati, S. Charfi-Kaddour, A.B. Ali, R. Bennaceur, M. Héritier, Phys. Rev. B **62**, 5957 (2000)
1019. A. Kobayashi, Y. Tanaka, M. Ogata, Y. Suzumura, J. Phys. Soc. Jpn. **73**, 1115 (2003)
1020. H. Fröhlich, Phys. Rev. **79**, 845 (1950)
1021. P. Auban-Senzier, C. Bourbonnais, D. Jérome, C. Lenoir, P. Batail, E. Canadell, J.P. Buisson, S. Lefrant, J. Phys. I (France) **3**, 871 (1993)
1022. K.D. Carlson, J.N. Williams, U. Geiser, A.M. Kini, H.H. Wang, R.A. Klemm, S.K. Kumar, J.A. Schlueter, J.R. Ferraro, K.R. Lykke, P. Wurz, D.H. Parker, J.D.M. Sutin, Inorg. Chem. **114**, 10069 (1992)
1023. A.M. Kini, K.D. Carlson, J.D. Dudek, U. Geiser, H.H. Wang, J.M. Williams, Synth. Met. **85**, 1617 (1997)
1024. Y. Watanabe, T. Shimazu, T. Sasaki, N. Toyota, Synth. Met. **86**, 1917 (1997)
1025. J.M. Williams, A.J. Schultz, U. Geiser, K.D. Carlson, A.M. Kini, H.H. Wang, W.K. Kwok, M.H. Whangbo, J.E. Schirber, Science **252**, 1501 (1991)
1026. A.M. Kini, J.A. Schlueter, B.H. Ward, U. Geiser, H.H. Wang, Synth. Met. **120**, 713 (2001)
1027. J.A. Schlueter, A.M. Kini, B.H. Ward, U. Geiser, H.H. Wang, J. Mohtasham, R.W. Winter, G.L. Gard, Physica C **351**, 261 (2001)
1028. J.D. Axe, G. Shirane, Phys. Rev. B **8**, 1965 (1973)
1029. S.M. Shapiro, G. Shirane, Phys. Rev. B **12**, 4899 (1975)
1030. N. Toyota, M. Lang, S. Ikeda, T. Kajitani, T. Shimazu, T. Sasaki, K. Shibata, Synth. Met. **86**, 2009 (1997)
1031. R. Zeyher, G. Zwicknagl, Z. Phys. B **78**, 175 (1990)
1032. Y.J. Uemura, G.M. Luke, B.J. Sternlieb, J.H. Brewer, J.F. Carolan, W.N. Hardy, R. Kadono, J.R. Kempton, R.F. Kiefl, S.R. Kreitzman, P. Mulhern, T.M. Riseman, D.L. Williams, B.X. Yang, S. Uchida, H. Takagi, J. Gopalakrishnan, A.W. Sleight, M.A. Subramanian, C.L. Chien, M.Z. Cieplak, G. Xiao, V.Y. Lee, B.W. Statt, C.E. Stronach, W.J. Kossler, X.H. Yu, Phys. Rev. Lett. **62**, 2317 (1989)

1033. Y.J. Uemura, Physica (Amsterdam) **282C-287C**, 194 (1997)
1034. V.J. Emery, S.A. Kivelson, Nature **374**, 434 (1995)
1035. C.C. Homes, S.V. Dordevic, M. Strongin, D.A. Bonn, R. Liang, W. Hardy, S. Komiya, Y. Ando, G. Yu, N. Kaneko, X. Zhao, M. Greven, D.N. Basov, T. Timusk, Nature **430**, 539 (2004)
1036. S. Hünig, A. Aumüller, P. Erk, H. Meixner, J.U. von Schütz, H.J. Groß, U. Langohr, H.P. Werner, H.C. Wolf, C. Burschka, G. Klebe, K. Peters, H.G. von Schnering, Synth. Met. **27**, B 181 (1988)
1037. R. Kato, H. Kobayashi, A. Kobayashi, J. Am. Chem. Soc. **111**, 5224 (1989)
1038. S. Hünig, P. Erk, *DCNQIs – New Electron Acceptors for Charge-Transfer Complexes and Highly Conducting Radical Anion Salts*, Adv. Mater. **3**, 225 (1991)
1039. S. Hünig, *N,N'-Dicyanoquinone Diimines (DCNQIs) – Unique Acceptors for Conducting Materials*, J. Mater. Chem. **5**, 1469 (1995)
1040. S. Hünig, M. Kemmer, H. Meixner, K. Sinzger, H. Wenner, T. Bauer, E. Tillmanns, F.R. Lux, M. Hollstein, H.G. Groß, U. Langohr, H.P. Werner, J.U. von Schütz, H.C. Wolf, Eur. J. Inorg. Chem. **1999**, 899 (1999)
1041. T. Mori, K. Imaeda, R. Kato, A. Kobayashi, H. Kobayashi, H. Inokuchi, J. Phys. Soc. Jpn. **56**, 3429 (1987)
1042. S. Tomić, D. Jérome, A. Aumüller, P. Erk, S. Hünig, J.U. von Schütz, Eur. Phys. Lett. **5**, 553 (1988)
1043. H. Kobayashi, R. Kato, A. Kobayashi, T. Mori, H. Inokuchi, Solid State Commun. **65**, 1351 (1988)
1044. H. Kobayashi, A. Miyamoto, R. Kato, A. Kobayashi, Y. Nishio, K. Kajita, W. Sasaki, Solid State Commun. **72**, 1 (1989)
1045. S. Aonuma, H. Sawa, R. Kato, J. Chem. Soc., Perkin Trans. 2 **2**, 1541 (1995)
1046. K. Sinzger, S. Hünig, M. Jopp, D. Bauer, W. Bietsch, J.U. v. Schütz, H.C. Wolf, R.K. Kremer, T. Metzenthin, R. Bau, S.I. Khan, A. Lindbaum, C.L. Lengauer, E. Tillmanns, J. Am. Chem. Soc. **115**, 7696 (1993)
1047. A. Kobayashi, R. Kato, H. Kobayashi, T. Mori, H. Inokuchi, Solid State Commun. **64**, 45 (1987)
1048. H. Kobayashi, H. Sawa, S. Aonuma, R. Kato, J. Am. Chem. Soc. **115**, 7870 (1993)
1049. H. Sawa, M. Tamura, S. Aonuma, R. Kato, M. Kinoshita, H. Kobayashi, J. Phys. Soc. Jpn. **62**, 2224 (1993)
1050. I.H. Inoue, A. Kakizaki, H. Namatame, A. Fijimori, A. Kobayashi, R. Kato, H. Kobayashi, Phys. Rev. B **45**, 5828 (1992)
1051. T. Takahashi, K. Kanoda, T. Tamura, K. Hiraki, K. Ikeda, R. Kato, H. Kobayashi, A. Kobayashi, Synth. Met. **56**, 2281 (1993)
1052. K. Hiraki, K. Takagawa, T. Nakamura, T. Takahashi, S. Aonuma, H. Sawa, R. Kato, Synth. Met. **70**, 1069 (1995)
1053. K. Ishida, Y. Kitaoka, H. Masuda, K. Asayama, T. Takahashi, A. Kobayashi, R. Kato, H. Kobayashi, J. Phys. Soc. Jpn. **64**, 2970 (1995)
1054. R. Kato, S. Aonuma, H. Sawa, Synth. Met. **70**, 1071 (1995)
1055. S. Uji, T. Terashima, H. Aoki, J.S. Brooks, R. Kato, H. Sawa, S. Aonuma, M. Tamura, M. Kinoshita, Phys. Rev. B **50**, 15597 (1994)
1056. H. Fukuyama, J. Phys. Soc. Jpn. **61**, 3452 (1992)
1057. Y. Suzumura, H. Fukuyama, J. Phys. Soc. Jpn. **61**, 3322 (1992)
1058. M. Tamura, H. Sawa, S. Aonuma, R. Kato, M. Kinoshita, H. Kobayashi, J. Phys. Soc. Jpn. **63**, 429 (1994)

1059. M. Tamura, H. Sawa, S. Aonuma, R. Kato, M. Kinoshita, H. Kobayashi, J. Phys. Soc. Jpn. **62**, 1470 (1993)
1060. M. Nakano, M. Kato, K. Yamada, Physica B **186-188**, 1077 (1993)
1061. Y. Nishio, M. Tamura, K. Kajita, S. Aonuma, H. Sawa, R. Kato, H. Kobayashi, J. Phys. Soc. Jpn. **69**, 1414 (2000)
1062. R. Kato, H. Kobayashi, A. Kobayashi, Synth. Met. **41 - 43**, 2093 (1991)
1063. M. Fujiwara, N. Tajima, T. Imakubo, M. Tamura, R. Kato, J. Solid State Chem. **168**, 396 (2002)
1064. H. Fujiwara, H. Kobayashi, Bull. Chem. Soc. Jpn. **78**, 181 (2005)
1065. N. Toyota, T. Suzuki, Comptes Rendus Chimie, in press
1066. H. Kobayashi, H. Tomita, T. Naito, A. Kobayashi, F. Sakai, T. Watanabe, P. Cassoux, J. Am. Chem. Soc. **118**, 368 (1996)
1067. T. Mori, M. Katsuhara, J. Phys. Soc. Jpn. **71**, 826 (2002)
1068. T. Mallah, C. Hollis, S. Bott, M. Kurmoo, P. Day, M. Allan, R.H. Friend, J. Chem. Soc. Dalton Trans. p. 859 (1990)
1069. M.A. Beno, D.D. Cox, J.M. Williams, J.F. Kwak, Acta Crystallogr. **C40**, 1334 (1984)
1070. M.A. Beno, D.D. Cox, J.M. Williams, J.F. Kwak, Acta Crystallogr. **C44**, 1544 (1988)
1071. H. Mori, T. Okano, M. Kamiya, M. Haemori, H. Suzuki, S. Tanaka, Y. Nishio, K. Kajita, H. Moriyama, Physica C **357-360**, 103 (2001)
1072. H. Tanaka, A. Kobayashi, A. Sato, H. Akutsu, H. Kobayashi, J. Am. Chem. Soc. **121**, 760 (1999)
1073. H. Fujiwara, E. Fujiwara, Y. Nakazawa, Z.B. Narymbetov, K. Kato, H. Kobayashi, A. Kobayashi, M. Tokumoto, P. Cassoux, J. Am. Chem. Soc. **123**, 306 (2001)
1074. T. Otsuka, A. Kobayashi, Y. Miyamoto, J. Kiuchi, S. Nakamura, N. Wada, E. Fujiwara, H. Fujiwara, H. Kobayashi, J. Solid State Chem. **159**, 407 (2001)
1075. B. Zhang, H. Tanaka, H. Fujiwara, H. Kobayashi, E. Fujiwara, A. Kobayashi, J. Am. Chem. Soc. **124**, 9982 (2002)
1076. F.L. Pratt, S.L. Lee, S.J. Blundell, I.M. Marshall, H. Uozaki, N. Toyota, Synth. Met. **133**, 489 (2003)
1077. H. Uozaki, *Superconductivity, Metal-Insulator Transition and Magnetism in (BEDT-TSF)*$_2$*MCl*$_4$ *(M =* $Fe, Ga, Ga_{1-x}Fe_x$*)*. Doctor thesis, Tohoku University, Sendai (2001)
1078. L.K. Montgomery, T. Burgin, J.C. Huffman, J. Ren, M.H. Whangbo, Physica C **219**, 490 (1994)
1079. H. Akutsu, E. Arai, H. Kobayashi, H. Tanaka, A. Kobayashi, P. Cassoux, J. Am. Chem. Soc. **119**, 12681 (1997)
1080. T. Suzuki, H. Matsui, H. Tsuchiya, E. Negishi, K. Koyama, N. Toyota, Phys. Rev. B **67**, R020408 (2003)
1081. L. Brossard, R. Clerac, C. Coulon, M. Tokumoto, T. Ziman, D.K. Petrov, V.N. Laukhin, M.J. Naughton, A. Audouard, F. Goze, A. Kobayashi, H. Kobayashi, P. Cassoux, Eur. Phys. J. B **1**, 439 (1998)
1082. I. Rutel, S. Okubo, J.S. Brooks, E. Jobiliong, H. Kobayashi, A. Kobayashi, H. Tanaka, Phys. Rev. B **68**, 144435 (2003)
1083. T. Sasaki, H. Uozaki, S. Endo, N. Toyota, Synth. Met. **120**, 759 (2001)
1084. D.B. Haviland, Y. Liu, A.M. Goldman, Phys. Rev. Lett. **62**, 2180 (1989)
1085. K. Penc, F. Mila, Phys. Rev. B **50**, 11429 (1994)

1086. T. Mori, J. Solid State Chem. **168**, 433 (2002)
1087. C. Hotta, H. Fukuyama, J. Phys. Soc. Jpn. **69**, 2577 (2000)
1088. M. Terao, Y. Ohashi, Physica C **412-414**, 324 (2004)
1089. T. Moriya, *Spin Fluctuations in Itinerant Electron Magnetism* (Springer, Berlin, 1985)
1090. S. Takagi, D. Maruta, H. Sasaki, H. Uozaki, H. Tsuchiya, Y. Abe, Y. Ishizaki, E. Negishi, H. Matsui, S. Endo, N. Toyota, J. Phys. Soc. Jpn. **72**, 483 (2003)
1091. S. Takagi, D. Maruta, H. Uozaki, H. Tsuchiya, Y. Abe, Y. Ishizaki, E. Negishi, H. Matsui, S. Endo, N. Toyota, J. Phys. Soc. Jpn. **72**, 3259 (2003)
1092. K. Maki, In *Superconductivity* (R.D. Parks, Ed., Marcel Dekker, New York, 1969), vol. II, chap. Gapless Superconductivity, pp. 1035 – 1106
1093. M.B. Maple, *Interplay between Superconductivity and Magnetism*, Physica B **215**, 110 (1995)
1094. H. Matsui, H. Tsuchiya, E. Negishi, H. Uozaki, Y. Ishizaki, Y. Abe, S. Endo, N. Toyota, J. Phys. Soc. Jpn. **70**, 2501 (2001)
1095. H. Matsui, H. Tsuchiya, T. Suzuki, E. Negishi, N. Toyota, Phys. Rev. B **68**, 155105 (2003)
1096. M.E. Lines, A.M. Glass, *Principles and Applications of Ferroelectrics and Related Materials* (Oxford Univ., 2001)
1097. B.A. Strukov, A.P. Levanyuk, *Ferroelectric Phenomena in Crystals* (Springer, Berlin Heidelberg, 1998)
1098. O. Klein, S. Donovan, M. Dressel, , G. Grüner, Int. J. Infrared Millimeter. Waves **14**, 2423 (1993)
1099. S. Komiyama, M. Watanabe, Y. Noda, E. Negishi, N. Toyota, J. Phys. Soc. Jpn. **73**, 2385 (2004)
1100. S. Endo, T. Goto, T. Fukase, H. Matsui, H. Uozaki, H. Tsuchiya, E. Negishi, Y. Ishizaki, Y. Abe, N. Toyota, J. Phys. Soc. Jpn. **71**, 732 (2002)
1101. K. Hiraki, H. Mayaffre, M. Horvatic, C. Berthier, H. Tanaka, A. Kobayashi, H. Kobayashi, T. Takahashi, J. Low Temp. Phys. **142**, 185 (2006)
1102. M. Watanabe, S. Komiyama, R. Kiyanagi, Y. Noda, E. Negishi, N. Toyota, J. Phys. Soc. Jpn. **72**, 452 (2003)
1103. N. Toyota, Y. Abe, H. Matsui, E. Negishi, Y. Ishizaki, H. Tsuchiya, H. Uozaki, S. Endo, Phys. Rev. B **66**, 033201 (2002)
1104. N. Toyota, Y. Abe, T. Kuwabara, E. Negishi, H. Matsui, J. Phys. Soc. Jpn. **72**, 2714 (2003)
1105. Y. Tokura, H. Okamoto, T. Koda, T. Mitani, G. Saito, Phys. Rev. B **38**, 2215 (1988)
1106. Y. Iwasa, T. Koda, S. Koshihara, Y. Tokura, N. Iwasawa, G. Saito, Phys. Rev. B **39**, 10441 (1989)
1107. R. Kumai, Y. Okimoto, Y. Tokura, Science **284**, 1645 (1999)
1108. E. Negishi, T. Kuwabara, S. Komiyama, M. Watanabe, Y. Noda, T. Mori, H. Matsui, N. Toyota, Phys. Rev. **71**, 012416 (2005)
1109. G.A. Smolenskii, I.E. Chupis, *Segnetomagnetics*, Sov. Phys. Usp. **25**, 475 (1982)
1110. G. Srinivasan, E.T. Rasmussen, B.J. Levin, R. Hayes, Phys. Rev. B **65**, 134402 (2002)
1111. T. Kimura, T. Goto, H. Shintani, K. Ishizuka, T. Arima, Y. Tokura, Nature **426**, 55 (2003)
1112. N. Hur, S. Park, P.A. Sharma, J.S. Ahn, S. Guha, S.W. Cheong, Nature **429**, 392 (2004)

1113. F. Goze, V.N. Laukhin, L. Brossard, A. Audouard, J.P. Ulmet, S. Askenazy, T. Naito, H. Kobayashi, A. Kobayashi, M. Tokumoto, P. Cassoux, Europhys. Lett. **28**, 427 (1994)
1114. S. Uji, H. Shinagawa, T. Terashima, T. Yakabe, Y. Terai, M. Tokumoto, A. Kobayashi, H. Tanaka, H. Kobayashi, Nature **410**, 908 (2001)
1115. L. Balicas, J.S. Brooks, K. Storr, S. Uji, M. Tokumoto, H. Tanaka, H. Kobayashi, A. Kobayashi, V. Barzykin, L.P. Gor'kov, Phys. Rev. Lett. **87**, 067002 (2001)
1116. S. Uji, T. Terashima, C. Terakura, T. Yakabe, Y. Terai, S. Yasuzuka, Y. Imanaka, M. Tokumoto, A. Kobayashi, F. Sakai, H. Tanaka, H. Kobayashi, L. Balicas, J.S. Brooks, J. Phys. Soc. Jpn. **72**, 369 (2003)
1117. L. Balicas, V. Barzykin, K. Storr, J.S. Brooks, M. Tokumoto, S. Uji, H. Tanaka, H. Kobayashi, A. Kobayashi, Phys. Rev. B **70**, 092508 (2004)
1118. H. Fujiwara, H. Kobayashi, E. Fujiwara, A. Kobayashi, J. Am. Chem. Soc. **124**, 6816 (2002)
1119. T. Konoike, S. Uji, T. Terashima, M. Nishimura, S. Yasuzuka, K. Enomoto, H. Fujiwara, B. Zhang, H. Kobayashi, Phys. Rev. B **70**, 094514 (2004)
1120. H. Uozaki, S. Kawamata, K. Okuda, K. Ueda, T. Sugimoto, N. Toyota, Synth. Met. **103**, 1981 (1999)
1121. Ø. Fischer, Helv. Phys. Acta **45**, 331 (1972)
1122. V. Jaccarino, M. Peter, Phys. Rev. Lett. **9**, 290 (1962)
1123. M. Isino, N. Kobayashi, Y. Muto, In *Ternary Superconductors* (G. K. Shenoy, B. D. Dunlap, F. Y. Fradin, eds., North-Holland, Amsterdam, 1981), p. 95
1124. H.W. Meul, C. Rossel, M. Decroux, Ø. Fischer, G. Remenyi, A. Briggs, Phys. Rev. Lett. **53**, 497 (1984)
1125. S. Uji, H. Shinagawa, C. Terakura, T. Terashima, T. Yakabe, Y. Terai, M. Tokumoto, A. Kobayashi, H. Tanaka, H. Kobayashi, Phys. Rev. B **64**, 024531 (2001)

Index

Springer Series in
SOLID-STATE SCIENCES

Series Editors:
M. Cardona P. Fulde K. von Klitzing R. Merlin H.-J. Queisser H. Störmer

Springer Series in

SOLID-STATE SCIENCES

Series Editors:
M. Cardona P. Fulde K. von Klitzing R. Merlin H.-J. Queisser H. Störmer

114 **Interatomic Potential and Structural Stability**
Editors: K. Terakura and H. Akai

115 **Ultrafast Spectroscopy of Semiconductors and Semiconductor Nanostructures**
By J. Shah

116 **Electron Spectrum of Gapless Semiconductors**
By J.M. Tsidilkovski

117 **Electronic Properties of Fullerenes**
Editors: H. Kuzmany, J. Fink, M. Mehring, and S. Roth

118 **Correlation Effects in Low-Dimensional Electron Systems**
Editors: A. Okiji and N. Kawakami

119 **Spectroscopy of Mott Insulators and Correlated Metals**
Editors: A. Fujimori and Y. Tokura

120 **Optical Properties of III–V Semiconductors**
The Influence of Multi-Valley Band Structures By H. Kalt

121 **Elementary Processes in Excitations and Reactions on Solid Surfaces**
Editors: A. Okiji, H. Kasai, and K. Makoshi

122 **Theory of Magnetism**
By K. Yosida

123 **Quantum Kinetics in Transport and Optics of Semiconductors**
By H. Haug and A.-P. Jauho

124 **Relaxations of Excited States and Photo-Induced Structural Phase Transitions**
Editor: K. Nasu

125 **Physics and Chemistry of Transition-Metal Oxides**
Editors: H. Fukuyama and N. Nagaosa

126 **Physical Properties of Quasicrystals**
Editor: Z.M. Stadnik

127 **Positron Annihilation in Semiconductors**
Defect Studies
By R. Krause-Rehberg and H.S. Leipner

128 **Magneto-Optics**
Editors: S. Sugano and N. Kojima

129 **Computational Materials Science**
From Ab Initio to Monte Carlo Methods. By K. Ohno, K. Esfarjani, and Y. Kawazoe

130 **Contact, Adhesion and Rupture of Elastic Solids**
By D. Maugis

131 **Field Theories for Low-Dimensional Condensed Matter Systems**
Spin Systems and Strongly Correlated Electrons
By G. Morandi, P. Sodano, A. Tagliacozzo, and V. Tognetti

132 **Vortices in Unconventional Superconductors and Superfluids**
Editors: R.P. Huebener, N. Schopohl, and G.E. Volovik

133 **The Quantum Hall Effect**
By D. Yoshioka

134 **Magnetism in the Solid State**
By P. Mohn

135 **Electrodynamics of Magnetoactive Media**
By I. Vagner, B.I. Lembrikov, and P. Wyder

136 **Nanoscale Phase Separation and Colossal Magnetoresistance**
The Physics of Manganites and Related Compounds
By E. Dagotto

137 **Quantum Transport in Submicron Devices**
A Theoretical Introduction
By W. Magnus and W. Schoenmaker